D1130436

Chain Reaction

Chain Reaction is a work of recent American political history. It seeks to explain how and why America came to depend so heavily on its experts after World War II, how those experts translated that authority into political clout, and why that authority and political discretion declined in the 1970s. Brian Balogh's pathbreaking research into the internal memoranda of the Atomic Energy Commission substantiates his argument in impressive historical detail. It was not the ravages of American anti-intellectualism, as so many scholars have argued, that brought the experts back down to earth. Rather, their decline can be traced to the very roots of their success after World War II. The need to overstate anticipated results in order to garner public support, incessant professional and bureaucratic specialization, and the sheer proliferation of expertise pushed arcane and insulated debates between experts into public forums at the same time that a broad cross section of political participants found it easier to gain access to their own expertise. These tendencies ultimately undermined the political clout of all experts.

Chain Reaction

Expert debate and public participation in
American commercial nuclear power, 1945–1975

BRIAN BALOGH

The right of the
University of Cambridge
to print and sell
all manner of books
was granted by
Henry VIII in 1534.
The University has printed
and published continuously
since 1584.

Cambridge University Press

Cambridge

New York   Port Chester   Melbourne   Sydney

Published by the Press Syndicate of the University of Cambridge
The Pitt Building, Trumpington Street, Cambridge CB2 1RP
40 West 20th Street, New York, NY 10011, USA
10 Stamford Road, Oakleigh, Melbourne 3166, Australia

First published 1991

Printed in the United States of America

Library of Congress Cataloging-in-Publication Data
Balogh, Brian.
Chain reaction : expert debate and public participation in
American commercial nuclear power, 1945–1975 / Brian Balogh.
p. cm.
Includes bibliographical references and index.
ISBN 0-521-37296-8
1. Nuclear industry – Government policy – United States. 2. Nuclear
industry – Government policy – United States – Citizen participation.
3. Consulting engineers – United States. 4. Nuclear engineering –
United States – Safety measures. I. Title.
HD9698.U52B35 1991
363.17'99'0973 – dc20   90-29093

British Library Cataloguing in Publication Data
Balogh, Brian
Chain reaction : expert debate and public participation in
American commercial nuclear power, 1945–1975.
1. United States. Nuclear power. Politics, history
I. Title
333.79240973

ISBN 0-521-37296-8

For KC and DAB

Contents

Acknowledgments

I have benefited from the generous financial assistance of The Johns Hopkins University Faculty of Arts and Sciences and Department of History, the Harvard University Faculty of Arts and Sciences, the Brookings Institution, the Herbert Hoover Presidential Library, and the American Association for State and Local History. I would like to thank these institutions for their encouragement and support. I would also like to thank the librarians and archivists at the Milton S. Eisenhower Library, The Papers of Dwight D. Eisenhower, The University of Southern California Manuscripts Collection, the Library of Congress Manuscripts Collection, the National Archives, the Herbert Hoover Presidential Library, the Commonwealth of Massachusetts State House Library, and the Nuclear Regulatory Commission Public Document Room. Several agency historians went out of their way to guide me to crucial documents and cut bureaucratic red tape in order to make those documents available. George Mazuzan and Samuel Walker, official historians of the Nuclear Regulatory Commission, epitomize the combination of scholarship, institutional memory, and collegial spirit that make public history such a valuable national resource. Abel and Reds Wolman were extremely generous with their time and gave me unlimited access to Abel Wolman's papers. I would also like to thank Senator Paul Sarbanes and the Office of the Senate Historian. Without their intervention, it is unlikely that I would have gained access to the thousands of United States Department of Energy documents that I ultimately was able to analyze.

My manuscript benefited tremendously from the comments and suggestions of a number of readers who were kind enough to plow through draft after draft. I want to thank members of The Seminar at The Johns Hopkins University, members of the Twentieth Century American History Workshop at Harvard University, and panel members and audience at the History of Science and Organization of American Historians sessions where portions of this work were presented. Samuel Beer, Ed Berkowitz, Alan Brinkley, Lawrence Brown, David Herbert Donald, John Mark Hansen, Hugh Heclo, Jim Hershberg, John Higham, Mark Kornbluh, Naomi Lamoreaux, Ernest May, Michael McGerr, Priscilla McMillan, Francis Rourke, Ron Walters, Steve Skowronek, Steve Thernstrom,

Daun Van Ee, Samuel Walker, and Spencer Weart all helped me sharpen the argument and improve the writing. Even before I conceived of this project, I had long benefited from the scholarship of Samuel Hays. His close reading and constructive criticism helped me expand my dissertation into this book. Nancy Landau, my copy editor, caught hundreds of mistakes and improved my prose. Research assistants Mark Bradley, David Condon, and Jessica Marshall helped me enormously. My editor at Cambridge University Press, Frank Smith, improved both the substance and style of this book. He did this professionally, and always with a sense of humor. Russell Hahn ably guided the manuscript through production without a hitch.

I owe my greatest intellectual and professional debts to my graduate adviser, Louis Galambos. Lou steered me toward post–World War II political history, introduced me to the "organizational literature" that frames my study, insisted that I articulate my ideas more forcefully at the same time that he relentlessly edited out hundreds of "to be sures." To be sure, we disagreed on many specific points. But his insistence that historians draw upon the patterns of the past to illuminate the present is a lesson I learned from him that I hope I will never forget.

Family and friends, though they probably never want to hear the word "expert" again, were unstinting in their enthusiasm for the project, their generosity, and their encouragement. My father didn't need a college degree to wade into historiography and bureaucratic politics. He even read the footnotes. I would particularly like to thank Clifton Crais, Jane Eliot Sewell, Caroline Ford, the Forsythes, Steve Golub, Mark Hansen, John Hielemann, Tom Hines, Cathy Kerr, the Kornbluhs, Bill Lanouette, Eva Moskowitz, John Muresianu, Riva Novey, Barbara Salisbury, Evelyne Schwaber, Pam Scully, the Sermiers, Ted Small, Vince Tompkins, Andrew Walsh, and John Yoo, for their coffee, wise counsel, humor, and patience.

For Kathy Craig and Dustin Anthony Balogh, to whom this book is dedicated, a whole big bunch of thanks can't even begin to express my gratitude.

Abbreviations

ACRS	Advisory Committee on Reactor Safeguards
AEC	Atomic Energy Commission
AWWA	American Water Works Association
BOB	Bureau of the Budget
CPUC	California Public Utilities Commission
DOE	Department of Energy
ECCS	Emergency Core Cooling System
EPA	Environmental Protection Agency
FDA	Food and Drug Administration
GAC	General Advisory Committee
GE	General Electric Co.
ICRLP	Industrial Committee on Reactor Location Problems
JCAE	Joint Committee on Atomic Energy
MPCA	Minnesota Pollution Control Agency
NACOR	National Advisory Committee on Radiation
NASA	National Aeronautics and Space Administration
NEWA	Nuclear Energy Writers Association
NSC	National Security Council
ORNL	Oak Ridge National Laboratory
OSRD	Office of Scientific Research and Development
OWMR	Office of War Mobilization and Reconversion
PDRP	Power Demonstration Reactor Program
PG&E	Pacific Gas and Electric Co.
PRDC	Power Reactor Development Company
RSC	Reactor Safeguard Committee
SCAE	Special Committee on Atomic Energy
SDI	Strategic Defense Initiative
SNAP	Systems for Nuclear Auxiliary Power
TVA	Tennessee Valley Authority
UCS	Union of Concerned Scientists
USGS	U.S. Geological Survey
USPHS	U.S. Public Health Service

1

From fission to fusion: professionalization and politics in twentieth-century America

Expertise has played an increasingly important role in modern American politics. The professions' ability to identify and resolve problems in our complex and interdependent world explains a large part of this phenomenon. Its roots reach back to the mid-nineteenth century. In his perceptive study of the emergence of social science as a profession, Thomas Haskell portrayed the demise of traditional deference to classical professions such as law, medicine, and the clergy in the Gilded Age. Professionals who recognized that social and economic disruptions stemmed from America's rapidly increasing interdependence – not, for instance, from character flaws – could reconstitute lost authority or grasp it for the first time. Recognizing that solutions to problems caused by immigration, urbanization, and industrialization depended on the management of complex factors that on the surface might seem unrelated to the problem, the modern professions developed powerful and appealing diagnoses: particularly in medicine, prescriptions soon followed. What disparate disciplines shared was their technique of problem solving – an approach grounded in specialization. Those professions sufficiently specialized to root out the hidden cause of a problem and treat its more explicit consequences earned back the authority lost with the demise of a more tradition-bound society.[1]

This response to interdependence stimulated more forceful professional organization, while the quest for order and control simultaneously spurred increasingly hierarchical and specialized organization in the public sector. Politics at the start of the twentieth century had already begun a long transition away from torchlight parades, 90 percent voter participation rates, and strong partisan ties of the

---

1 Thomas L. Haskell, *The Emergence of Professional Social Science: The American Social Science Association and the Nineteenth-Century Crisis of Authority* (Urbana: University of Illinois Press, 1977), pp. 1–47; Paul Starr, *The Social Transformation of American Medicine* (New York: Basic Books, 1982), pp. 17–21; John Higham, "The Matrix of Specialization," in *The Organization of Knowledge in Modern America, 1860–1920*, ed. John Voss and Alexandra Oleson (Baltimore: Johns Hopkins University Press, 1979), pp. 3–18; Louis Galambos, "By Way of Introduction," in Galambos, ed., *The New American State: Bureaucracies and Policies Since World War II* (Baltimore: Johns Hopkins University Press, 1987), p. 7.

previous century. In their place slowly emerged an administrative state.[2] "Continuous management" – Robert Wiebe's label for the never-ending adjustments required to maintain stability in this newly interdependent world – lay at the heart of bureaucratization.[3] Ideally, experts staffed the command posts of "continuous management": economists guided rate-setting commissions; doctors directed the crusade for public health improvements; Department of Agriculture scientists promoted more efficient farming techniques. As Samuel P. Hays demonstrated in his pioneering study *Conservation and the Gospel of Efficiency*, expertise was an integral part of the Progressive Era bureau.[4]

The impact of interdependence, bureaucratization, and their relationship to expertise captivated several scholars writing more than two decades ago. Labeled the "organizational synthesis" by Louis Galambos, that scholarship is today a mainstay of historical interpretation.[5] Implicitly, the organizational synthesis was built on two theoretical foundations: Weberian sociology informed its perception of organizations as rational and efficient actors, and Parsonian sociology shaped its treatment of professionals as independent and objective experts who brought unique skills to the problems they faced. The most influential individual work in this synthesis was Robert Wiebe's *Search for Order*.[6]

The hidden hands of Weber and Parsons gave the original organizational synthesis a distinctly functionalist cast. Experts were granted authority because

---

2 On the proliferation of professional organizations, see Louis Galambos, *America at Middle Age: A New History of the United States in the Twentieth Century* (New York: McGraw-Hill, 1982), pp. 43–7. On the decline of electoral, partisan politics, see Mark Kornbluh, "From Participatory to Administrative Politics: A Social History of American Political Behavior, 1880–1920" (Ph.D. diss., The Johns Hopkins University, 1987), ch. 1; Michael E. McGerr, *The Decline of Popular Politics: The American North, 1865–1928* (New York: Oxford University Press, 1986); Richard L. McCormick, *The Party Period and Public Policy: American Politics from the Age of Jackson to the Progressive Era* (Oxford: Oxford University Press, 1986), ch. 5.

3 Robert H. Wiebe, *The Search for Order, 1877–1920* (New York: Hill & Wang, 1967), p. 145, for definition of "continuous management."

4 Wiebe, *Search*, pp. 170–5; Samuel P. Hays, *Conservation and the Gospel of Efficiency: The Progressive Conservation Movement, 1890–1920* (Cambridge, Mass.: Harvard University Press, 1959). Of course, the Progressive Era was not the first time Americans turned to the public use of experts. On issues where interdependence proved particularly disruptive, the pattern had begun much earlier. In railroad regulation at the state level shortly after the Civil War, for instance, Charles Francis Adams called for mechanisms foreshadowing continuous management by experts (Thomas K. McCraw, *Prophets of Regulation: Charles Francis Adams, Louis D. Brandeis, James M. Landis, Alfred E. Kahn* [Cambridge, Mass.: Harvard University Press, 1984], pp. 7–15).

5 Louis Galambos, "The Emerging Organizational Synthesis in Modern American History," *Business History Review* 44 (Autumn 1970): 279–90; and "Technology, Political Economy, and Professionalization: Central Themes of the Organizational Synthesis," *Business History Review* 57 (Winter 1983): 471–93.

   For a critical perspective on the organizational synthesis, see Alan Brinkley, "Writing the History of Contemporary America," *Daedalus* 113, no. 3 (Summer 1984): 121–41; and Michael McGerr, "Organization, Individualism, and Class Conflict: Redefining Early Twentieth-Century America," paper presented to The Seminar, October 20, 1986, The Johns Hopkins University.

6 Other central works include Alfred D. Chandler, *The Visible Hand: The Managerial Revolution in American Business* (Cambridge, Mass.: Harvard University Press, 1977), and Samuel P. Hays, *The Response to Industrialism, 1885–1914* (Pittsburgh: University of Pittsburgh Press, 1957).

a complex and interdependent society required it, the argument went. Relatively little attention was paid to the professions' active role in acquiring authority.[7] To the extent that the pioneers of the original organizational synthesis added a historical gloss to this otherwise relatively static picture of expertise, they concentrated on tracing the historical developments that led to this social need, that is, industrialization, urbanization, and the like. All causal paths led to interdependence.[8]

Like a literary aircraft carrier, *Search for Order* dwarfed smaller vessels that had helped chart its course and provided a platform for new exploratory sorties.[9] Since the christening, however, the empirical data that a second generation of organizational recruits amassed and a sea change in the theoretical underpinnings of the organizational synthesis have altered the course of that scholarship and with it our understanding of the relationship between experts and administrators, and of the way each has exercised authority and power. Their work recasts the relationship between experts and the federal government in a far less deterministic framework. Incorporating experts into hierarchical organizations hardly ended political struggles, it simply transferred them to new forums. Nor did it mean that the professions embraced the bureaucracies that housed them, or vice versa.[10] What ultimately emerged was a symbiotic relationship fraught with tensions, not a union of apolitical rationalizers.[11]

To argue that the professions in the Progressive Era were just beginning to integrate a new scientific base of knowledge and that the federal government had not developed much administrative capacity is not to say that there were no

7 The exception was Samuel Hays's account of the rise of scientific conservation (in *Conservation and the Gospel of Efficiency*). Although the profession that Pinchot pioneered ultimately achieved a degree of expertise, credentials, and autonomy, it was hardly born professional, as Hays displays in impressive detail. Its influence was constructed through a series of political, organizational, and intellectual struggles during which the very nature of the profession was altered.

   In treating professionalization and the application of professional values and knowledge as a struggle with the outcome far from determined, Hays anticipated the thrust of much recent scholarship. Along these lines, see "Gifford Pinchot Creates a Forest Service," in Jameson W. Doig and Erwin C. Hargrove, eds., *Leadership and Innovation: A Biographical Perspective on Entrepreneurs in Government* (Baltimore: Johns Hopkins University Press, 1987), pp. 63–95.

8 It is ironic that the revisionist historians who so impressively debunked accounts that accepted the Progressives' own rhetoric about representing the people against giant interests themselves accepted virtually in its entirety Progressive rhetoric regarding the apolitical nature of experts and the administrative forum. For a discussion of the need to politicize the administrative forum in order to "democratize" expertise, see Brian Balogh, "Democratizing Expertise: State Building and the Progressive Legacy" (paper presented to the 1989 annual meeting of the American Political Science Association).

9 Samuel P. Hays, for instance, anticipated many of the central themes in *Search for Order* in his *Response to Industrialism*, published ten years earlier.

10 The tension between professionals and bureaucratic organization is explored by Wayne K. Hobson in "Professionals, Progressives and Bureaucratization: A Reassessment," *The Historian* 39 (1977): 639–58.

11 A summary of the more recent literature spawned by the organizational approach and a detailed assessment of its impact on the relationship between professionals and the federal government is presented in Brian Balogh, "Reorganizing the Organizational Synthesis: Reconsidering Modern American Federal–Professional Relations," *Studies in American Political Development* 5, no. 1 (1991).

experts serving well-developed administrative mechanisms. In fact, A. Hunter Dupree has labeled the Department of Agriculture during this period a "predominant" agency of science. It was built around a system, not a personality; the scope of its various branches ranged from land-grant colleges to state experimentation stations; and the variety of functions it performed ran the gamut, coupling basic research to the application of science by individual farmers.[12] Even before the Progressive Era, the Department of Agriculture was sufficiently developed to have spawned a specialized group of agricultural administrators. Almost all agricultural scientists were full-time government employees. Federal funding was no doubt a contributor to the "Golden Age" that this field experienced in the last two decades of the nineteenth century and the first two of the twentieth century.[13] Yet Agriculture was the clear exception, and even in this "predominant" agency of science, it appears that the full extent of the professional expert's repertoire was not always engaged.

There were two critical differences between virtually all of the federal agencies' relationships to expertise in the Progressive Era and the kinds of relationships that emerged after World War II. Before that war the federal government had little to do with the production of professional experts and as a result exercised little influence over the cognitive base of the professions it employed or the number of professionals produced. Second, and even more significantly, professionals were employed in Progressive Era federal bureaus primarily to apply skills already tested and routinized. As such, the agendas dictated by the cutting edge of each profession's research – and, for that matter, basic research itself – remained outside the scope of the federal–professional relationship.

The federal government's enlarged regulatory role accounted for much of the growth of federally funded science during this period. Scientists employed by the agencies concentrated on the routine tasks of inspection and testing. Although the federal government's Hygienic Laboratory began to develop an independent research laboratory in the first two decades of the twentieth century, a large part of its political legitimacy was derived from its enforcement of the Biologics Act of 1902.[14] The same pattern obtained in the Agriculture Department's enforcement of the Food and Drug Act of 1906.[15] Despite the fact that federal expenditures on research in 1900 totaled $11 million – a staggering amount compared to other sources – federal spending did not dominate science because

12 A. Hunter Dupree, "Central Scientific Organization in the United States Government" *Minerva* (Summer 1963): 457–8. The definitive account of those scientific fields in which the federal government was active before World War II and the analysis of the nature of federal support is Dupree's *Science in the Federal Government: A History of Policies and Activities to 1940* (Cambridge, Mass.: Harvard University Press, 1957).

13 Margaret W. Rossiter, "The Organization of Agricultural Sciences" in Voss and Oleson, eds., *The Organization of Knowledge,* p. 215.

14 Victoria A. Harden, *Inventing the NIH: Federal Biomedical Research Policy, 1887–1937* (Baltimore: Johns Hopkins University Press, 1986), p. 42.

15 Ibid., p. 3.

so much of it was devoted to practical day-to-day applications as opposed to basic research.[16] By the end of the Progressive Era, science as a whole had become distinctly specialized: basic research was soundly grounded in the new research universities, while applied research remained the domain of government and increasingly was to be found in industry as well.[17]

Leading physical scientists, housed in universities or funded by voluntary agencies, resisted federal funding. In 1934 the well-known astronomer and astrophysicist George Ellery Hale warned his colleagues that a National Research Administration might injure the voluntary organizations that housed science, such as the National Academy of Sciences (NAS) and the National Research Council. "I do not believe," Hale warned, "that under a Government system the University of Chicago and other institutions of which I have personal knowledge would be what they are today."[18] His colleague Frank Jewett of the American Telephone and Telegraph Company was more blunt: "Outside of a restoration of scientific activities within the legitimate functions of the governmental departments, appropriation of Federal funds is a very grave question which goes to the root of the whole matter of state participation and control of functions which hitherto we have jealously guarded as the affairs of individual or non-political cooperative effort." Sooner or later, Jewett continued, "there would be a large measure of bureaucratic control or attempts to control by the Federal government."[19] Basic science's elite continued to resist federal funding throughout the thirties.[20]

Social scientists also resisted proposals that might have brought federal funding at the expense of federal control. Economists, while relying heavily on the government's statistics-gathering ability, remained aloof from what they feared would be government control. They were housed in voluntary structures such as the National Bureau of Economic Research or the Brookings Institution. The vision they offered relied on voluntary cooperation in order to build a national countercyclical economic strategy or to improve economic and public administration's efficiency.[21] In explaining why so many professionals resisted the urge

16 Roger L. Geiger, *To Advance Knowledge: The Growth of American Research Universities, 1900–1940* (New York: Oxford University Press, 1986), pp. 29, 59.

17 Dupree, "Central Scientific Organization," p. 462.

18 Robert Kargon and Elizabeth Hodes, "Karl Compton, Isaiah Bowman, and the Politics of Science in the Great Depression," *Isis* 76 (1985): 316. For an account of some of the funding pressures facing the sciences and the failed effort to organize industrial support, see Lance E. Davis and Daniel J. Kevles, "The National Research Fund: A Case Study in the Industrial Support of Academic Science," *Minerva* 12, no. 2 (April 1974): 207–20.

19 Kargon and Hodes, "Karl Compton," ibid.

20 Kargon and Hodes, "Karl Compton"; Lewis E. Auerbach, "Scientists in the New Deal: A Prewar Episode in Relations Between Science and Government in the United States," *Minerva* 3, no. 4 (Summer 1965); Daniel J. Kevles, *The Physicists: The History of a Scientific Community in Modern America* (New York: Knopf, 1978), ch. 17.

21 See, for instance, Guy Alchon, *The Invisible Hand of Planning: Capitalism, Social Science and the State in the 1920s* (Princeton, N.J.: Princeton University Press, 1985), and Donald T. Critchlow, *The Brookings Institution, 1916–1952* (DeKalb: Northern Illinois University Press, 1985).

to seek federal funds, we must go beyond their concern for what federal funding might do to the professions and consider broader cultural reasons for their position. These men and women were products of the same antistatist political culture as their nonprofessional neighbors: they resisted expanding the federal government for the same reasons that millions of other Americans did.[22]

Elected officials were also hesitant, particularly in regard to research and its application. That politicians were not eager to fund research where tangible results might be years off or nonexistent is not surprising. Nor did politicians leap at the opportunity to pass on the political discretion and resources that might accompany the application of expertise to problems. Because early-twentieth-century American public administration was so underdeveloped – at least by European standards – the most enduring relationship between expert and public sector, and the most representative, was licensing.[23] Through this relationship states contributed to the authority of professionals, and in many instances reduced competition from other occupations. Professionals, in turn, solved some of society's problems without disrupting the private or voluntary framework already in place.

Rather than fuse expertise with a more centralized state, Americans viewed it as an alternative to an extended state. From the perspective of those steeped in America's antistatist culture, the hierarchy and self-governing mechanisms of the myriad professionalizing organizations were not unlike state and local govern-ments: they dealt with problems without requiring the expansion of the centralized state.

By the 1920s there were four kinds of institutions that formed crucial bridges between professionals and the national government and that thrived precisely because of professional, public, and political reticence about the fusion of these two powerful sources. The first such institution, the research university, actually produced the new professionals and conducted much of the nation's basic research. The second was the voluntary coordinating agency, such as the Social Science Research Council or National Research Council already mentioned. The third was the foundation – such as those established by Rockefeller and Carnegie or in memory of Laura Spelman – which provided funding and sought to direct the development of both universities and coordinating agencies. Finally, there were the trade associations and professional societies that President Herbert Hoover hoped would guide his "associative state." The unique combination of charity and faith in science that created foundations like the Laura Spelman Rockefeller Memorial filled a void created by the tradition of federalism at the same time that it hastened the process of professionalization. Foundations existed, Barry Karl and Stanley Katz argue, because Americans were unwilling "to give their

22 On antistatist nature of American political culture, see Ellis Hawley, "Social Policy and the Liberal State in Twentieth Century America," in Hawley and Donald T. Critchlow, eds., *Federal Social Policy: The Historical Dimension* (University Park: Pennsylvania State University Press, 1988).
23 Galambos, *America at Middle Age*, pp. 42–7.

national government the authority to set national standards of social well-being, let alone to enforce them."[24]

Though Progressives initially branded it unthinkable, it was war, more than reform, that expanded administrative capacity and pushed professionals into these new positions on a national scale. Progressives like John Dewey executed an abrupt about-face in their thinking about America's role in World War I. Dewey felt that the crisis of war might lead to more extensive use of science for communal purposes, illuminate the public aspect of every social enterprise, create new mechanisms for "enforcing the public interest in all agencies of production and exchange," and temper America's excessive individualism.[25]

The new wartime managers discovered, however, that the science-based professions were just in their infancy. The war made it the federal government's business to force-feed development. Barry Karl summed up the problem in regard to the social sciences, writing that "in virtually every field the lesson was the same. From the economists who worked for the War Industries Board [WIB] to the historians and political scientists who advised the president at Versailles, the issue boiled down to one basic problem: American specialization in such fields was essentially in its infancy."[26]

While the challenge for the social sciences was to demonstrate that their cognitive base was in fact scientific, the challenge for physical sciences was to apply basic knowledge quickly enough to affect the war's outcome. That chemists and chemical engineers were able to produce in short order optical glass, nitrates, and poison gases justified the label "chemist's war" that was subsequently applied to the conflict.[27]

Physicists, who before the war were viewed by the press as impractical scientists – good subjects for the staff humorists but not particularly newsworthy – had a hard time convincing the federal government that it needed their help.[28] Only through the persistent lobbying of the National Research Council – itself a product of the war – were physicists allowed to join the race to develop a submarine detection device. The physicists were allowed to set up a small lab at New London, Connecticut, where under the direction of Max Mason they developed such a device. They also played an instrumental role in the work of the National Advisory Committee for Aeronautics, and developed a range of signaling devices for the infantry.[29]

24 Barry D. Karl and Stanley N. Katz, "The American Private Philanthropic Foundation and the Public Sphere, 1890–1930," *Minerva* 19, no. 2 (Summer 1981): 238; see also Barry D. Karl, "Philanthropy and the Social Sciences," and Robert E. Kohler, "Philanthropy and Science," *Proceedings of the American Philosophical Society* 129, no. 1 (1985): 14–19 and 9–13.

25 David M. Kennedy, *Over Here: The First World War and American Society* (Oxford: Oxford University Press, 1982), p. 50.

26 Barry D. Karl, *The Uneasy State: The United States from 1915 to 1945* (Chicago: University of Chicago Press, 1983), p. 46.

27 Kevles, *Physicists*, p. 137.

28 Ibid., p. 95.   29 Ibid., ch. 9.

Though limited, and surely inefficient, the turn to professionals and the expansion of the federal government's administrative capacity proved sufficient to win the war. Despite the compromised nature of both developments, they proved far too controversial to be retained once the shooting stopped. Better than anybody, Bernard Baruch recognized it was the wartime emergency that had provided the fragile base from which he operated. If the WIB was not dissolved quickly, it would crack apart.[30] Congress was so quick to dismantle the nation's war machinery that it left some Washington office workers searching for funds to get home when their salaries were abruptly cut off.[31] For professionals who had served the federal government there was also a quick return to the private or voluntary institutions in which they previously had been housed.

Yet the military crisis did have some long-term implications for what had previously been the parallel paths of state building and professionalization. If nothing else, the successes would be remembered. Science had proven it could be applied rapidly and practically.[32] The experience of actually working for the federal government undoubtedly revised some of the traditional assumptions about the dangers of government-sponsored research.[33] The war also introduced social scientists to corporate managers. In some instances, that resulted in the direct adaptation of social science techniques to industry, such as using psychological testing to categorize manpower.[34] The war also mobilized economists and demonstrated the potential of economics and the importance of establishing a good data base for economic projections. It was during the war that economists were able to challenge older deductive paradigms. The war – as it did for the other social sciences – also revealed just how precarious a stage of development the profession of economics was in. Most significantly, the war introduced professional economists to policy and managerial elites. Stripped of the mobilization machinery, social scientists built a postwar equivalent around Hoover's vision of an associative state using apolitical institutions like the National Bureau of Economic Research and the foundations as a crucial source of funding.[35]

As with the first decade of twentieth-century state and profession building, it was a crisis – this time economic crisis brought on by the Great Depression – not a reform-minded social movement, that pushed experts and the state closer together. Nevertheless, the path towards convergence remained strewn with

30 Robet Cuff, *The War Industries Board: Business Government Relations During World War I* (Baltimore: Johns Hopkins University Press, 1973), p. 241.

31 Karl, *Uneasy State*, p. 46.

32 The war created an ideology of science: it demonstrated the scientist's ability to bring basic research to bear on practical problems and the government's ability to solve the difficulties of coordinating research. See Geiger, *To Advance Knowledge*, p. 98. Success, however, was tempered by another new public image for science and technology – one that associated it with death and destruction. See Kevles, *Physicists*, p. 181.

33 Harden, *Inventing the NIH*, p. 3.

34 Loren Baritz, *The Servants of Power: A History of the Use of Social Science in American Industry* (New York: Wiley, 1965).

35 Alchon, *Invisible Hand of Planning*, pp. 68–75.

obstacles. Given much of the credit for the technological breakthroughs such as radios and other consumer goods mass-marketed during the twenties, scientists were now often associated with some of the new technology's negative consequences – particularly "technological unemployment."[36] Rather than increasing their inroads into the federal government during the Great Depression, the physical sciences came under increasing pressure to demonstrate their relevance. President Franklin D. Roosevelt was persuaded to establish a Science Advisory Board, but the board received no funds and lasted only two years. The Roosevelt administration rejected its costly proposals. Meanwhile, the budget director, Lewis Douglas, slashed scientists from the Bureau of Standards just as he cut jobs in other federal agencies in response to Roosevelt's campaign pledges of belt tightening.[37]

Scientists responded to the new demand for relevance. In physics, nobody understood the changed conditions of fund-raising better than Ernest Lawrence. He adapted research on the newly developed cyclotron to meet the needs of medical science – producing radioactive isotopes for the treatment of cancer. On the lecture circuit, Lawrence served radio-sodium cocktails, Geiger counter in hand, tracing the isotope's progress through a startled volunteer's bloodstream.[38] Although such vaudeville techniques were criticized by some of Lawrence's colleagues, his ability to demonstrate practical applications for the cyclotron helped pay the bills for this costly research tool. Both the foundations and the federal government demanded such a demonstration. At the Rockefeller Foundation, for instance, Warren Weaver steered his resources increasingly toward the field of biomedicine, where, he correctly predicted, it could have the most immediate impact on human suffering.[39] Meanwhile, the Bankhead–Jones Act in agricultural research and the National Cancer Institute in the field of biomedical research increased federal funding for basic research in areas that promised to lead to direct social applications.[40]

While scientists demonstrated their relevance, administrators extended Hoover's concept of a society built around organized interests to far less organized sectors of the polity. At the Tennessee Valley Authority (TVA), David Lilienthal brought the benefits of technical expertise to an entire region, mobilizing citizens through the mechanism of grass-roots democracy. What started out as an appeal to the grass roots ended up looking like politics-as-usual through a series of compromises with firmly established interest groups like the American Farm Bureau Federation.[41] A more radical commitment to organizing at the grass

36 Kevles, *Physicists*, ch. 16.   37 Ibid., pp. 250–8.   38 Ibid., p. 272.

39 Robert E. Kohler, "The Management of Science: The Experience of Warren Weaver and the Rockefeller Foundation Programme in Molecular Biology," *Minerva* 14, no. 3 (Autumn 1976): 285.

40 Kevles, *Physicists*, pp. 260, 265.

41 Philip Selznick, *TVA and the Grass Roots: A Study of Politics and Organization* (Berkeley: University of California Press, 1984).

roots, however, was carried out by administrators of the Farm Security Administration.[42] Whether appealing to established interest groups or beginning to create new constituencies, federal administrators recognized that access to expertise was an invaluable asset. Experts could produce concrete benefits for the agency's constituency. Experts also carried weight in Congress, where there was increasing competition for resources.

It was the demand for resources, more than any other factor, that pushed experts and federal administrators together. Lawrence's new cyclotron – estimated to cost more than $1.5 million – epitomized the need for greater resources. By the 1930s the cost of research and training was threatening to outstrip private and voluntary sources of funding.[43] In Lawrence's case, the Rockefeller Foundation shouldered the increasing burden. As early as the mid-1920s, however, the Rockefeller Foundation's General Education Board had discovered just how costly its commitments to building a scientific base for medical education could be. If the board was to have a national impact, it would have to mingle its funds with those of state-funded institutions – a procedure it had scrupulously avoided until 1923.[44] It was the rising costs of research that ultimately forced foundations to narrow what had been broad support of medical education. The foundations had succeeded in stimulating a more research-oriented approach on the nation's leading campuses. That very success, both in producing new scholars and scientists and in the growing specialization of their work, outstripped the resources of the voluntary sector by the 1930s.[45]

The federal government, however, was not yet prepared to fill that gap in a comprehensive way. Nor were professional experts entirely ready to accept the central government's embrace. Given the severe social upheaval the Great Depression caused, the development of administrative capacity in areas that brought the expertise of social scientists to bear on these problems might have been expected. There certainly were some early indications that this might happen. Again, the field of agriculture led the way. In the Department of Agriculture, Richard Kirkendall has argued, proposals generated by economists and planners actually altered the behavior of key farm groups. Throughout the decade, social scientists were employed at top levels of the department, and influenced policy. Ultimately their influence depended on political support. When a deteriorating relationship with the Secretary of Agriculture exacerbated by Rexford Tugwell's confrontational style eroded that support shortly before World War II, social scientists lost much of their influence.[46]

42 Sidney Baldwin, *Poverty and Politics: The Rise and Decline of the Farm Security Administration* (Chapel Hill: University of North Carolina Press, 1968).
43 Kevles, *Physicists*, p. 285.
44 E. Richard Brown, *Rockefeller Medicine: Medicine and Capitalism in America* (Berkeley: University of California Press, 1979), p. 175.
45 Ibid., p. 406.
46 Richard S. Kirkendall, *Social Scientists and Farm Politics in the Age of Roosevelt* (Columbia: University of Missouri Press, 1966).

The National Resources Planning Board was another potentially powerful mechanism for fusing the social sciences and federal administrative capacity. Besides the political opposition that ultimately hobbled the NRPB, social scientists continued to question a closer relationship with the federal government. The counterpart of the National Research Council – the Social Science Research Council – when requested by the NRPB to study the relationship of the federal government to social research "emphatically disapproved any government subsidization of research."[47] The SSRC committee was concerned that social research might deteriorate into "merely government propaganda" if federal dollars became the main source of support.[48] As with the physical sciences, the watershed for federal funding was World War II.

Within the White House itself, of course, much was made of Roosevelt's "brain trust." Relying on academicians interested in economic and industrial planning such as A. A. Berle, Raymond Moley, and Rexford Tugwell, Roosevelt established new links between the presidency and the academic community.[49] Yet for all of the attention they drew, it is not clear how much of an impact their expertise had on Roosevelt's program. Barry Karl has argued that working relationships between the president and his experts were "personal and basically political. [T]he edge was usually given to experience," Karl concluded, "and the newly established experts simply didn't have it. Roosevelt looked to academics for ideas, but his programs seem more traceable to the experienced political leaders and the community of managers he had always worked with."[50]

Even when expert advice was housed in more formalized structures such as the Temporary Economic Committee and the National Resources Planning Board, it seems to have had a limited impact. Yet, by expanding employment and allowing social scientists to apply their skills on high-priority problems, the New Deal experience expanded knowledge and technique in the disciplines themselves, even if these advances were not always applied.[51] Government service, as it had in World War I, also helped overcome some of the fears about the politicization that professionals harbored along with the broader public. Although prestigious organizations like the Brookings Institution voiced concern about the excessive independence of a reorganized executive branch and cautioned about the limits of expert-driven social policy, increasing numbers of prestigious social scientists were willing to take the plunge into government service.[52]

Despite the crisis of the Great Depression, Roosevelt continued to struggle with the restrictions imposed by the political culture that any president had to

---

47 Martin Bulmer, "Philanthropy and Social Science in the 1920s: Beardsley Ruml and the Laura Spelman Rockefeller Memorial, 1922–29," *Minerva* 19, no. 3 (Autumn 1981): 347; Samuel Z. Klausner and Victor M. Lidz, eds., *The Nationalization of the Social Sciences* (Philadelphia: University of Pennsylvania Press, 1986), p. 5.
48 Klausner and Lidz, ibid.   49 Karl, *Uneasy State*, p. 113.
50 Ibid.   51 Ibid., p. 163.
52 Critchlow, *Brookings*, pp. 108–9, 131–6.

operate within. Wherever possible, FDR sought to use existing administrative mechanisms to implement his new programs. Thus all of the titles of the Social Security Act, with the exception of social insurance, were administered by state and local governments. Roosevelt was not happy about national administration of social insurance either, but was persuaded that it simply was not technically feasible to administer this contributory retirement plan in any other fashion.[53] The National Industrial Recovery Act harkened back to Hoover's vision of an associative state, relying on voluntary cooperation between industry and labor, coordinated by the federal government.[54] Although the New Deal changed the scope of government activity, its methods were familiar. Given the conservative reaction that swept away some of the New Deal's more radical reforms – such as the Farm Security Administration or the National Resources Planning Board – it is not clear how much of the administrative capacity would have survived if America had not been plunged into yet another crisis.[55]

Viewed historically, it was the parallel development, not the merger, of professionals and the federal government that characterized relations between the two in the first half of the twentieth century. Incorporating this revised analysis of the relationship between professionals and the state into a refurbished organizational synthesis shifts the date of their union from the Progressive Era to World War II. Only after the triple crises of the Great Depression, World War II, and the ensuing cold war would that shotgun marriage take place, bringing with it a new kind of politics. Epitomized by the development of commercial nuclear power after the war, the new politics eventually pervaded virtually all policy areas. In fact, agenda setting by professionals directing federal resources and relying on federal administration for implementation in part explains the dramatic expansion of the federal government's reach following the war. At the core of this new kind of politics lay a symbiotic relationship between professional experts and the nation's public bureaucracies. A full seventy-five years after the American state and professions began their separate responses to the crises generated by increased interdependence, the "prominstrative state" – a term I have coined to signify the merger of professionals and federal administrative capacity – finally arrived.

World War II forged a more permanent link at the national level between expertise and most federal institutions. The federal government bankrolled the production of record numbers of experts who then moved into government-

53 Brian Balogh, "Securing Support: The Emergence of the Social Security Board as a Political Actor, 1935–1939," in Hawley and Critchlow, eds., *Federal Social Policy*.

54 Peri E. Arnold, "Herbert Hoover and the Continuity of American Public Policy," *Public Policy* 20 (Fall 1972): 541.

55 For the challenge to institutionalization of many of the New Deal programs that had relied on social science expertise, see James T. Patterson, *Congressional Conservatism and the New Deal* (Lexington: University of Kentucky Press, 1967), pp. 288–324; and Baldwin, *Poverty and Politics*, pp. 365–94.

funded positions.[56] Agencies courted experts: they were now an essential political resource.[57] The war demonstrated in dramatic fashion that expertise – directed and funded by the national government – could produce in a crisis.

As policymakers framed pleas for additional experts in the jargon of national priorities, meeting the expert shortage became a major postwar concern – at times even a "crisis."[58] This concerted effort by the federal government goes far toward explaining the diffusion of expertise and the growth of major research universities. Using the dramatic expansion of Ph.D. programs in the physical sciences as a model, other disciplines imitated their better-funded counterparts: they, too, turned to the federal government for resources.

An elaborate federal network supported these newly minted experts. To ensure an uninterrupted flow of accomplishments and to control their direction better, more permanent administrative arrangements were established – ranging from the National Science Foundation to the Council of Economic Advisers. The infusion of experts permanently altered the traditional composition of administrative personnel. This, in turn, changed the way programs were administered.

56 On the professionalization of the postwar state, see Frederick C. Mosher, *Democracy and the Public Service* (New York: Oxford University Press, 1968), pp. 99–133. Don K. Price comments on the trend of scientists moving into administrative positions in *The Scientific Estate* (Cambridge, Mass.: Harvard University Press, 1965), pp. 66–71.

57 On expertise as an agency resource, see Francis E. Rourke, *Bureaucracy, Politics and Public Policy* (Boston: Little, Brown, 1969), pp. 39–61. A good summary of how a particular profession can make its policy advocacy more effective in the bureaucratic setting is contained in Robert H. Nelson, "The Economics Profession and the Making of Public Policy," *Journal of Economic Literature* 25 (March 1987): 49–91.

One of Magali Sarfatti Larsen's greatest contributions in *The Rise of Professionalism: A Sociological Analysis* (Berkeley: University of California Press, 1977) has been to recognize the growing significance of the bureaucratic framework in which professionals often work. She contends that, far from interfering with professionalization, this bureaucratic phenomenon has been the primary catalyst behind many of the more recent professionalization projects (pp. 145–7). New forms of bureaucratic specialization, Larsen argues, spawned new claims to expert status by the would-be specialist (pp. 148, 178–85).

Although this sequence may occur in some instances – particularly in those occupations where claims to professional status are most hotly challenged – my research in two very different policy areas suggests just the opposite. In its early years, the Social Security Board was desperate to staff state welfare offices with trained social workers. The board hoped that the professional standing of these workers – no matter how marginal it was – would help to offset the pork-barrel stigma long attached to local assistance programs (Balogh, "Securing Support," in Hawley and Critchlow, eds., *Federal Social Policy*). In the case of the Atomic Energy Commission, it is clear it was the agency that enhanced its status by bringing into its fold highly respected physicists. The desire to enhance agency status seems to be one important reason why so many postwar agencies turned to "independent" experts for policy agendas and implementation.

58 The best work on the "crisis" attributed to the shortage of highly trained experts, the major role played by the federal government in both publicizing and responding to this shortage, and the ultimate inflation of categories of expertise considered crucial to the state is an outstanding Ph.D. dissertation by Frank J. Newman, "The Era of Expertise: The Growth, the Spread and Ultimately the Decline of the National Commitment to the Concept of the Highly Trained Expert: 1945 to 1970," Stanford University, 1981.

Increasingly, policy agendas were expert-driven; support building monopolized more of the experts' time.[59]

Just as the first burst of growth began to slow in the mid-1950s, Sputnik provided a new stimulus. The implications for the physical sciences were obvious. Winning the competition with the Soviets for the hearts and minds of the third world, a host of policymakers argued, required more than a quick fix in the basic sciences. Whether teaching Johnny to read, or ameliorating some of America's more embarrassing social problems, the appeal to wounded American pride and fears of strategic blackmail or worse stimulated funding for policies as disparate as education, welfare, and medical care. To ensure that adequate supplies of experts were on hand to guide policy in such a wide variety of fields, the National Defense Education Act funded expansion in all areas of higher education.[60]

Perhaps the best indication of how far professionalization had gone in postwar America was the literature on the "New Class." Writing during the heyday of enthusiasm for this trend – the mid-1960s – Daniel Bell noted that society was moving from blue collar to white collar. The new dominant institutions were intellectual, Bell proclaimed, and the political implications of this massive change were among its most significant. Politics would not disappear entirely, but political decisions increasingly would be circumscribed by the experts of the New Class, thus limiting the power of ideology to influence politics. The governing attitude in the Kennedy White House, according to Arthur Schlesinger, Jr., was "that public policy is no longer a matter of ideology but of technocratic management."[61] Even Don K. Price, who, having struggled to hold scientists accountable to the Truman administration, was wary of their apolitical image, did not doubt the deference they commanded in politics. Scientific institutions were the only ones where funds were appropriated "almost on faith," Price wrote in *The Scientific Estate*.[62]

The development of commercial nuclear power is an ideal leading case with

59 For a good description of the various factors that go into agenda setting, see John W. Kingdon, *Agendas, Alternatives, and Public Policies* (Boston: Little, Brown, 1984).
60 Walter A. McDougall, "Technocracy and Statecraft in the Space Age – Toward the History of a Saltation," *American Historical Review* 87, no. 4 (October 1982): 1017, 1029–36; Richard Hofstadter, *Anti-intellectualism in American Life* (New York: Vintage, 1962), pp. 4–5; Newman, "Era of Expertise," pp. 65–72; Hugh Davis Graham, *The Uncertain Triumph: Federal Education Policy in the Kennedy and Johnson Years* (Chapel Hill: University of North Carolina Press, 1984).
61 Quoted in David Dickson, *The New Politics of Science* (Chicago: University of Chicago Press, 1984), pp. 265–6.
62 Daniel Bell, "Notes on Post-Industrial Society," *The Public Interest*, no. 6 (Winter 1967): 24–35. The richest source of commentary on the New Class is *The Public Interest*, from its inception in the fall of 1965 through the late 1960s. Besides Bell, see articles by Daniel Patrick Moynihan, particularly "The Professionalization of Reform," *The Public Interest*, no. 1 (Fall 1965): 6–16. See also Daniel Bell, *The Coming of Post-Industrial Society: A Venture in Social Forecasting* (New York: Basic Books, 1973), chs. 3 and 6; Moynihan, *Maximum Feasible Misunderstanding: Community Action and the War on Poverty* (New York: Free Press, 1969), pp. 21–36. See Peter Steinfels, *The Neoconservatives: The Men Who Are Changing America's Politics* (New York: Simon & Schuster, 1979), pp. 250–60, for a description of Moynihan's declining confidence in policy professionals. Price, *Scientific Estate*, p. 12.

which to explore the trends discerned by 'New Class' theorists. No honeymoon was sweeter than that of the Atomic Energy Commission in its first days, nor were the politics of the prominstrative state any purer than in this policy area.[63] Chapters 2 through 7 recount this period, culminating in the mid-1960s. By our probing inside administrative and professional forums, as 'New Class' theorists

63 Very little on nuclear power has been written from primary sources. Fortunately, the first volumes produced by official historians of the Atomic Energy Commission and the Nuclear Regulatory Commission have been excellent. Unfortunately, they cover only a small portion of the history of those agencies. See Richard G. Hewlett and Oscar E. Anderson, *A History of the United States Atomic Energy Commission*, Vol. 1: *The New World, 1939–1946* (University Park: Pennsylvania State University Press, 1962), and Hewlett and Francis Duncan, *A History of the United States Atomic Energy Commission*, vol. 2: *Atomic Shield, 1947–1952* (University Park: Pennsylvania State University Press, 1969). Volume III of the official history authored by Richard G. Hewlett and Jack M. Holl, was recently released. See *Atoms for Peace and War, 1953–1961* (Berkeley: University of California Press, 1989). A manuscript of Volume III was completed several years prior to publication. Despite numerous requests, coauthor and Department of Energy chief historian Jack Holl has refused to make it, footnotes to it, or documents cited in footnotes, available to me. In addition to the official history of the AEC, agency historians have produced several outstanding works on issues related to the development of nuclear power. See Richard G. Hewlett and Francis Duncan, *Nuclear Navy, 1946–1962* (Chicago: University of Chicago Press, 1974); George T. Mazuzan and J. Samuel Walker, *Controlling the Atom: The Beginnings of Nuclear Regulation, 1946–1962* (Berkeley: University of California Press, 1984). Though not a professional historian, David Okrent, a member of the AEC's Advisory Committee on Reactor Safeguards, has written a carefully documented history of some of that committee's more important decisions, entitled *Nuclear Reactor Safety: On the History of the Regulatory Process* (Madison: University of Wisconsin Press, 1981). See also the agency-commissioned RAND study by Arturo Gandara, *Electric Utility Decision Making and the Nuclear Option* (Santa Monica, Calif.: RAND, 1977).

Even though they do not rely on agency or congressional internal documents, a number of scholars have provided useful and balanced accounts of commercial power's development. The definitive work on the relationship between the Atomic Energy Commission and the Joint Committee on Atomic Energy remains Harold P. Green and Alan Rosenthal, *Government of the Atom: The Integration of Powers* (New York: Atherton Press, 1963). Since that work was published, most accounts have been organized around the questions of why nuclear power was developed in the manner that it was, or why nuclear power has failed.

Published a decade ago, Steven L. Del Sesto's *Science, Politics and Controversy: Civilian Nuclear Power in the United States, 1946–1974* (Boulder, Colo.: Westview Press, 1979) and Irwin C. Bupp and Jean-Claude Derian, *The Failed Promise of Nuclear Power: The Story of Light Water* (New York: Basic Books, 1978), were two of the most balanced surveys of the field published as the controversy heated up. Del Sesto provides a particularly thorough account of the origins of opposition to nuclear power. The best monograph on the origins of opposition at one site is Dorothy Nelkin's *Nuclear Power and Its Critics* (Ithaca, N.Y.: Cornell University Press, 1971).

More recently, political scientists, sociologists, and cultural historians have probed these questions from a number of valuable perspectives. See, for instance, Lee Clarke, "The Origins of Nuclear Power: A Case of Institutional Conflict," *Social Problems* 32, no. 5 (June 1985): 474–87; Rebecca S. Lowen, "Entering the Atomic Power Race: Science, Industry, and Government," *Political Science Quarterly* 102, no. 3 (Fall 1987): 459–79; James R. Temples, "The Politics of Nuclear Power: A Subgovernment in Transition," *Political Science Quarterly* 95, no. 2 (Summer 1980); Joseph G. Morone and Edward J. Woodhouse, *The Demise of Nuclear Energy? Lessons for Democratic Control of Technology* (New Haven, Conn.: Yale University Press, 1989); John L. Campbell, *Collapse of an Industry: Nuclear Power and the Contradictions of U.S. Policy* (Ithaca, N.Y.: Cornell University Press, 1988); Elizabeth Rolph, *Nuclear Power and the Public Safety: A Study of Regulation* (Lexington, Mass.: Lexington Books, 1979); Paul Boyer, *By the Bomb's Early Light: American Thought and Culture at the Dawn of the Atomic Age* (New York: Pantheon, 1985); Spencer R. Weart, *Nuclear Fear: A History of Images* (Cambridge, Mass.: Harvard University Press, 1988).

have neglected to do, a far different picture emerges than the politically circumscribed portrait painted by Bell and others. From the start, experts and administrators were forced to confront the political landscape in which they operated. This required keen political skills. Initially, the uneasy alliance between administrator and expert held. But its very success soon generated a new set of problems. The two most difficult problems faced by the Atomic Energy Commission were in fact a product of the changing nature of politics in the proministrative state.

Ultimately, commercial nuclear power foundered on the twin hurdles of insufficient demand for the product that its experts produced and the loss of the experts' public authority. The former was always a problem, plaguing even the first generation of nuclear experts and administrators. It was the direct product of expert agenda-setting. Support-building pushed experts and administrators into politics. Not surprisingly, it politicized the proministrative forum. It was this crucial support-building function – essential to esoteric programs seeking funding in a pluralist democracy – that the New Class theorists initially missed.[64]

David Lilienthal, the Atomic Energy Commission's first chair, understood the support-building problem well. Traditionally, powerful economic interest groups had initiated policies – working closely with congressional committees and administrative agencies to weld a wedge of support so powerful that political scientists referred to these relationships as "iron triangles."[65] Lilienthal had presided over the institutionalization of support at the Tennessee Valley Authority that altered this traditional pattern: the chief initiator of policy was the agency itself – constituency building its major subsequent task.[66]

Again, at the Atomic Energy Commission, Lilienthal sought to create what in the past had generally been one of the preconditions for establishing a program in the first place – interest group support. After broader appeals failed to mobilize a public that didn't particularly care how its electricity was generated, Lilienthal and his successors labored to build an iron-triangle arrangement. That iron triangle would anchor the commission to the nuclear industry and the Joint Committee on Atomic Energy. Although a facsimile of such an arrangement had indeed been manufactured by the early 1960s, its post hoc construction and

---

64 Ironically, it was the discovery that experts did harbor political agendas, and that those agendas often were at odds with New Class theorists, that was a crucial catalyst in the transition to neoconservatism for many of these scholars. Unfortunately, many neoconservatives attributed these agendas to conspiracy-like power grabs as opposed to the political adaptations necessary to pursue policy objectives in the proministrative state that they were. For a good example of the former, see Carolyn Weaver, *The Crisis in Social Security: Economic and Political Origins* (Durham, N.C.: Duke University Press, 1982).

65 For a good summary of the most recent scholarship on iron triangles and issue networks, see Thomas L. Gais, Mark A. Peterson and Jack L. Walker, "Interest Groups, Iron Triangles and Representative Institutions in American National Government," *British Journal of Political Science* 14 (April 1984): 161–85.

66 The classic work on Lilienthal's support-building activities at the TVA is Philip Selznick, *TVA and the Grass Roots*.

history of expert-driven agendas were betrayed by one telltale characteristic: There was still no demand for its product in the civilian market. Utilities, not to mention rate payers, found little reason to go nuclear. If creating programs that found few clients was a legacy the state inherited from its experts, publicly promoting programs was a skill the experts honed while working for the state. But promotion sometimes led to inherently contradictory objectives that eroded the consensus nuclear experts had hammered out for public consumption.

Just as nuclear experts finally began to establish some visible demand for their programs they confronted a still more serious challenge: the loss of authority. As with the rise of experts in the federal government, the trend extended well beyond nuclear power. By the mid-1970s, a good number of professional journals were asking what had become of the authority of experts.[67] The answers they and a number of scholars offered referred to recurring bouts of deep-rooted American anti-intellectualism (the theme Richard Hofstadter had popularized in his Pulitzer prize–winning *Anti-intellectualism in America*) or more specific political events, such as Vietnam and Watergate, that eroded the authority of anyone associated with the "establishment."[68] At the same time, powerful social movements such as civil rights and the New Left pressed claims to democratic participation.[69]

The sharpest confrontation between experts and democratic participation was over the environment. *Beauty, Health, and Permanence: Environmental Politics in the United States, 1955–1985* is the most thorough history of this wide-ranging debate. One of the "central dramas" of Hays's powerful account is the struggle between "one set of perspectives based on the occurrences of daily life in distinctive geographical settings" and another set of perspectives "arising out of professional expertise." Hays has produced a masterful account, analyzing each side's perspective well, elaborating on the changing values of a wealthier consumption-oriented

---

67 See, for instance, "Decline in Science Support," *Science and Public Policy* 2, no. 8 (September 1975): 378–9; F. J. Ingelfinger, "Deprofessionalizing the Profession," *New England Journal of Medicine* 294, no. 6 (February 5, 1976): 334–5; Amitai Etzioni, "Public Views of Scientists," *Science* 181 (September 1973): 1123. A good summary of the declining faith in science literature is contained in Kenneth Prewitt, "The Public and Science Policy," *Science, Technology and Human Values* 7, no. 39 (1982): 5–7.

68 Even a sophisticated and recent analysis of technical debates like Dianna B. Dutton's *Worse than the Disease: Pitfalls of Medical Progress* (Cambridge: Cambridge University Press, 1988) continues to place the primary emphasis on the Watergate/Vietnam explanation. For an equally thoughtful account of the decline of expert authority, which also relies on the Vietnam/Watergate explanation, see Henry Aaron, *Politics and the Professors: The Great Society in Perspective*: (Washington, D.C.: Brookings, 1978).

69 Even Daniel Bell acknowledged the significance of participatory democracy and the tension between it and expert models of decision making. As he put it, "What is evident, everywhere, is a society-side uprising against bureaucracy and a desire for participation....To a considerable extent, the participation revolution is one of the forms of reaction against 'professionalization' of society and the emergent technocratic decision-making of a post-industrial society" (*Coming of Post-Industrial Society*, pp. 365–6). For a representative synthetic interpretation of postwar history that emphasizes these sorts of broad social movements, see William H. Chafe, *The Unfinished Journey: America since World War II* (New York: Oxford University Press, 1986).

postwar society concerned with issues affecting quality of life, and recognizing the specialized nature of expertise.[70]

Yet, so polarizing is the people-versus-experts drama, rarely do these two protagonists share so much as a conversation. In *Beauty, Health and Permanence* the environmental experts share neither social spheres nor values with the rest of the community. This, according to Hays, "led to their detachment from the public and their alignment with other scientific, technical, and managerial institutions in their strategies for action and defense."[71]

This bipolar framework – shared by virtually all scholarship on expert policymaking – obscures some of the most crucial elements in the political process.[72] Just as Hays opened up new vistas by questioning the liberal dichotomy that polarized businessmen and Progressive reformers, I hope to revise the scholarly framework used to consider relations between experts and the public.[73] Nuclear power provides rich material for reconsidering the source of experts' declining authority.

Chapter 8 argues that it was the evolution of prominent administrative politics that best explains the declining authority of nuclear experts.[74] Specialization – both bureaucratic and professional – was one significant contributor to the decline of authority.[75] The scope of debate was also inexorably broadened as the experts

70 Samuel P. Hays, *Beauty, Health, and Permanence: Environmental Politics in the United States, 1955–1985* (New York: Cambridge University Press, 1987), pp. 1–12, 13–39, 329–62, and 458–90; quotation from p. 7. See also Hays, "Political Choice in Regulatory Administration," in Thomas K. McCraw, ed., *Regulation in Perspective: Historical Essays* (Cambridge, Mass.: Harvard University Press, 1981), pp. 124–54; and "The Structure of Environmental Politics since World War II," *Journal of Social History* 14, no. 4 (Summer 1981): 719–38.

71 Hays, *Beauty*, p. 490.

72 A good example of and summary of this literature is Guy Benveniste, *The Politics of Expertise*, 2nd ed. (Berkeley: University of California Press, 1977). For a more recent example of this all-pervasive bipolar framework, see David Dickson, *The New Politics of Science*, 2nd ed. (Chicago: University of Chicago Press, 1988), which analyzes debate over technical issues in a "technocratic" versus "democratic" framework. (See particularly ch. 5.)

73 Hays challenged the Progressive business dichotomy in *Conservation* and in "The Politics of Reform in Municipal Government in the Progressive Era," *Pacific Northwest Quarterly* (October 1964): 157–69.

74 Clearly, shocks to the political system such as Vietnam and Watergate should not be ignored when seeking to explain the declining authority of experts housed in government institutions. David Halberstam, *The Best and the Brightest* (New York: Random House, 1972), for instance, is a classic example of how Vietnam damaged the authority of both experts and government institutions. Scholars, however, have failed to consider factors endemic to the symbiotic relationship between expertise and the administrative state established after World War II. These are the factors that provide the most serious lasting challenge to the authority of experts.

75 There is a significant literature that discusses the breakdown of public consensus within the professions. See, for instance, Thomas Haskell, ed., The *Authority of Experts: Studies in History and Theory* (Bloomington: University of Indiana Press, 1984), pp. xiii–iv; Aaron, *Politics*, pp. 155–9; Walter Heller, "What's Right with Economics," *American Economic Review* 65 (March 1975): 1–3; Barbara Culliton, "Science's Restive Public," *Daedalus* 107 (Spring 1978): 149; Christopher Hohenemser, Roger Kasperson, and Robert Kates, "The Distrust of Nuclear Power," *Science* 196 (April 1977): 25–34. As Harvey Brooks has pointed out, part of the problem has been caused by the powerful new techniques that the professions have developed to identify previously undetected problems (Brooks, "Technology: Hope or Catastrophe?" *Technology in Society* 1 [1979]: 7).

developing nuclear power tackled complex agendas that cut across professional disciplines.[76] In organizational terms, these issues spanned agency and federal jurisdictions as well. As each new discipline was drawn into the discussion, its representatives – now fused to an institutional base – brought with them their agency as well as state, local, or federal allegiance. They also brought the varied perspectives that such amalgams of discipline and political institutions tended to produce. Committed to particular issues, these networks of officials and consultants further broadened the debate, and extended it to the full range of forums available in America's decentralized, pluralist political system, eventually engaging the broad social movement spawned by environmentalists.[77]

With the triumph and elaboration of the prominstrative state, political cleavages no longer cut neat swaths between the people and the experts. Rather, an intricate web of issue networks linked specialized experts in a host of subdisciplines to a variegated cross section of public representatives, ranging from state and local agencies to mass-based citizens groups. That the highly exclusive and insulated conditions in which commercial nuclear power began following World War II ultimately blossomed into a policymaking arena characterized by such

---

76 The concept of expanding the scope of debate is based on E. E. Schattschneider's discussion of "managing the scope of conflict" (*The Semisovereign People: A Realist's View of Democracy in America* [New York: Holt, Rinehart & Winston, 1960]; see especially ch. 1, "The Contagiousness of Conflict"). As Schattschneider puts it: "So great is the change in the nature of any conflict likely to be as a consequence of widening involvement of people in it that the original participants are apt to lose control of the conflict altogether" (p. 3).

John L. Campbell adds a valuable refinement to Schattschneider, noting that the policymaking stage in America tends to be far more insulated than the implementation stage (*Collapse*, ch. 5). This certainly applies in the case of nuclear power, as Campbell deftly demonstrates. Like other scholars, however, Campbell frames his analysis in terms of the people versus the experts, thus missing the structural reasons behind the expanded scope of debate and conflict within the expert community.

77 On issue networks, see "Issue Networks and the Executive Establishment," in Anthony King, ed., *The New American Political System* (Washington D.C.: American Enterprise Institute, 1978). Contrasting issue networks – far more fluid, often ideologically driven alliances that cut across numerous programs – to the older, iron-triangles notion that presumes "small circles of participants who have succeeded in becoming largely autonomous," Hugh Heclo notes: "Issue networks, on the other hand, comprise a large number of participants with quite variable degrees of mutual commitment or dependence on others in their environment; in fact it is almost impossible to say where a network leaves off and its environment begins" (pp. 102–3). He adds: "More than mere technical experts, network people are policy activists who know each other through the issues." And: "It is the issue network that ties together what would otherwise be the contradictory tendencies of, on the one hand, more widespread organizational participation in public policy and, on the other, more narrow technocratic specialization in complex modern policies" (pp. 119–21). As Heclo has noted, issue networks, because they make policy more complex, because they make consensus more difficult, and because they provide few incentives for compromise while attracting "true believers" not prone to compromise, increase the likelihood of political stalemate.

The best comparative studies of the relative impact of political decentralization on the course of commercial nuclear power are Dorothy Nelkin and Michael Pollak, *The Atom Besieged: Antinuclear Movements in France and Germany* (Cambridge, Mass.: MIT Press, 1981); and Campbell, *Collapse*. The best description of the pluralist science policymaking mechanism in the United States is A. Hunter Dupree, "Central Scientific Organization in the United States Government," pp. 453–69.

kaleidoscopic variety, suggests that relations between the experts and the people in other domestic programs hardly proceeded along bipolar paths.

Chapter 9 applies the pattern that I have detected in the case of nuclear power to a broad range of domestic policies in postwar America. It confirms that insulated expertise was crucial to building political support for prominministrative agendas. But it also demonstrates the universal tendency of those insulated communities to splinter and of their memberships to realign with a variety of more populist political bases, ultimately changing the trajectory of policy development and eroding the political clout of expertise. In policy areas as disparate as the Strategic Defense Initiative and national programs to reduce drunk driving, the pattern obtained.

The inexorable expansion of expert debate in America's porous political environment was perhaps the most crucial prerequisite to public participation. No matter how insulated the origins of policy, nor how self-conscious the effort to cement expert consensus, access to the crucial ingredients in prominministrative politics – professionals and administrative bases – was simply too easy to procure by the late 1960s to prevent the trend toward public participation. How ironic that the unrivaled (and unexamined) political success of America's first generation of prominministrators, who for the most part were committed to restricting public debate, virtually ensured this dramatic change in the nature of expert debate and public participation.

2

The promise of the proministrative state: nuclear experts and national politics, 1945–1947

That World War II and the ensuing cold war heated up the courtship between professionals and the federal government does not mean that external factors alone accounted for the new symbiotic relationship that emerged by 1950 – what I have called the "proministrative state." The foundation had been laid in the interwar years. In the Progressive Era, four barriers prevented the flirtation from being consummated. By the eve of World War II, two of these obstacles had almost been surmounted.

Immaturity was the first to fall. In organizational terms, maturity could be measured by "state capacity." In profession-building terms, it could be measured by professional autonomy. The interwar years were a crucial period of parallel development for both the professions and the federal government. The millions of dollars poured into professional development by the voluntary sector and the growth of American research universities nurtured the professions in both the physical and social sciences. The crises of World War I, the Great Depression, and America's role as an economic world power enhanced the federal government's capacity to administer complex programs. Both the professions and the federal government matured in fits and starts. Particularly in the public sector, rapid expansion was often followed by even swifter demobilization. The professions developed more unevenly: those grounded in the physical sciences advanced faster than those drawing on the social sciences. Both the professions and the state were far more mature on the eve of World War II than they had been forty years earlier when Progressive rhetoric called upon experts to serve an expanded state.[1]

A second barrier to the marriage of professionals and the state that was partially surmounted during the interwar years might best be described as "fear of commitment." Even in the heart of the Great Depression basic science's elite

1 For a good example of how "state capacity" can be measured, see Theda Skocpol and Kenneth Finegold, "State Capacity and Economic Intervention in the Early New Deal," *Political Science Quarterly* 97 (Summer 1982): 255–74. On professional autonomy, see Eliot Friedson, "Are Professions Necessary?" in Haskell, ed., *Authority*, p. 10; and for general discussion of sources of professional autonomy, Friedson, *Professional Powers: A Study of the Institutionalization of Formal Knowledge* (Chicago: University of Chicago Press, 1986).

continued to resist federal funding: they hoped to preserve the set of national voluntary institutions that antedated the central government's eventual role in the production and development of professionals. For their part, federal administrators hesitated to commit their agency's future to professionals insensitive to the practical political concessions required to maintain congressional and interest group support.[2]

Ultimately, it was the resources of the state – both financial and managerial – that the professionals could not do without; it was the prestige and the problem-solving capability of the professionals that tempted the state. Again, World War I and the Great Depression provided tastes of these advantages for each would-be partner.

By the eve of World War II these parallel paths had begun to converge. The professions, aided by foundations, had established strong beachheads in research universities. Even after retrenchment in the late 1930s, Americans embraced an enhanced central government. Both experts and administrators inched toward each other as the cost of research skyrocketed and the expectations of the public grew. At the same time, however, Roosevelt's 'brain trust' hardly solved the problems of overproduction or unemployment. On the eve of World War II, and in the very field where fusion after the war would become most pervasive (both literally and figuratively) – military research and development (R and D) – the paths of experts and the government looked as though they would never meet. The army reduced its research budget, claiming that "the crisis was too serious to wait for research."[3]

World War II pushed the maturing, increasingly committed partners together. As we will see in the case of the physicists, the Manhattan Project extended the federal government and its use of expertise far beyond the wildest dreams of Progressive state builders. It drew upon a well-developed profession, attracting some of the leading physicists in the world. The crucial nature of their mission, the esoteric basis of their knowledge, and the secrecy of their quest brought them unparalleled resources. The federal government received in return the total commitment of some of the world's most respected experts. While some scientists bridled at wartime controls and secrecy, given the astounding rise in the cost of "big science," very few felt that voluntary techniques after the war would suffice.

The dramatic breakthroughs produced by expert-federal fusion was the war's greatest legacy, demonstrating in highly public fashion the potential that this partnership offered to postwar America. The war proved that expertise – directed and funded by the federal government – could produce in a crisis. In the fields of military security and economic and social planning, experts put their theories into action, and events convinced many Americans that experts could solve large

2 Kargon and Hodes, "Karl Compton," pp. 301–18; Davis and Kevles, "The National Research Fund, pp. 207–20; Auerbach, "Scientists in the New Deal"; Kevles, *Physicists*, ch. 17.

3 Dupree, *Science in the Federal Government*, quoted in Dickson, *The New Politics of Science*, p. 117.

problems. Massive deficit spending during the war, for instance, accomplished what all of the New Deal programs had been unable to achieve: It lowered unemployment to politically acceptable levels, just as Keynesian economists had argued it would. By far the most decisive shift in the federal government's role following the war was its sponsorship of science and technology. Here, the war effort produced such highly publicized triumphs as the atomic bomb, radar, and penicillin.[4]

The mobilization model remained attractive long after the fighting in Europe and Japan stopped. Writing to Dwight D. Eisenhower shortly before the 1952 election, a longtime friend forwarded the general a plan that, if implemented, would head off the appeal of socialized medicine. Entitled "War on Untimely Death," the plan sought to eliminate the "Great Killer" diseases. Designed to "smash the atoms" of disease, Edward Everett Hazlett wrote, the bomb project would serve as a model for this war on disease. "The attack on disease seems no more likely to fail than did that on the atom. It has, in addition, the spiritual advantage of being a campaign to save life and not to take it."[5]

Nevertheless, the fusion of central state and professional expert might well have dissolved once again following World War II had a fourth condition failed to materialize – demand for the policies that this new alliance generated. In the past, experts housed in the federal government had worked together most successfully when a narrowly drawn set of economic interests demanded services. Whether railroads and eventually truckers via the Interstate Commerce Commission, or the financial community through the Securities and Exchange Commission or Federal Reserve Bank, interest group demand had been crucial to the institutionalization of expertise in the federal government. The proministrative state

---

4 The best work on the "crisis" attributed to the shortage of highly trained experts, the major role played by the federal government in both publicizing and responding to this shortage, and the ultimate inflation of categories of expertise considered crucial to the state, is Newman's dissertation "The Era of Expertise."

On the professionalization of the postwar state, see Mosher, *Democracy and the Public Service*, pp. 99–133. Don K. Price comments on the trend of scientists moving into administrative positions in *The Scientific Estate*, pp. 66–71.

For the influence of World War II on the acceptance of Keynesian economics, see Herbert Stein, *The Fiscal Revolution in America* (Chicago: University of Chicago Press, 1969), pp. 169–70. Robert M. Collins, in *The Business Response to Keynes, 1929–1964* (New York: Columbia University Press, 1981), pp. 115–72, finds a more divided response, in the immediate wake of the war. But by the mid-1950s, Keynesians had carried the day.

On the wartime mobilization model, see Kevles, *Physicists*, ch. 20; J. Stefan Dupre and Sanford Lakoff, *Science and the Nation: Policy and Politics* (Englewood Cliffs, N.J.: Prentice Hall, 1962), pp. 9–11, and Price, *Scientific Estate*, pp. 32–8, document wartime scientific breakthroughs. On the Manhattan Project's impact on funding for science, see Dickson, *The New Politics of Science*, pp. 5–6. As Robert Gilpin put it in *American Scientists and Nuclear Weapons Policy* (Princeton, N.J.: Princeton University Press, 1962), p. 25, "those wartime projects initiated a military revolution which changed irrevocably the relationship of science and war." For a good comparison to the impact of science on World War I, see Gilpin, pp. 10–11.

5 Edward Everett Hazlett to Eisenhower, September 24, 1952, Dwight D. Eisenhower Personal Papers (1916–53), Presidential Library, Abilene, Kans.

– whether offering electricity too cheap to meter, or the eradication of poverty – promised services that outstripped organized demand for them.

In the absence of such well-articulated demand following World War II (or, in the case of military expenditures, even in the face of it) the powerful tendency in American politics toward decentralization and a weak central state triumphed when Truman slashed spending for the army. Defense expenditures plummeted from $81.6 billion in fiscal 1945 to $13.1 billion for fiscal 1947 before begin-, ning to rise again. Eisenhower conducted a rearguard battle, ultimately railing against the military–industrial complex in his frustration. Both presidents, however, failed in their quest to limit the scope of the prominstrative state.[6]

A permanent crisis – the cold war – cemented demand for the kinds of services provided by the prominstrative state. Truman and Eisenhower took actions to ensure that America's professionals remained permanently mobilized – pronouncing the "Truman Doctrine," providing virtually unlimited funding for it by implementing NSC 68, and responding to Sputnik. These actions spoke far louder than both presidents' genuine longings for a simpler, more traditional central government. Funding for federal research – which had totaled approximately $97 million in 1940 – jumped to $1.6 billion by 1945 and, after leveling off in 1948 at $865 million, climbed to $2.1 billion by 1953. Though frequently remembered now for his 1961 farewell address warning that the nation's public policy might become the captive to a "scientific–technological elite," General Eisenhower was more concerned in 1946 about "Scientific and Technological Resources as Military Assets." Distilling these views in an April 30 memo with that title, Ike argued that the nation's security required civilian resources that were crucial in an emergency "be associated closely with the activities of the Army in time of peace. [The army] must establish definite policies and administrative leadership which will make possible even greater contributions from science, technology, and management than during the last war."[7]

James Conant, president of Harvard University, and formerly the second-ranking civilian overseeing the Manhattan Project, commented on the startling change in a secret address to military and government officials in 1952. Before the atomic bomb, Conant noted, technological conservatism was the military's chief stumbling block. Military officials had been "perhaps unduly slow in some cases to take up new ideas developed by civilian scientists." But in the wake of

---

6 On military spending, see John Lewis Gaddis, *Strategies of Containment: A Critical Appraisal of Postwar American National Security policy* (Oxford: Oxford University Press, 1982), p. 23. On Eisenhower's battle, see Charles C. Alexander, *Holding the Line: The Eisenhower Era, 1952–1961* (Bloomington: Indiana University Press, 1975).

7 U.S. National Science Foundation, *Federal Funds for Science: III. The Federal Research and Development Budget Fiscal Years 1953, 1954, and 1955* (Washington D.C.: U.S. Government Printing Office), p. 15. For Eisenhower's farewell address, see *New York Times*, January 18, 1961. Quotation from Eisenhower to Directors and Chiefs of War Department, April 30, 1946, in Louis Galambos, ed., *The Papers of Dwight David Eisenhower*. Vol. 7: *Chief of Staff* (Baltimore: Johns Hopkins Press, 1970–9), pp. 1046–7.

the bomb and the cold war something akin to "the old religious phenomenon of conversion" had struck. "As I see it now," Conant continued, "the military, if anything, have become vastly too much impressed with the abilities of research and development." Underscoring the professional–federal union that atomic and cold wars had sealed, Conant noted wryly that "some of your colleagues have become infected with the virus that is so well known in academic circles, the virus of enthusiasm of the scientist and the inventor."[8]

The cold war stimulated demand for centralized, expert responses to America's threatened national security. It was now the experts who often identified specific problems that required federal action. As the cold war spread from military to economic and social fronts, experts in the social sciences and even the humanities linked security to solutions offered by a broader range of disciplines. In the wake of Sputnik, policies as far afield from military security as civil rights and highway construction benefited from America's mass passion for national security. Even before Sputnik, however, educators claimed an increasing share of the national security budget by joining the drive for civil defense. Bert the Turtle warned schoolchildren that they not only had to learn to cross the street safely and know what to do in case of fire – they now had to know what to do in case of atomic attack. Americans turned directly to experts to soothe jittery nerves: approximately one out of six white middle-class Americans consulted a professional for emotional or marriage problems by the mid-1950s. As one attitudinal survey put it, "Experts took over the role of psychic healer.... They would provide advice and counsel about raising and responding to children, how to behave in marriage, and what to see in that relationship.... Science moved in because people needed and wanted guidance."[9]

8 James B. Conant, "The Problems of Evaluation of Scientific Research and Development for Military Planning," speech to the National War College, February 1, 1952, quoted in James G. Hershberg, "Over My Dead Body: James B. Conant and the Hydrogen Bomb," unpublished paper presented to The Conference on Science, Military and Technology, Harvard/MIT, June 1987, revised draft, p. 50.

9 For case studies on how the cold war stimulated demand for centralized expertise in space and weaponry, see Walter A. McDougall, *The Heavens and the Earth: A Political History of the Space Age* (New York: Basic Books, 1985); Gregg Herken, *Counsels of War* (Oxford: Oxford University Press, 1987), and *The Winning Weapon: The Atomic Bomb in the Cold War, 1945–1950* (Princeton, N.J.: Princeton University Press, 1988); Herbert York, *The Advisors: Oppenheimer, Teller and the Superbomb* (San Francisco: Freeman, 1976). James G. Hershberg's "James B. Conant, Nuclear Weapons and the Cold War, 1945–1950," unpublished Ph.D. diss., Tufts University, 1989, is an excellent study of one of the nation's leading scientist administrators and his response to cold war weaponry. See also Dickson, *The New Politics of Science*, pp. 26, 119–23. Don K. Price's *Scientific Estate* is a testimonial to the increased authority and political clout of scientists in postwar America. Price correctly attributes to the new role of science in the federal government several major changes, including the narrowing gap between the public and private sectors, greater autonomy for and more initiative by executive agencies as a result of this expertise, and the adaptation of the research contract, leading to federal support for open-ended research (pp. 15–21, 36–40, 46–51). Robert Gilpin contrasts the revolution in scientific participation in weapons development during World War II and after, to the case of World War I in *American Scientists and Nuclear Weapons Policy*, pp. 9–12.

The marriage between the professions and the federal government, announced as early as the Progressive Era, had finally been consummated. Drawing upon a far more sophisticated cognitive base, confident, convinced that their very survival required federal assistance, respected for their dramatic accomplishments, and in demand everywhere, experts embraced federal support. Federal administrators, eager to capitalize on the prestige of experts, and anxious to tap expert skills to solve a host of problems that, particularly in a crisis context and given the government's wartime success, were seen as federal responsibilities, reciprocated. The newlyweds turned their attention to building their prominstrative nest.

Their union, like millions of others after the war – and for some of the same reasons, including prosperity, wartime abstinence, and great faith in the future – soon produced heirs in larger quantities than ever before. The permanent crisis dwarfed the New Deal's contribution to state building. Federal expenditures more than doubled during the 1950s, increasing from $42 billion to $92 billion by 1960. Federal expenditures also consumed a larger share of the nation's gross national product. Federal expenditures accounted for 16.1 percent of GNP in 1950, and 18.5 percent of GNP by 1960. The explosion of federal administration and oversight was even more dramatic than the rise in expenditures. Agencies directly related to national security proliferated. The National Security Act of 1947 created a new Department of Defense as well as a National Security Council and Central Intelligence Agency. The federal government established competing bureaucracies ranging from the Agency for International Development to an Office for International Security Affairs within the Department of Defense. The prominstrative state moved most forcefully to create an organizational infrastructure for science in postwar America. The Office of Naval Research, the Atomic Energy Commission, the National Science Foundation, and a greatly expanded National Institutes of Health soon housed and funded thousands of America's most sophisticated experts. Nor were the social sciences left behind. Established by the Employment Act of 1946, the Council of Economic Advisers, for instance, institutionalized economics within the White House.[10]

The best survey of the wide range of responses across disciplines is Paul Boyer, *Bomb's Light*; see also Newman, "Era of Expertise." On the impact of Sputnik, see McDougall, *Heavens*, chs. 6 and 7, as well as "NASA, Prestige, and Total Cold War: The Expanded Purview of National Defense," paper delivered to the History of Science Society, Chicago, December 1984, and "Technocracy and Statecraft in the Space Age – Toward the History of a Saltation." For Bert the Turtle, see Federal Civil Defense Administration, "Bert the Turtle Says 'Duck and Cover'" (Washington, D.C., 1950), n.p., quoted in JoAnne Brown, "'A Is for Atom, B Is for Bomb': Civil Defense in American Public Education, 1948–1963," *Journal of American History*, 75 (no. 1, June 1988): 84. The quotation on experts is from Joseph Veroff, Richard A. Kulka, and Elizabeth Douvan, *The Inner American: A Self-Portrait from 1957 to 1976* (New York: Basic Books, 1981), p. 194, quoted in Elaine Tyler May, *Homeward Bound: American Families in the Cold War Era* (New York: Basic Books, 1988), p. 27; for statistics on counseling, see ibid.

10 Statistics on federal growth are from Matthew A. Crenson and Francis E. Rourke, "By Way of Conclusion: American Bureaucracy since World War II," in Louis Galambos, ed., *The New American State*, p. 147. On military–foreign policy bureaucracy, see Charles E. Neu, "The Rise of the National Security Bureaucracy," in Galambos, ed., *New American State*, pp. 85–108; Anna

The professions grew at an equally extraordinary rate. Universities responded to the demand for highly skilled experts by minting record numbers of doctorates. The number of graduate degrees awarded between 1940 and 1959 doubled and quadrupled between 1950 and 1970, with degree programs in the most specialized fields leading the way. American universities conferred 3,277 doctorates in 1940, 3,940 in 1948, and 6,535 by 1950. Nearly 10,000 doctorates were awarded in 1960. With this burst of higher learning, the theoretical drove out the applied. Math and science began to dominate engineering curricula, for instance. Training for research replaced preparation for patient-care at most prestigious medical schools. As Frank Newman put it in his Ph.D. dissertation, "Increasingly, faculty saw their most important role as creating new researchers – a sort of institutionalized cloning." The newly minted professionals led the drive toward federal support for graduate training. Within the prominustrative state, it was these research interests that also powered the drive toward a host of new federal policies.[11]

At the heart of the new 'prominustrative' state lay a symbiotic relationship between expert and the federal government. The federal government now actively produced and developed experts. Experts, on the other hand, just as in the early days of foundation funding, actively defined what their benefactor's policy agenda should be. By 1950, however, there was a significant difference between foundations in the interwar years and the federal government. Whereas the management of grants was relatively centralized within the major foundations, the equivalent mechanisms in the federal government were scattered among hundrds of agencies and dozens of oversight committees, and were at least potentially subject to pressure from thousands of interest groups. Under these circumstances, it was impossible to re-create the kind of direction that Warren Weaver was able to achieve in the field of biomedicine at the Rockefeller Foundation. Nor was it so easy to distinguish between grantor and grantee: the federal administrators responsible for such decisions increasingly were professionals produced by this new relationship.

The relationship between federal administrators and professionals was hardly

Kasten Nelson, "President Truman and the Evolution of the National Security Council," *Journal of American History* 72, no. 2 (September 1985): 360–78; Bert A. Rockman, "America's Departments of State: Irregular and Regular Syndromes of Policy Making," *American Political Science Review* 75 (1981); and Crenson and Rourke, "Conclusion," p. 142. On new and expanded institutions that housed science, see Dickson, *The New Politics of Science*, intro. and ch. 1; Price, *The Scientific Estate*; Daniel S. Greenberg, *The Politics of Pure Science* (New York: New American Library, 1967); Detlev W. Bronk, "Science Advice in the White House: The Genesis of the President's Advisers and the National Science Foundation," *Technology in Society* 2, nos. 1 and 2 (1980): 245–56; Margaret W. Rossiter, "Science and Public Policy since World War II," *Osiris* 1 (1985): 273–94; Kevin Michael Saltzman, "Countdown to Sputnik: The Institutionalization of Scientific Expertise in the White House, 1945–1957," undergraduate honors thesis, Department of History, Harvard University, 1988, ch. 3. On the social sciences, see Samuel Z. Klausner and Victor M. Lidz, eds., *The Nationalization of the Social Sciences* (Philadelphia: University of Pennsylvania Press, 1986).

11 Newman, "Era of Expertise," p. 55 for statistics, p. 59 for quotation, and ch. 3 for transformation of academic community vis-à-vis federal support.

static. Over time, funding agencies increased their control over the research agenda. We must understand this often overlooked tension between organizational mission and professional agenda if we are accurately to portray postwar policy development.[12]

Beneath the veneer of expert–agency relations, another powerful dynamic – spurred by specialization – operated. Massive infusions of federal dollars, expanded research universities, and the emergence of a knowledge-based economy accelerated professional specialization. Subdisciplines proliferated. Whether seeking or saddled with new responsibilities, the organizational infrastructure also specialized at a breathtaking pace. Nor was organizational articulation confined to the federal bureaucracy. Eventually Congress, state governments, and even local administration specialized to keep pace. As each political actor acquired access to its own expertise a tripartite matrix built on a specialized cognitive base, organizational perspective – now divided even within the same agency – and jurisdiction emerged. The interplay of expertise and these structural factors created its own dynamic – a process that ultimately shattered the nation's confidence in its experts. By the mid-1970s, expertise had been relegated to a necessary, but no longer decisive, condition for political success.

As surely as Progressive reform and rhetoric altered the distributive style of politics that it in part supplanted, proministrative politics revised interest group politics in post–World War II America. This is not to argue that distributive or interest group politics disappeared. Just as the distributive style of politics that pervaded the nineteenth-century system of "courts and parties" was built into the administrative state that emerged by the 1920s, elements of both distributive and interest group politics were built into the very foundation of the proministrative state that was constructed in the wake of World War II. Like their predecessors in the Progressive Era, post–World War II administrators and professionals had to build political support in order to acquire the autonomy necessary to pursue their distinct approach to problem solving. Precisely because the kind of politics they tended toward excluded public participation, general public support for and belief in expertise was crucial to their success.

Had that support not existed in the wake of World War II, the marriage of experts and large-scale federal organization would have been annulled before the honeymoon was over. Nor could massive public support – funded by steep increases in the income tax – be engineered by clever conspirators. Far from conspiratorial, the marriage of federal administrative capacity to professional expertise could only occur with tremendous popular support. Support poured in because the professions had matured, and during World War II demonstrated in dramatic fashion the powerful results that expert–Federal fusion – which both partners were now ready to embrace – could deliver. When the crisis continued

---

12 The best sketch of this dynamic is Don K. Price's "Endless Frontier or Bureaucratic Morass?" *Daedalus* 107, no. 2 (Spring 1978): 75–92.

into the cold war, the American public gladly embraced the prominstrative state: the nation's very survival depended on it. No group was more heralded than the nuclear physicists, and no honeymoon any sweeter than the early days of the Atomic Energy Commission's life. The best indicator of the esteem in which experts and administrative agency were held was the extraordinary degree of autonomy that the Atomic Energy Commission initially was granted.

A DRAMATIC INTRODUCTION

"Sixteen hours ago an American airplane dropped one bomb on Hiroshima, an important Japanese army base. That bomb had more power than 20,000 tons of T.N.T. It had more than two thousand times the blast power of the British 'Grand Slam' which is the largest bomb ever yet used in the history of warfare." These were the opening lines of Truman's press release of August 6, 1945. Through news reports of this statement and thousands of stories broadcast and printed soon after, experts entered the living rooms of American households.[13]

Over the ensuing forty years, the nuclear industry, utilities using commercial power, and the Atomic Energy Commission would spend millions of dollars on efforts to separate the image of peaceful nuclear power from nuclear weapons. This task proved difficult. The roots of commercial nuclear power, the men who developed it, and the public's earliest and most significant perceptions of it were integrally linked to the bomb and to the circumstances in which it was unleashed.[14]

Americans approved of using the bomb against Japan, and credited it favorably with ending the war. Although historians have argued recently that surrender may have come without an invasion or the bomb, most Americans at the time did not question Truman's decision to bomb Hiroshima and Nagasaki. To them, the bomb was the decisive factor in ending the war without risking further American lives. The September 17 issue of *Life* summed things up with the banner headline next to its picture of the mushroom cloud: "What Ended the War?: The Atomic Bomb."[15]

13 *New York Times*, August 7, 1945, p. 4.
14 On the need to replace the image of weapons with that of peaceful applications in the 1950s and 1960s, see Michael L. Smith, "The Physicist's New Clothes: The Atom in Postwar Culture," paper presented to the Society of the History of Technology, November 3, 1984; on the AEC's earliest efforts, see Boyer, *Bomb's Light*, pp. 293–8.
15 "What Ended the War," *Life*, September 17, 1945, p. 37. For a discussion of American options for ending the war, see Barton J. Bernstein, "The Atomic Bomb and American Foreign Policy, 1941–1945: An Historiographical Controversy," *Peace and Change* 11, no. 1 (Spring 1974): 1–16. Asked by Gallup in August 1945 whether they "approve or disapprove of using the new atomic bomb on Japanese cities," 85 percent of those questioned expressed approval (George H. Gallup, *The Gallup Poll: Public Opinion 1935–1971*, vol. 1 [New York: Random House, 1972]). The best summary of public reaction to the use of atomic weapons on Hiroshima and Nagasaki is Michael J. Yavenditti, "The American People and the Use of Atomic Bombs on Japan: The 1940s," *The Historian* 36 (February 1974): 224–47.

Everything about this new force seemed larger than life. In presenting it for the first time to the American public, Truman stated that "the basic power of the universe ... [t]he force from which the sun draws its power has been loosed against those who brought war to the Far East." Truman's description sounded like the product of a committee, which in fact it was, particularly compared to accounts that soon followed in the press. With so little information to go on, yet such a highly charged topic, journalists relied on a familiar stock of images of sudden death and apocalyptic power. One representative newsreel reported that Hiroshima was annihilated by a "cosmic power...hell-fire...described by eyewitnesses as Doomsday itself!"[16]

What most Americans thought about the atomic future was less clear. To judge from the voluminous correspondence received by the Senate Special Committee on Atomic Energy, individuals – following the lead of journalists – were struggling to fit this event into a framework that reflected their own past experience and beliefs. Thus, one man viewed the bomb as God trying to speak to the people. Another considered it as not much different from adolescents playing with fireworks. Public opinion polls suggest that individuals often held contradictory views. One poll, for instance, reported that 83 percent of those surveyed in September 1945 thought there was a real danger that most of the world's population would be annihilated in an atomic war. Of that same group, however, 52 percent felt that atomic fission would improve the welfare of mankind.[17]

In *By the Bomb's Early Light*, historian Paul Boyer has reviewed American responses to Hiroshima and Nagasaki. He concludes that the dominant response "was confusion and disorientation" coupled with a determination not to surrender to fear. Boyer also points out that with the passage of time, doubts about the advantages of nuclear energy crept into people's responses. In the course of a year, the number of respondents who felt the world would be a better place had declined from 52 percent to 37 percent. The most common justifications for a bleak outlook emphasized the destruction that the bomb had already caused and the possibility that the Soviet Union would soon develop a bomb of its own. Those who were more sanguine about atomic power pointed to its potential to revolutionize international relations and, to a lesser extent, its potential peaceful applications.[18]

16 *New York Times*, August 7, 1945, p. 4.; Spencer R. Weart, *Nuclear Fear*, p. 103; newsreel quotation from Weart, p. 104.

17 Belle Davis Cheatham to Senator McMahon, January 6, 1946, and "American" to "Senator," April 4, 1946, Record Group 46, "Records of the Special Committee of the Senate on Atomic Energy," Misc.–General folder, Box 2, National Archives, Washington D.C.; National Opinion Research Center poll, September 1945, in Hazel Gaudet Erskine, "The Polls: Atomic Weapons and Nuclear Energy," *Public Opinion Quarterly* 27 (Summer 1963): 156, 179.

18 Boyer, *Bomb's Light*, pp. 24–5; NORC public opinion poll, September 1946, in Erskine, "The Polls," p. 179. Elizabeth Douvan and Stephen Withey, "Public Reaction to Non-Military Aspects of Atomic Energy," *Science* 119, no. 3079 (January 1, 1954): 1.

Without sensationalist journalism, it is unlikely that the public would have shown much interest in peaceful uses of the atom. Whereas Manhattan Project officials were quite sober in their assessment of the possibilities, some journalists and authors rushed into print with speculations about peaceful uses of atomic power ranging from weather control to atomically powered automobiles. "Instead of filling the gasoline tank of your automobile two or three times a week, you will travel for a year on a pellet of atomic energy the size of a vitamin pill," wrote David Dietz, science editor of the Scripps-Howard newspapers. "No baseball game will be called off on account of rain in the Era of Atomic Energy." That era was scheduled to arrive within ten to twenty-five years, according to Dietz: "At the moment I am sufficiently optimistic to say that it may prove to be closer to 10 than 25."[19]

In a flash, the bomb seemed to bring together the public's great faith in science and technology epitomized by the consumer goods revolution of the twenties and the fears raised by the destructive use of science in World War I fueled by charges of "technological unemployment" during the thirties. John O'Neill, science editor of the *New York Herald Tribune*, claimed that atomic energy might make it possible for "the human race to create a close approach to an earthly Paradise." Others promised a potential paradise if the world could choose wisely between atomic power's awesome destructive potential and its alluring peaceful applications. William Laurence, a correspondent for the *New York Times* – and the only journalist to witness the first atomic explosion in New Mexico – put the choice to his readers: "Today we are standing at a major crossroads. One fork of the road has a signpost inscribed with the magic word '*Paradise*'; the other fork also has a signpost bearing the word '*Doomsday*.'"[20]

Whether a force for evil or good, atomic fission was not to be separated from the scientists who introduced it; nor were those scientists lone heroes toiling away in musty laboratories, scrambling for resources. Rather, they were members of a vast organization, tightly linked to the military and funded by overflowing federal coffers. Sharing their limelight after the war was General Leslie Groves – the organizational genius behind the Manhattan Project. Groves was a career army man who, as deputy chief of construction for the entire U.S. Army, had built the Pentagon before accepting the promotion to brigadier general in 1942

19 David Dietz, *Atomic Energy in the Coming Era* (New York: Dodd, Mead, 1945), pp. 13, 16, 23. Boyer, *Bomb's Light*, pp. 112, 111, 109. Even with sensationalist journalism, the postwar public could name few specific constructive uses for atomic energy. As one survey of public opinion put it in 1946, "To the general public atomic energy means the atomic bomb" (quoted in Weart, *Nuclear Fear*, p. 162). One indicator of the relative importance of national security concerns as opposed to interest in peaceful applications was the type of question asked by the public pollsters. In the first three years following the war, military-use questions outnumbered peaceful-application questions eighteen to one.

20 O'Neill is quoted in Donald Porter Geddes, ed., *The Atomic Age Opens* (New York: Pocket Books, 1945), pp. 185–6; See also John O'Neill, *Almighty Atom: The Real Story of Atomic Energy* (New York: Ives Washburn, 1945); "Paradise or Doomsday," *Women's Home Companion*, May 1948, p. 33.

in order to head up the secret bomb project. Standing less than six feet tall but weighing more than two hundred and fifty pounds, Groves had learned how to throw his weight around in order to make things happen in Washington. Not a terribly expressive man, Groves responded to his new assignment and the suggestion that he might win the war if it succeeded with the profound comment, "Oh."[21]

Before the war was over, the army and the scientists used almost unbelievable resources – whether measured by the $2 billion spent for the entire project or 5 million bricks laid at Oak Ridge alone – to accomplish their breakthrough. Despite the magnitude of the operation, Groves – who understood construction management, but not nuclear physics – was beholden to a small group of experts for virtually all of the project's major substantive decisions. The Manhattan Project scientists, on the other hand, could not have begun to assemble the vast organizational network required to sustain this secret operation without the federal government's organizational capacity and resources.[22]

Once exposed, the drama and magnitude of these events left few doubters: America's very survival depended on integrating its scientific and technological expertise into the military and adjusting to the latest scientific developments on a continuous basis. The overwhelming importance of science and technology had been demonstrated dramatically. The complexity of issues like nuclear power dictated that questions regarding its ultimate use – questions that were largely political – be answered at least in part by the experts who specialized in understanding such complexities. As one editorial put it, "Mention of atomic energy makes any other noun in the same sentence seem a minor matter.... It was not the army but the civilian scientific laboratory full of 4-fs that sparked this war's revolution in weapons." A leading student of postwar popular culture, Elaine Tyler May, summarized the attitudes of millions following the war, labeling that period the "era of the expert." Americans looked to the experts, May concludes, "to make the unmanageable, manageable."[23]

Scientists, who had played a major role in directing atomic policy behind closed doors during the war, suddenly enjoyed a brief period of direct communication with the mass public. After all, they had a momentary monopoly on the scientific facts surrounding the bomb's development. The news was filled with stories about and interviews with Vannevar Bush, director of the Office of Scientific Research and Development (OSRD), James Conant, chairman of the National Defense Research Committee, and Robert Oppenheimer, the head of Los Alamos Laboratory, where the bomb was built. Truman's terse statement on the bomb was followed shortly by a technical report authored by Princeton

21 Richard Rhodes, *The Making of the Atomic Bomb* (New York: Simon & Schuster, 1986), pp. 425–6.
22 Hewlett and Anderson, *The New World*, p. 153. A more readable account of the bomb project that places it in a diplomatic context is Martin J. Sherwin, *A World Destroyed: Hiroshima and the Origins of the Arms Race* (New York: Vintage, 1987).
23 *Life*, September 17, 1945, p. 40. May, *Homeward Bound*, p. 26.

physicist and future AEC commissioner Henry Smyth. Despite its dense content and title (*A General Account of the Development of Methods of Using Atomic Energy for Military Purposes under the Auspices of the United States Government, 1940–1945*), the Smyth Report quickly became a best seller.[24]

Months before the public debate began, Vannevar Bush and James Conant had pressed Secretary of War Henry Stimson to consider the matter of atomic energy, its future, and the likely public reaction to its introduction. Bush and Conant, in turn, were feeling pressure from scientists in the Chicago laboratory, eager to investigate new applications and consider the postwar role of the atom. Stimson responded by establishing, in May 1945, the Interim Committee to consider plans for the future of atomic energy. Topics included international control and the shape of the postwar atomic development, the need for public statements, and draft legislation. Chaired by Stimson, the committee also included Bush, Conant, Karl Compton, and representatives from the Department of State, the military, and the president's office. At the urging of Conant, the committee named a Scientific Panel consisting of Arthur Compton, Ernest Lawrence, Robert Oppenheimer, and Enrico Fermi to express the scientists' views on the issues. As the official historians of the Atomic Energy Commission recount, Conant recognized that consensus among the experts was crucial: "The Government needed full support from the scientific community. There should be no public bickering among the experts."[25]

This means of officially channeling scientific advice was adequate for the moment. But with the decision to drop the bomb at the end of the war, it became apparent that this small group did not speak for many of the scientists who had labored to build the bomb. In October 1945 three physicists from the Manhattan Project's Chicago laboratory published an article entitled "Atomic Scientists Speak Up." They insisted that scientists did not aspire to political leadership but had to become politically involved with the consequences of atomic power because they had "helped man to make the first step into this new world." The article appealed for international control of the bomb; it argued that the Soviet Union would soon develop atomic power, that there was no defense against the bomb, and that a strategy of defense based on the threat of retaliation would lead to a world dominated by fear and suspicion.[26]

In December 1945, economist Stuart Chase summed up just how visible the experts had become, writing: "Thorstein Veblen in his grave must now be permitting himself a sardonic smile. His technicians and scientists have come

24 On new role in policy for science, see Gilpin, *American Scientists*, pp. 9–10; on sudden notoriety, see Rhodes, *Making*, p. 751. Henry D. Smyth, *A General Account of the Development of Methods of Using Atomic Energy for Military Purposes Under the Auspices of the United States Government, 1940–1945* (Washington D.C.: U.S. Government Printing Office, 1945).

25 Hewlett and Anderson, *New World*, pp. 322–31, 344–5, 353–60; quotation from p. 345.

26 David Hill, Eugene Rabinowitch, and John Simpson, "Atomic Scientists Speak Up," *Life*, October 29, 1945, pp. 45–47. For accounts of the atomic scientist's wartime role, see Alice Kimball Smith, *A Peril and a Hope: The Scientist's Movement in America: 1945–1947* (Chicago: University of Chicago Press, 1965), and Sherwin, *A World Destroyed*.

roaring into their own as the acknowledged and undisputed arbiters of human destiny....For the first time, scientists are developing a social conscience, and even daring to dispute the generals and politicians."[27]

How did America feel about these experts? In the immediate wake of the war, the Manhattan Project scientists enjoyed extraordinary personal and collective prestige. In Washington they were in great demand as everybody sought to learn something about the atom. Yet by the early 1950s, America's most prominent wartime scientist – J. Robert Oppenheimer – had been stripped of his security clearance and humiliated. A number of scholars have used Richard Hofstadter's framework of America's anti-intellectualism to interpret these attacks on science in the fifties – and have argued that the implications of this anti-intellectualism extended well beyond the fifties. As Boyer concludes, postwar attitudes toward the experts were ambivalent at best, "deteriorating into a mood of anti-intellectualism by the early fifties...a legacy that has continued to resonate powerfully in the American culture down to the present." As early as 1957, sociologist Edward Shils anticipated Hofstadter's and Boyer's conclusions, observing that in the past decade "the uproar of antiintellectualism and distrust for scientists was louder than it has ever been in America."[28]

Yet Shils also points to contradictory evidence, evidence often ignored or simply taken for granted. It is the most conclusive evidence suggesting that Americans overwhelmingly embraced their experts – particularly in the fifties. As Shils puts it, the period from 1945 to 1957 "was also the decade of the greatly enhanced influence of scientists within public bodies and of a moderate but nonetheless unprecedented effectiveness of scientists working outside the government seeking to influence opinion and policy. The federal government bankrolled research and development after the war: The federal government was now the nation's chief supporter of 'Big Science.' Nobody benefited from this revolution in public support more than the physicists."[29]

Starting with a 1950 quotation from *Physics Today* proclaiming that the "springtime of Big Physics has arrived," Daniel Kevles has documented the explosive growth in this profession and its postwar appetite for new projects funded by the federal government. Physicists became accustomed to seemingly unlimited resources. "The Los Alamos generation thinks big and expensively," concluded Kevles. In March 1948 the Atomic Energy Commission funded gigantic particle accelerators at opposite ends of the country: a 2- to 3-billion-volt accelerator was authorized at Brookhaven, New York; at Berkeley, an accelerator

27 Stuart Chase, "Atoms and Industry: The New Energy," *The Nation*, December 22, 1945, p. 709.
28 Boyer, *Bomb's Light*, ch. 22; quotation from p. 274; Smith, *A Peril*, p. 528; Hewlett and Anderson, *New World*, p. 446; Margaret Smith Stahl, "Splits and Schisms, Nuclear and Social," Ph.D. diss. University of Wisconsin, 1946, p. 126; Donald A. Strickland, *Scientists in Politics: The Atomic Scientists Movement, 1945–46* (Purdue University Studies, 1968), pp. 105–11; Edward Shils, "Freedom and Influence: Observations on the Scientists Movement in the United States," *Bulletin of the Atomic Scientists* [hereafter *BAS*] 13 (January, 1957): 15.
29 Ibid.

was designed to operate at more than 6 billion volts. Projects in the physical sciences funded by the Atomic Energy Commission and the Defense Department received the bulk of the nation's rapidly expanding research and development funds. By the outbreak of the Korean War, federal R and D topped the $1 billion level, climbing to more than $3 billion by 1956.[30]

The federal government's new efforts toward profession building were as bold as its investment in bricks and mortar. By the late 1940s there were twice as many physics graduate students as there had been before the war. American universities minted 275 doctorates in physics in 1949 – more than 500 by 1953. Many of the new students were funded by federal agencies. In 1953, for instance, sixty students received National Science Foundation fellowships – the Atomic Energy Commission and Office of Naval Research funded another six hundred.[31]

Though the Los Alamos generation was optimistic – "Isn't physics wonderful?" Isidor Rabi bubbled in 1948 – Americans had more confidence in the experts' ability to handle the problems posed by atomic power than the experts themselves had. For instance, the American public believed that scientists and engineers would design a foolproof defense against atomic weapons. Sixty percent of the American public believed that scientists would be able to develop a defense against the bomb in the next ten years, despite insistence by experts that no such defense could be designed in the foreseeable future. The belief in a defense was so pervasive that Oppenheimer – summarizing the views of other scientists – began his testimony before the Senate Special Committee on Atomic Energy (SCAE) with the blunt statement, "there are, and there will be, no specific countermeasures to atomic weapons."[32]

One reason for these high expectations was the mood of confidence that victory in the war inspired. If progress, implicit in the American experience in the nineteenth century, could no longer be assumed, the "physicists' war" proved that it could be "discovered," designed, and organized. The war also showed that such breakthroughs as the bomb came suddenly – or so it seemed to a public who had no knowledge of this massive effort until Truman's statement. Thus it was reasonable to expect more dramatic breakthroughs regardless of what the experts were saying.

Occasional dramatic statements by the experts themselves encouraged the public to expect startling new accomplishments. An article in the December 1945 *Nation* pointed out that in the immediate aftermath of the bomb, the few statesmen and scientists who ventured remarks about peaceful uses "were in a rather negative mood and predicted that we needed more than a generation ... before we could see any possibilities for the peaceful application of atomic power. But what a

---

30 Kevles, *Physicists*, pp. 367–9.   31 Ibid., p. 370.

32 Ibid., p. 369 for Rabi quotation. Only 19 percent of those polled thought that a defense could not be developed. American Institute of Public Opinion poll, October 3 1945, in Erskine, "The Polls," p. 161. Special Committee on Atomic Energy [hereafter SCAE], "S. Res. 179," p. 187.

change has occurred in the last few weeks!" The author referred to none other than J. Robert Oppenheimer to demonstrate this turnabout. "It is quite feasible," Oppenheimer was quoted as saying, "that a city the size of Seattle should be completely heated from an atomic energy source in less than five years."[33]

As the months passed, public interest declined noticeably. A series of articles by William Laurence in the *New York Times* placed the responsibility in part with the Bikini tests. That the explosions at Bikini left many of their targets still afloat encouraged the public to treat the atomic bomb as just another weapon. Since Bikini, which was widely covered by the press, a sense of awe had been supplanted by a sense of relief, Laurence wrote. America's citizens were trying to regain some peace of mind. The bomb had briefly engaged the hopes and fears of a broad spectrum of the American public, but within a year of Hiroshima, many seemed to have integrated the new atomic force into their daily pattern of life. They found that the atom made little difference, at least at a conscious level.[34]

To Senator Brien McMahon, first chair of the newly created Joint Committee on Atomic Energy (JCAE), it made quite a difference. As atomic energy became the focal point of his political career, he had good reason to commemorate the first anniversary of the "Atomic Age," as he called it. But as Robert Bacher, a Manhattan Project scientist who was equally absorbed in the development of atomic energy, pointed out, by the first anniversary of the bombing of Hiroshima, complacency had already set in. Atomic energy was the concern increasingly of specialized publics. Foremost among these were the scientists in whom the vast majority of the public had placed their confidence after the war.[35]

ATOMIC SCIENTISTS REMOBILIZE

The perspective of the scientists who worked in the Manhattan Project was far different from that of the general public. Having initiated the project when Leo Szilard and several other émigré physicists prevailed upon Albert Einstein to write to President Roosevelt, and having participated in the project's development, these scientists were well aware of the magnitude of the effort they were involved in. As the project moved from the theoretical to the engineered, they became caught up in its momentum and observed the mobilization of more and more of their colleagues: they were only too well acquainted with the project's elaborate organization – which they often bridled at – and equally conscious of the unprecedented level of funding lavished on it. Like parents who had prepared a child for a concert career from birth, they viewed their prodigy's public debut

---

33 Boris Pregel, "Power and Progress," *The Nation*, December 22, 1945, p. 710.

34 *New York Times*, August 4, 1946. Although the explosions were less than spectacular, the massive doses of radiation they created would have crippled the crew on this hypothetical enemy fleet.

35 National Committee on Atomic Information press release, August 6, 1946, "National Liaison Committee on Atomic Information" folder, Box 3, SCAE Files.

with more than the temporary awe of the rest of the audience: the performance was framed in the context of years of arduous practice, many missed notes, and, perhaps most importantly, visions of possibilities for the future.[36]

By the end of the war there was no longer any debate among scientific leadership about the role of the federal government: federal funding was crucial to the future of American science. Nor was there any hesitance by federal administrators to spend freely on research. As Edward Condon, director of the Bureau of Standards, commented in 1952, "every new agency feels it must do research in order to have status in the world of bureaucracy." James Newman, counsel to the Special Committee on Atomic Energy, wrote, "The war demonstrated the miracles which science can perform when fully mobilized and directed to the production of weapons. Now for the first time in our history, we are making an experiment to see what science can do when organized and geared to a constructive peacetime objective."[37]

Most scientists apparently agreed on three other issues in the wake of the war. They recognized that the gap between scientific discovery and social application had been drastically narrowed. While acknowledging the potential horrors that might result from their creations, they remained firmly optimistic that over the long term positive benefits would outweigh such negative consequences. And finally, to a degree never experienced before, they felt personally responsible for ensuring that outcome.

The Manhattan Project demonstrated that research at the edge of theoretical understanding could be translated into dramatic applications in a relatively short period of time. The speed with which theoretical principles were translated into a powerful practical application through the Manhattan Project staggered even the scientists directly involved. George Weil, a young research assistant to Enrico Fermi who personally removed the control rods to start the world's first controlled chain reaction and who went on to head up the Atomic Energy Commission's reactor development unit, wrote to a friend one month after the first atomic bomb was tested in New Mexico, "I don't believe there has ever been a development passed so quickly from the experimental stage to 'practical' uses." Basic researchers could not remain removed from the consequences of their work when within a matter of months, or at most years, they participated directly – with the aid of massive federal funding and organization – in translating that basic research into hardware.[38]

---

36 For a detailed account of the physicists' early efforts to initiate the project, see Rhodes, *Making*, ch. 10.

37 Kargon and Hodes, "Karl Compton"; Edward Condon, "Some Thoughts on Science in the Federal Government," *Physics Today* 5 (April 1952): 12; Daniel Kevles, "The National Science Foundation and the Debate over Postwar Research Policy, 1942–1945: A Political Interpretation of Science the Endless Frontier," *Isis* 68, no. 241 (March 1977) 16; James R. Newman, "America's Most Radical Law: The Atomic Revolution Begins," *Harper's Magazine* 194 (May 1947): 442.

38 George L. Weil, "Nuclear Power: From Promises to Problems," unpublished manuscript, January 28, 1988, p. 49. On Weil background, see Weil, "Nuclear Energy: Promises, Promises" (Washington D.C., 1971).

Although the scientists' movement, in its bid for international control of the atom, did dramatize some of the horrors of the bomb, the movement on the whole was profoundly optimistic – perhaps too optimistic. Some leading scientists expected to modify political institutions in accordance with the ideals of science. Niels Bohr, the Danish physicist, exemplified this perspective. Bohr was all too aware of the potential destructive consequences of atomic energy. Yet he saw this awesome power as ultimately leading to positive results. Scientific progress, rationality, and peace went hand in hand. Ultimately, Bohr believed, these ideals would be transferred to the political sphere. Bohr exemplified a perspective shared by many scientists in the wake of World War II. As the *Bulletin of the Atomic Scientists* put it, "the development of science and technology is rapidly changing the realities of human existence," destroying "historical concepts of international struggle for power."[39]

Leading scientists were also optimistic about the domestic implications of the Manhattan Project. For the wartime director of the University of Chicago's Metallurgical Laboratory, Arthur H. Compton, the Manhattan Project – with its fusion of science, engineering, and large-scale management – was an important model for mobilizing domestic resources. Shortly after the war Compton wrote, "In the success of the atomic bomb project the United States has perhaps caught a new view of its titanic strength. It is a strength that comes when a compelling objective draws the co-ordinated effort of trained and educated citizens."[40]

Even the "hardheaded" scientist administrator most experienced in academia, the world of business, and government service, Vannevar Bush, believed that science and technology would dramatically improve the lives of most Americans. In *Science, the Endless Frontier* – published a month before the first bomb was dropped – Bush argued that economic growth, centralization, and interdependence were essential to America. Science-based technology, Bush maintained, was the way to achieve these objectives. After noting such direct benefits of scientific research as penicillin, radar, radio, air-conditioning, and synthetic fibers, *Science, the Endless Frontier* proclaimed that when applied, advances in science meant, "more jobs, higher wages, shorter hours, more abundant crops, more leisure.... Advances in science will also bring high standards of living, will lead to the prevention or cure of diseases, will promote conservation of our limited national resources and will assure means of defense against aggression."[41]

Scientists were determined to find a silver lining even in the destructive mushroom cloud of the atom. James Conant, who from the beginning had serious

39 Boyer emphasizes the scientists' campaign of fear in *Bomb's Light*, chs. 3–9. The material on Niels Bohr is from Sherwin, *World Destroyed*, pp. 92–4. On optimistic scientific perspective, see Gilpin, *American Scientists*; quotation from "The Dawn of a New Decade," *BAS* 16, no. 1 (January 1960): 5–6, quoted in Gilpin.
40 American Academy of Political and Social Science, *Annals* 249 (January 1947): 9–19.
41 U.S. Office of Scientific Research and Development, *Science, the Endless Frontier* (Washington, D.C.: U.S. Government Printing Office, 1945), p. 5.

doubts about the peaceful applications of atomic energy, struggled to find some positive benefit. The Harvard president quoted from Ralph Waldo Emerson's "Law of Compensation," reminding listeners, "There is a crack in everything God has made. It would seem there is always this vindictive backstroke, this kick of the gun.... [I]n nature nothing can be given, all things are sold." Conant urged Americans to think of the bomb's destruction as the price paid for health and comfort in the scientific age.[42]

Despite the explicit concern about proliferation that dampened enthusiasm for the immediate development of peaceful uses, most scientists directly involved in the development of the bomb believed that in the long run, the peaceful application of the atom – particularly as a new source of electricity – would benefit humanity. According to Farrington Daniels – a pioneer in reactor research at the Metallurgical Laboratory – it was not the bomb that was the war's most important scientific development. As he told the American Physical Society: "The most important thing which has come out of scientific research during the war period is the pile, consisting of a structure of uranium and graphite or equivalent materials. Simple, silent, powerful, these potential giants are waiting for man to decide between abundant life and race suicide."[43]

A fourth broad area of consensus united most scientists: Increasingly, they felt responsible for the social consequences of their work. This followed directly from their recognition of the speed with which theory could now be translated into practical applications and their optimism that such applications could improve the lot of humanity. John Simpson, a physicist instrumental in organizing the Federation of Atomic Scientists (FAS) after the war, summed up this change in the *Bulletin of the Atomic Scientists* – a journal that in its very inception reflected the scientists' new concern with social policy. "What was different about scientists' role in the bomb," Simpson argued, "is that no sooner had they verified the theory of fission, the same scientists applied it, consequently they were brought face to face with the problems of its application." Guilt undoubtedly contributed to some of this growing sense of responsibility, but as Robert Gilpin has pointed out, the deeper roots of the scientists' social responsibility lay in a radically changed relationship between science and society. With science altering society daily, scientists sought greater control over the fruits of their labors; with government policy altering science daily, scientists sought to retain their professional autonomy.[44]

Broad consensus about objectives, however, did not guarantee a united front when it came to devising a political strategy for achieving these ends. Having accepted federal support as a fact of life, scientists split over the most effective

42 Hershberg, "Conant," p. 254
43 Farrington Daniels, "Plans and Problems in Nuclear Research," *Science* 104, no. 26922 (August 2, 1946): 91.
44 John A. Simpson, "The Scientists as Public Educators: A Two Year Summary," *BAS* 3: 243. Gilpin, *American Scientists*, pp. 23–8.

technique for ensuring that as few strings as possible would be attached to those government dollars.

Wartime experience shaped scientists' views on this matter as well. Because of the hierarchical organization of the Manhattan Project and its highly compartmentalized operations, that experience often depended on the scientists' access to policymaking. Those scientists at the top, who had carved out satisfactory arrangements with the army, sought to extend these arrangements following the war. To them, broader public participation – particularly congressional intrusion – was a greater potential evil than military control. Thus Bush, Conant, and Oppenheimer stressed quiet diplomacy with existing organizations – mainly military. At least these organizations, because of their wartime track records, were known quantities. These scientist administrators gambled on competition for scarce expertise strengthening their hand in future negotiations with well-established organizations. Though difficult to document, it is likely that those scientists who took on administrative responsibility in the Manhattan Project were a self-selected group, more prone to accept hierarchical control and bureaucratic compromise. As administrators, these scientists long ago had learned to accept a less than perfect world. Recognizing that some sort of control over science would be required, they preferred to maintain their existing relationships, even though these were not ideal.

Scientists with less access to policymaking – the "laboratory scientists" – preferred more explicit means of ensuring their future autonomy. In one regard, their approach was an extension of the profession-building strategy of the early twentieth century. What most distinguished them from their professional predecessors, however, was the consensus these scientists shared with the administrative scientists about the necessity for federal funding and organization on the one hand, and their strategy of mass public appeals on the other. Out of the organizations formed at various Manhattan Project laboratories during the war – Chicago, Oak Ridge, Los Alamos – these scientists formed the Federation of Atomic Scientists. Along with the quest for international control, this organization tried to ensure that scientific values were not compromised by political or bureaucratic expedients. Continued military control embodied just such a threat. The FAS and its supporters questioned whether the centralized, limited-access, and highly secretive structure of the Manhattan Project had been the wisest administrative arrangement even though it ultimately achieved its wartime objective. That structure, they argued, could not have been farther removed from the pluralist federal political system that Americans had fought the war to preserve. It certainly was a radical departure from the voluntary, and local, administrative mechanisms that housed expertise before the war.

Even Alice Kimball Smith, who is at pains to portray the scientific community as united, acknowledges that there was a "gulf ... between those who thought that scientists' influence upon policy could best be exerted through a small group of administrators and advisers and those who saw the usefulness of a broadly based organization explicitly recognizing political obligations." The scientist administrators sought to contain the debate to the nation's organizational trenches, where they felt that they had the upper hand. But unlike the wartime organization of the Manhattan Project, the final peacetime apparatus hinged on congressional action – highly visible and open to public scrutiny. Scientist administrators, siding with the military, viewed this as a problem, and hoped to skirt it by pushing through, as quickly as possible, legislation that would limit congressional and, for that matter, presidential control. Laboratory scientists, on the other hand, organized public protests and sought broad public discussion to head off this end run. In the standoff that ensued there were two important intermediaries: congressional entrepreneurs and general administrators. Both were eager to regularize the politics of atomic energy and, not coincidentally, extend their own institutional control over it.[45]

The standoff also foreshadowed a pattern that reshaped nuclear politics over the next forty years. Nuclear insiders – first scientists, but soon administrators and legislators – pushed nuclear politics toward its insular limits. In the early years of development, a monopoly on expertise and consensus among those experts contained debate within that narrow community. Over time, however, organizations, jurisdictions, and disciplines with different perspectives addressed issues that cut across nuclear politics. They soon began to acquire their own nuclear expertise. Inside the nuclear community, as experts and organizations specialized, it became more difficult to contain debate.

The chief spokesperson for an insulated approach was Office of Scientific Research and Development director Vannevar Bush. Planning in the fall of 1944, Bush envisioned a part-time commission, virtually independent of government controls, that would include technical specialists selected by the military, the president, and the National Academy of Sciences. His plan left little room for broader political control. As Bush envisioned it, this group of experts not only would advise the president on atomic development but would also be responsible for its operation and control. A newly established National Research Foundation would fund independent research.[46]

In July 1945, Bush reported his proposal to the Interim Committee, which also considered legislation drafted by the army's lawyers. The army legislation also proposed a part-time commission and a strong administrator to whom sweeping powers were delegated. Both proposals envisioned commissions that included direct representation for the armed services. Both framed the issue of atomic energy almost exclusively in the military context in which it had been initially developed.[47]

Bush and Conant were, of course, aware of the growing concerns of scientists in the Manhattan Project's laboratories who had reacted to the military's sometimes

45 Smith, *A Peril*, p. 72. For interpretations of the scientists' movement that emphasize the differences between groups of scientists within the laboratories, see Strickland, *Scientists in Politics* and Stahl, "Splits and Schisms."
46 Hewlett and Anderson, *New World*, pp. 409–11.
47 Ibid., p. 412.

heavyhanded administration. In response to the army's proposed legislation, Bush and Conant suggested that the commission should be made up entirely of civilians. Also, Bush was concerned that the powers of the commission were simply too broad, a feature always likely to attract congressional attention. The army compromised on the make-up of the commission – only four members would be active or retired officers – but stood firm on sweeping powers for it.[48]

Because the Truman administration was slow to resolve its approach to the international control of atomic energy, the May–Johnson bill – as the army legislation became known – was not introduced in the House until October. House Military Affairs Committee chair Andrew May hoped to send the bill to the floor after only one day of hearings during which pro-administration witnesses ranging from Bush to General Groves appeared. In his testimony endorsing the bill, Bush played down the possibilities for peaceful applications. There were no great commercial applications just around the corner: the primary purpose of the legislation was to promote the national defense, according to the author of *Science, the Endless Frontier*. Conant reinforced that testimony. Currently, he said, the only proven use for atomic energy was as a bomb (although the future held the promise of medical by-products). In Conant's opinion, industrial uses were at least ten years away. Both Bush and Conant stressed the need for quick passage of the May–Johnson legislation. Conant warned that the entire atomic project was losing momentum.[49]

More delays, however, were in store. At the labs, concern over the atom's future had been mounting. Eugene Rabinowitch, who had already called for a scientists' organization in July, chaired a committee of Chicago scientists in August that emphasized the importance of international control. But the most visible target of the laboratory scientists' concern was the army. Policies such as compartmentalization and restrictions on publication that had been barely tolerated during the war were challenged openly during peacetime. On September 1, Samuel Allison, a senior physicist at Los Alamos and the newly appointed director of the University of Chicago's Institute of Nuclear Studies, voiced publicly what many scientists had already discussed informally: Scientists, he warned, would not tolerate continued military control. If military regulation prohibited the free exchange of scientific information, Allison quipped, atomic researchers would devote themselves to studying the color of butterfly wings instead. This piece of advice earned the headline in the next day's *Chicago Tribune*: "Scientist Drops A-Bomb: Blasts Army Shackles."[50]

Chicago scientists organized more formally in September, establishing the

---

48 Ibid., pp. 413–15.

49 U.S. Congress, House, 79th Cong., 1st sess., H.R. 4280, introduced October 3, 1945. U.S. Congress, House Committee on Military Affairs, "Hearings on H.R. 4280," 79th Cong., 1st sess., 1945, pp. 34, 56–7.

50 For a lively account of one scientist's frustration, see Richard P. Feynman, *Surely You're Joking, Mr. Feynman: Adventures of a Curious Character* (New York: Bantam, 1986), part 3. Smith, *A Peril*, pp. 86–9; quotation from *Chicago Tribune*, September 2, 1945, p. 5, quoted in Smith, p. 89.

Atomic Scientists of Chicago. The group's general purpose was to clarify and consolidate scientific opinion on the responsibility of science; to influence the national administration on matters concerning atomic power; and to educate the public.[51]

Although the Chicago group was the most influential – its roots reaching back to the Frank Report, which had warned the Interim Committee in June 1945 that an unannounced atomic attack against Japan would precipitate an arms race and poison the opportunity for international control – it was clearly not alone in its concerns. Scientists also organized at Los Alamos, Columbia, and Oak Ridge. The Oak Ridge scientists were particularly frustrated by the fact that General Groves – the man they held responsible for restricting their rights to communicate – was constantly in the headlines, disparaging estimates that Russia would have the bomb in two to five years. By stretching out the projected arrival date for atomic competition, Groves undercut one of the primary reasons that some scientists had advocated international control. International control was crucial, these scientists argued, precisely because the Soviet Union would soon have the bomb, making the world far more dangerous unless some form of international control was in place. The Oak Ridge scientists expressed their frustration at being muzzled on this matter of technical judgment that also carried such important political implications. "Up to this time," the scientists pointed out in a letter to the Interim Committee, "security regulations and interpretations based on them have largely prevented us from making our convictions known to the public.... The statement by General Groves, in view of his official position, tends to discount in advance the opinion of the scientific workers of the Project, the only large group of the population having intimate knowledge of its details and implications."[52]

Congressman May's attempt to push through legislation in October galvanized the laboratory scientists: they formed a national organization. While scientist administrators Bush and Conant testified at those hearings in favor of the May–Johnson legislation, senior scientists such as Leo Szilard and Edward Condon worked furiously to head off the legislation. Statements, telegrams, and finally representatives from the Los Alamos, Oak Ridge, and Chicago atomic scientists' organizations flooded Washington. They argued that the legislation failed to offset military applications and military representation on the commission with measures to encourage the dissemination of information for research purposes and the promotion of international control. The scientists also objected to the bill's drastic security provisions and criticized its provision for part-time rather than full-time commissioners.[53]

51 Smith, *A Peril*, pp. 87–93.
52 Ibid. pp. 87–125; Stahl, "Splits and Schisms," pp. 257–8; Stahl argues that a similar organization was *not* formed at Hanford because the environment there was far closer to industrial science as opposed to pure science; September 22 letter from Oak Ridge scientists to Interim Committee, quoted in Smith, p. 107.
53 Smith, *A Peril*, ch. 3.

May's effort to limit debate and obtain quick passage of the War Department legislation jump-started the laboratory scientists' effort to organize. It gave them an immediate, tangible target to shoot at. It also was an issue that could be presented in simple and familiar terms to the American public. After the Atomic Scientists of Chicago and the Association of Oak Ridge Scientists issued press releases warning against hasty passage, the *Chicago Sun* ran banner headlines charging that the scientists had been muzzled. Thus the laboratory scientists learned first hand a lesson that aided future generations of political activists seeking to open up the politics of technical decision making to a broader constituency: in the United States, undue secrecy or bypassing concerned participants in the political process could become powerful issues in their own right. By the late sixties, "access" was as powerful a rallying cry for some participants in nuclear politics as "national security" had been for their adversaries in the fifties.[54]

To counter this negative publicity, Oppenheimer persuaded two other members of the Interim Committee's Scientific Panel, Enrico Fermi and Ernest Lawrence, to urge prompt passage of the May–Johnson bill. But when the full Scientific Panel met to discuss the matter on October 17, 1945, all its members, including Oppenheimer, Fermi, and Lawrence, acknowledged that they had reservations, particularly about the bill's security provisions. By this time, May had agreed to reopen hearings, giving a broader cross section of scientists their opportunity to testify. By the end of the month, the separate site organizations had combined into a national organization – the Federation of Atomic Scientists.[55]

Testimony at the reopened hearings before the Military Affairs Committee underscored the difference in strategy between the administrative scientists and the laboratory scientists. Leo Szilard, perhaps the most vocal critic of the May–Johnson legislation, scored the proposed commission's independence. Szilard feared that continued dominance of the military on such a commission would handicap scientists. To underscore this point, the Feisty Szilard recounted horror stories about the Manhattan Project's compartmentalization policy. The policy was so crippling to lab work that scientists simply followed their common sense and disregarded the rules, Szilard revealed. For the Association of Oak Ridge Scientists and the Atomic Scientists of Chicago, physicist Herbert Anderson strongly opposed the legislation because it delegated responsibility to a commission that was not responsive to the electorate, the president, or any other authority. Anderson added that commission members were "virtually immune from outside criticism or review" because security regulations prevented the disclosure of the actions or policies that might provoke such criticism.[56]

---

54 House Military Affairs Committee, "Hearings," 1945, p. 71; Smith, *A Peril*, pp. 128, 133, 200; *Chicago Sun*, October 11, 1945, quoted in Hewlett and Anderson, *New World*, p. 431.
55 House Military Affairs Committee, "Hearings," 1945, p. 107; Hewlett and Anderson, *New World* p. 432; Smith *A Peril*, pp. 203–4.
56 House Military Affairs Committee, "Hearings," 1945, pp. 79–81; quotation from p. 99.

Presenting the opposing view, Oppenheimer urged administrative discretion. Only those close to the scene and intimately familiar with rapidly changing technical developments should guide atomic policy. In light of the rapidly changing technology, Oppenheimer testified, "it is fair to say that the May bill has been written largely from the point of view that we must have confidence in the Commission – it's the best we can do for the moment."[57]

The scientists' organizations slowed the momentum of the May–Johnson legislation. On substantive issues like security provisions, international control, and relatively unrestricted research, the views of the laboratory scientists did not differ greatly from the perspectives of the official representatives on the Scientific Panel of the Interim Committee. Where they did differ was in their willingness to trust that acceptable accommodations with the army could be quietly and pragmatically negotiated. Whereas the scientist administrators on the Scientific Panel – like the members of the Interim Committee – sought to limit the scope of debate, the laboratory scientists organized to broaden it. For a brief moment, that is precisely what they achieved. A greater challenge, however, lay ahead: they had to sustain the momentum they had helped to create.

THE INSTITUTIONAL RESPONSE: CONGRESSIONAL ENTREPRENEURS AND ADMINISTRATIVE CONTROL

After May's bill failed to gain speedy passage, Senator Brien McMahon made the issue his own, ultimately garnering deference from his colleagues for his expertise – or at least access to expertise. His legislation established a new joint committee in Congress to oversee atomic energy. McMahon's new political turf was carved directly from tracts that arguably belonged to established committees, such as Military Affairs or Foreign Relations.

While McMahon staked a claim for himself and Congress, administrators including Budget Director Harold Smith and Office of War Mobilization deputy James Newman surveyed the landscape on behalf of the White House. These administrators had to coordinate policy in a vastly expanded federal government. They sought to ensure presidential control over both atomic experts and the army while retaining as much executive discretion as possible vis-à-vis the Congress.

The new presidential and legislative contenders all found it politically expedient to stress the potential impact of the peaceful (as opposed to military) applications of atomic power. Although McMahon worked closely with the scientists' organizations, which, fearing proliferation, downplayed the potential benefits of atomically generated electricity, the senator was less constrained by such considerations than the scientists. Increasingly, McMahon emphasized the exciting possibilities for commercial nuclear power. He did so for explicitly

57 Ibid., p. 128.

political reasons. If civilian control were to persuade, impressive civilian uses were required. The case for civilian control was built in part on the promise of cheap electricity produced by nuclear reactors.

McMahon was a Democratic liberal – soon to be cold war liberal – from Connecticut. Schooled at Yale, the best-dressed or at least most flamboyantly dressed senator on the floor, he was a tough political brawler as well. This freshman senator with the solid jaw and heavy eyebrows was determined to make a name for himself, and equally determined to influence world affairs. McMahon eyed the chairmanship of the Senate Foreign Relations Committee as a stepping stone to the White House. A firm supporter of extending the New Deal at home and a tough talker in international affairs, McMahon was often compared to John F. Kennedy.[58]

An aggressive legislator, a kind of congressional entrepreneur, McMahon seized the opportunity to frame atomic power in a new context. Like any entrepreneur, McMahon was taking a risk when he wrapped himself in this issue – eventually earning the sobriquet of "Mr. Atom." Because of its complex and technical subject matter and volatile emotional associations, senior legislators were extremely cautious in their approach to atomic energy. Millard Tydings, for instance, often prefaced even his questions about atomic energy with the qualifier "I'm only thinking out loud."[59]

McMahon's path to atomic oversight rested initially on a technical parliamentary claim that, in his case, bucked the far more politically powerful tradition of selecting committee leadership by seniority. On the same day as Congressman May's disputed hearings, McMahon introduced a resolution calling for a special Senate committee responsible for conducting hearings and drafting legislation on atomic energy. Senator Arthur Vandenberg's resolution to establish a joint congressional committee had been blocked in the House. Vandenberg, in turn, blocked Senator Edwin Johnson's effort to introduce the administration's bill in the Senate, claiming that it should go to the Foreign Relations Committee. The Senate passed McMahon's resolution on October 22 as a way of breaking the deadlock. When Johnson argued that the chair of the new committee should be selected by seniority, McMahon reminded the Senate that the senator introducing legislation establishing a special committee (as opposed to a standing committee) had been named chair in the last seventeen instances. Parliamentary precedent prevailed: on October 26 the Senate named McMahon to chair the Special Committee on Atomic Energy.[60]

Although McMahon's colleagues let him run with the politically untested atomic issue, they surrounded him with some of the Senate's leading lights. Joining McMahon on the Special Committee were ten other senators including

---

58 Weart, *Nuclear Fear*, p. 141; Herken, *Counsels of War*, pp. 39–40.
59 See SCAE, "Hearings on S. 179." for Tydings' cross-examination style.
60 Senate Resolution 179, in U.S. Congress, *Congressional Record*, 79th Cong., 1st sess., p. 9472; see also pp. 9889–90, 9898; *Congressional Quarterly* 1, no. 4 (October–December 1945): 674–5.

Johnson and Vandenberg, as well as Senator Tom Connally, chair of the Foreign Relations Committee.[61]

McMahon faced a situation that would become increasingly common in the postwar world of government initiatives: He had in hand a program that existed before it had a constituency. Advocates of the May–Johnson bill had a built-in constituency – the army. But it had been precisely this cozy relationship between the military at nuclear developers that had provoked much of the opposition to the legislation. Unfortunately for McMahon, however, opposition to military control did not fill any familiar organizational mold. "Several unique factors combined to deprive the legislator of his comfortable patterns for reaching policy decisions," wrote Byron Miller, a legislative drafter. Legislators were not dealing with a conventional controversy, Miller pointed out, "a labor versus management or debt reduction versus public spending issue, on which his attitudes had long been fixed.... Facing this dilemma," Miller continued, most members of Congress "felt lost in a morass of technology." The first task at hand for McMahon was to establish a constituency of his own.[62]

He found one in the scientists' movement. The movement provided McMahon with a cast of experts "whose skin had been worn thin by their unhappy experiences in the Manhattan project. Sensitive as their own Geiger counters, they registered an immediate and powerful reaction to the May–Johnson bill, which threatened to make permanent the military control of atomic energy," observed James Newman. "It may be that the American scientists were grossly naive in politics," Newman continued, "but there is no denying that they showed more capacity for improvisation and prompt action than any of the economic groups whose interests were vitally involved."[63]

From its inception, commercial nuclear power was pushed by the very groups that provided the service rather than pulled toward those seeking to consume it. It was the scientists, of course, who had put all the uses of atomic energy – starting with the bomb, and eventually encompassing peaceful applications – on the political agenda in the first place. These "producers" of atomic energy remained an important constituency for peaceful applications. Organized consumers of atomically generated electricity, however, were nowhere to be found in the early years of atomic development.

McMahon developed a close working relationship with John Simpson of the Atomic Scientists of Chicago. Soon after meeting Simpson, McMahon asked him to write a brief statement about the Special Committee's objectives. Simpson did this, and when McMahon read it on the radio the next day, virtually unaltered, Simpson thought, "If it is this easy, I might as well stick around."[64]

61 U.S. Congress, *Congressional Record*, 79th Cong., 1st sess., p. 10068.
62 Quotation from Byron Miller, "A Law Is Passed – The Atomic Energy Act of 1946," *University of Chicago Law Review* 15, no. 4: 799.
63 Newman, "America's Most Radical Law," p. 439.
64 Smith, *A Peril*, pp. 262–3; quotation is from Donald Strickland interview, in Strickland, *Scientists in Politics*, p. 103.

The atomic scientists (along with the Federation of American Scientists) were meanwhile trying to broaden their base of support. The scientists formed the National Committee on Atomic Information specifically for this purpose. It informed citizens "through channels which people must trust, namely their churches, clubs, unions, schools." The scientists also consolidated their legislative activities; they established the Emergency Conference for the Civilian Control of Atomic Energy to coordinate lobbying. To increase their visibility and stature, the scientists brought together a group of prominent individuals in the Emergency Committee for the Civilian Control of Atomic Energy. McMahon used this web of overlapping public interest groups to broaden his base of support. Thus technical decisions were of relatively little concern to this lobby that addressed far more accessible issues, such as world peace and civil liberties.[65]

The relationship between McMahon and the scientists was symbiotic. While the ever-broadening organizations that radiated outward from the atomic scientists offered McMahon and his congressional allies an important base of support and means of communication, the hearings conducted by the Special Committee gave the opponents of May–Johnson their first friendly congressional forum. There was little doubt that the power of science had impressed even the conservative members of Congress. As Millard Tydings put it:

If in 1939 we had been conducting a hearing like we are conducting today, and men like yourself had come before our committee and projected the possible development of the bomb....I imagine they would have been called a lot of crackpots....I certainly would have not had the receptivity that I have today, to say the least.

Senator Johnson was more succinct: "I believe our security lies in the laboratory and in scientific development. That is my religion."[66]

McMahon promoted his ties to this highly visible constituency in the fall of 1946 by bombarding the Special Committee on Atomic Energy with a stream of scientists promoting the views of the Federation of Atomic Scientists. This was in part a response to experts excluded from Congressman May's hearings earlier that year. The Special Committee probed the War Department's policy of compartmentalization, options for international control, the likelihood and timing of atomic proliferation, and the degree of damage actually inflicted by the bombs dropped on Hiroshima and Nagasaki.

The potential for peaceful uses – though hardly the issue most frequently discussed – came up on a number of occasions. Virtually all of the scientists downplayed the short-term impact of commercial atomic power. Irving Langmuir, associate director of the General Electric Research Laboratory, was the least enthusiastic. Even if atomic energy were to replace coal and oil as sources of industrial power, he testified, this would be "a relatively trivial matter." Many

---

65 National Liaison Committee on Atomic Information Newsletter #18 (April 3, 1946), "National Liaison Committee" folder, Box 3, Records of the SCAE; Miller, "A Law," p. 808.
66 SCAE, "Hearings on S. 179," pp. 290, 121.

of the other scientists who testified – particularly Alvin Weinberg – envisioned a great deal of promise in the distant future, but consistent with the position of the Federation of Atomic Scientists they insisted that domestic applications should be restrained until an agreement for international control was in place. At times McMahon seemed to endorse this position.[67]

Yet, McMahon was determined to find some encouragement for his position that atomic energy had civilian uses that required civilian control. Scientists may have placed atomic energy on the nation's agenda, but this congressional entrepreneur had a lot to say about the timing of that agenda. Despite the pessimistic projections of his hand-picked experts, McMahon concluded that the benefits to be derived from the peaceful application of atomic energy were too important to delay. During John Simpson's testimony, for instance, McMahon voiced his minority rationale. Senator Tydings had been pushing his pet approach: simply arresting the development of atomic energy – locking it up. McMahon insisted that "it is not fair to sit on power, just because the United States doesn't need it right away." He added later that the surest path to peace lay in developing atomic power in order to diminish the gap between the haves and have-nots. These were themes that promoters of commercial nuclear power returned to again and again over the next thirty years, even in the face of expert testimony suggesting less cause for optimism. Simpson, like most of his scientific colleagues, replied that he was skeptical of what atomic power would do for the world during the next generation or two.[68]

McMahon was a promoter by profession. If the existing military context for controlling nuclear power was to be altered – and if a freshman senator was to emerge as its leading congressional expert – a great deal of promoting was required. Diamond stickpin gleaming, McMahon built his case on the most optimistic assessments of atomic power. He gradually made the potential of peaceful uses crucial to his emerging argument for broadly based political control. McMahon, it turned out, was a producer as well. By wresting congressional control from standing military or foreign policy committees, McMahon produced a powerful new congressional domain. It provided an influential platform for McMahon until his death several years later. Better than the scientists, McMahon understood that in politics, potential had better mean soon.

Turf was as crucial as timing in politics. In the new field of domestic control of the atom, McMahon could plan to become – and in fact did become – the Senate's leading authority. Ironically, once firmly ensconced as the chair of the Joint Committee on Atomic Energy, McMahon devoted most of his attention to the military and strategic aspects of atomic energy. But the emphasis on peaceful uses of atomic energy had been a crucial factor in securing that platform in the

---

67 Ibid., p. 110. See also the testimony of Harold Urey, Leo Szilard, John Simpson, and Alvin Weinberg.
68 Ibid., pp. 359, 311, 313. Simpson was chair of the Executive Committee of the Atomic Scientists of Chicago.

first place. As the first paragraph of S. 1717, introduced by McMahon in December 1945, emphasized:

It is reasonable to anticipate ... that tapping this new source of energy will cause profound changes in our present way of life. Accordingly, it is hereby declared to be the policy of the people of the United States that the development and utilization of atomic energy shall be directed toward improving the public welfare, increasing the standard of living, strengthening free competition among private enterprises so far as practicable, and cementing world peace.[69]

After introducing this legislation, McMahon pushed the power-generating applications of atomic energy even harder. Anticipating another recurring theme among nuclear power's promoters over the next thirty years, McMahon framed the peaceful applications of atomic energy in the same internationally competitive context as the debate over weapons. At the hearings he chaired in the winter of 1946, McMahon stressed the potential military advantages that power generated by the atom might bring or the economic damage that might ensue should America lose the race for commercial nuclear power. McMahon adapted the domestic use of atomic power to the tense international situation, asking mathematician John von Neumann whether the development of peacetime industrial uses by another country before the United States "might ruin our economy." McMahon also pointed out during the hearings that America did not leave its national policy for either electricity or oil up to the military.[70]

Given the make-up of the committee, McMahon had no chance to get his way on everything in S. 1717. The provisions for limiting the exchange of information – although not as restrictive as the May–Johnson legislation – still concerned many scientists. And the military was still represented through a military liaison committee – an amendment proposed by Senator Vandenberg.[71]

But McMahon had no trouble mustering support for another provision in his bill. SCAE members at both ends of the ideological spectrum agreed on the need for more congressional control. Where May–Johnson would have dispensed with even the trappings of congressional control, S. 1717, as amended by the Special Committee, called for a joint committee to comprise nine members of the Senate and nine members of the House. The legislation required the Atomic Energy Commission to keep the Joint Committee on Atomic Energy "fully and currently informed with respect to the Commission's activities." Herbert Marks, the first general counsel to the Atomic Energy Commission, marveled at the power invested in the JCAE. Never before had Congress concentrated such broad statutory powers in a joint committee. Though a joint committee, the JCAE had the

69 S. 1717, A Bill for the Development and Control of Atomic Energy in, U.S. Congress, Senate Special Committee on Atomic Energy, "Hearings on S. 1717," 79th Cong., 2nd sess., 1946, p. 1.
70 SCAE, "Hearings on S. 1717," pp. 219, 144.
71 U.S. Congress, Senate Special Committee on Atomic Energy "Report to Accompany S. 1717," 79th Cong., 2nd sess., Report no. 1211, pp. 22–4, 11–13; Miller, "A Law," pp. 810–13.

"permanence of a standing committee; jurisdiction over legislation in a broad field; permanent powers of inquiry over a broad subject; and the authority to keep itself 'fully and currently informed' about – and therefore intimately involved in – the daily operations of an important executive agency." By handling all legislative matters related to the development, use, and control of atomic energy, the Joint Committee on Atomic Energy laid the groundwork for treating nuclear power as an exceptional issue within Congress – precisely the perspective McMahon had fought for.[72]

McMahon was not alone in seeking to regularize control of the atom: Truman administration officials charged with the responsibility of coordinating postwar policy sought to rein in the atom as well. Two offices vied for the opportunity to coordinate policy across the burgeoning bureaucracy in the wake of World War II – the Bureau of the Budget (BOB) and the Office of War Mobilization and Reconversion (OWMR). Vannevar Bush had already felt the impact of both these coordinators in the summer of 1945 when the budget director, Harold Smith, wondered officially "whether endless frontier implied endless expenditure," and when the Reconversion director, Fred Vinson, warned President Truman against endorsing Bush's report. Because its consequences were potentially so far-reaching, both BOB and OWMR scrambled to fold atomic energy into the larger executive structure that they sought to construct following the war.[73]

Harold Smith took over the Bureau of the Budget in 1939 shortly after the President's Committee on Administrative Management (commonly known as the Brownlow Committee) recommended greatly enhanced powers for that office. With his background in public administration and experience as Michigan's budget director, Smith guided BOB to its new role as the president's chief policy coordinator. He recruited a professional staff that valued organizational loyalty over fleeting political objectives. This and Smith's ability to protect the long-term interests of the presidency – as opposed to the shorter-term interests of its occupant – institutionalized BOB in the Executive Office of the White House.[74]

Smith framed the atomic question in administrative, not technical, terms. "A knowledge of nuclear physics and engineering alone...cannot solve this problem

---

72 For a liberal critique of May–Johnson's lack of congressional control, see Congressmen Chet Holifield and Melvin Price's "Dissenting Views of the Military Affairs Committee on H.R. 4566," in U.S. Congress, House, "Report to Accompany H.R. 4566," 79th Cong., 1st sess., Report no. 1186, p. 17; for a similar critique from Senator Millikin, see SCAE, "Hearings on S. 1717," p. 126; the powers of the JCAE are enumerated in the SCAE, "Report on S. 1717," pp. 29–30; quotation is from Herbert S. Marks, "Congress and the Atom," *Stanford Law Review* 1 (November 1948): 27–8. On the unusual powers of the Joint Committee, see also Morgan Thomas, *Atomic Energy and Congress* (Ann Arbor: University of Michigan Press, 1956), p. 9, and Green and Rosenthal, *Government of the Atom*, pp. 266–72.

73 Quoted in Saltzman, "Countdown," p. 33. On BOB and OWMR, see, Larry Berman, *The Office of Management and Budget and the Presidency* (Princeton, N.J.: Princeton University Press, 1979), pp. 20–36.

74 U.S. President's Committee on Administrative Management, *Report of the Committee, with Studies of Administrative Management in the Federal Government* (Washington D.C.: U.S. Government Printing Office, 1937). Berman, *Office of Management and Budget*, pp. 20–4.

that faces the Nation and the world," Smith testified in January 1946. "At the moment, our national safety and welfare depend more immediately on our legislative and administrative knowledge." Testifying on S. 1717, Smith proposed changes in the legislation that would make the administrative apparatus of the commission more accountable to the executive branch. He suggested a new presidentially appointed general administrative officer for the atomic commission. On one hand, Smith was concerned about military domination – thus his support for the principles of McMahon's legislation. But as an experienced administrator, Smith was also wary of a commission dominated by independent experts. As he confided to the Special Committee, "I think engineers and scientists ought to be advising on the technical side, and not become involved in major policy. I would say the technical people, in the main, are the worst people to deal with large policy issues."[75]

Harold Ickes – the Secretary of the Interior, who was too accountable for some freer-spending New Dealers – went even farther than Smith in demanding executive branch control. Ickes testified before the same committee that atomic fission "is too big a matter to be set off in a field by itself." He urged that a single administrator report to a committee of cabinet officers so that no single existing department would dominate the development and control of atomic energy. The Ickes plan proposed for the executive what Senator Johnson's commission of legislators would have provided for the Congress: direct political accountability, and a mixture of differing perspectives on how atomic energy was to be developed. Neither plan got off the ground, however. Atomic energy was set off in "a field by itself," vis-à-vis the executive as well as the legislative branch. This hardly eliminated politics from the field; it merely ensured that the primary political arena would be inside the commission itself. Over time, the Joint Committee on Atomic Energy would also force its way into that insulated arena. This did little to broaden the agency's perspective: the JCAE was even less accountable to the Congress as a whole than the Atomic Energy Commission was to the executive branch.[76]

If the Ickes plan was unrealistic, there was a more routine means of ensuring broad executive oversight: Bureau of the Budget policy coordination. BOB director Smith was determined to place the Bureau of the Budget at the heart of the president's policy-coordinating mechanism, and from that perspective the May–Johnson legislation was a step in the wrong direction. The bill, Smith reported, would make the commission "virtually independent of executive control." Truman – not a man to take the erosion of presidential authority lightly – immediately informed Secretary of War James Patterson that the May–Johnson legislation did not yet carry a presidential endorsement. Truman also asked Smith to arrange a

75 SCAE, "Hearings on S. 1717," pp. 31, 36; quotation is from p. 44.
76 Quotation from ibid., pp. 90–1.

meeting between the War Department and other administration critics of the bill.[77]

Although not so highly placed as Smith, the most influential of these critics was James Newman. Newman had been an assistant to Secretery of War Patterson and now assisted the director of the Office of War Mobilization and Reconversion, James Snyder. Newman thought May–Johnson overemphasized military applications. Like McMahon, he wanted legislation that would emphasize the great potential for peaceful applications of atomic power. And like Smith, he saw the need for greater executive control. Newman's ideas reached the president at about the same time as Smith's, and undoubtedly contributed to the president's reevaluation of May–Johnson. Moreover, Newman got the opportunity to follow through on his interests when Truman designated responsibility for atomic control legislation to the OWMR. The president's decision reflected his growing tendency to shift science policy away from Vannevar Bush, the wartime leader, toward those advocating that science advice be subordinated to the broader principles of sound public administration.[78]

In the field of atomic energy, as with the National Science Foundation legislation, Bush was unable to recapture presidential support. Newman, on the other hand, shortly moved into an even more strategic position to implement his approach when he became special counsel to the Senate Special Committee on Atomic Energy. At the same time, Newman maintained his ties to the OWMR, working with associate general counsel Byron Miller to draft legislation embodying Newman's criticism of May–Johnson. Newman's role in the development of the Atomic Energy Act reflects the degree to which advocates of broader public control subordinated traditional checks and balances between the legislative and executive branches in an effort to gain some control over this new force.[79]

The OWMR lobbied for its position even beyond the Special Committee. Byron Miller, for instance, drafted Congressman Chet Holifield's dissenting report on the May–Johnson legislation. Like McMahon had earlier, it warned

77 William G. Wells, Jr., "Science Advice and the Presidency: An Overview from Roosevelt to Ford," *Technology in Society* 2, nos. 1 and 2 (1980): 198; Smith to Truman, October 22, 1945, Records of the Bureau of the Budget, Washington D.C., quoted in Hewlett and Anderson, *New World*, p. 438; Truman to Smith, October 30, 1945, Records of BOB, quoted in ibid. Smith had already written to the president on October 17, 1945, expressing his doubts about the administrative arrangement called for by the May–Johnson bill.

78 Snyder to the president, October 15, 1945, "S. 1717 – Prior to Introduction, 12/20/45" folder, Box 1, The Papers of Byron S. Miller, Library of Congress, Washington, D.C.; Hewlett and Anderson, *New World*, pp. 437–8; Wells, "Science Advice," p. 198; The most renowned example of this was the dispute over the development of the National Science Foundation. Reacting to Truman's new position, an angry Bush lashed out at OWMR director Snyder: "I find that in general primary responsibility for the ... legislative program is assigned to the agency of government which may be presumed to have the most experience and knowledge of the subject. A conspicuous exception occurs in the case of the proposal ... for the establishment of a single federal research agency." (Quotation is from Vannevar Bush to John Snyder, November, 23, 1945, "National Science Foundation Act" folder, Box 3, Miller Papers.) On decline of Bush, see Saltzman, "Countdown," ch. 2.

79 Miller, "A Law," p. 806.

against overlooking the tremendous possibilities of atomic power, citing scientific testimony that harnessing atomic energy for civilian purposes could "produce a degree of human comfort the likes of which the world has never seen or even imagined."[80]

The Office of War Mobilization and Reconversion, like the Special Committee on Atomic Energy, developed ties to the scientists' lobby. While the military maintained that May–Johnson was supported by the country's most distinguished scientists – and that its opposition was merely an organized minority – Newman (through OWMR director Snyder) countered with a tabulation of hundreds of scientists who opposed the legislation, listing the more prominent names, among them Harold Urey, Irving Langmuir, and Henry DeWolf Smyth. On his way to discuss legislation with a group of scientists, another administrator temporarily serving the legislative branch due to the scarcity of expertise, Edward Condon, urged both Newman and Miller to attend, warning that if they did not, the scientists might get some "queer ideas about what the Committee was up to."[81]

Atomic power offered more than a new scientific and military frontier: the struggle over control revealed new political, legal, and administrative vistas as well. That pioneers such as McMahon and Newman so readily ignored traditional political barriers between executive and legislative underscores the tremendous premium placed on establishing political control over the experts and the military. Whether that control was exercised through the president or Congress was for the moment immaterial.

The new law sought to prevent either the military or the scientists from gaining the upper hand. Newman and Miller were certainly influenced, while drafting legislation for McMahon, by the laboratory scientists' opposition to May–Johnson. Byron Miller met several times with Edward Levi, who was coordinating the Chicago scientists' version of a bill. The Newman–Miller draft's relaxation of security provisions and its emphasis on the importance of research echoed concerns of organized scientists. But McMahon and Newman needed more than that: they were struggling against impending monopoly by the military – atomic energy's most established consumer – on the one hand, and scientists – atomic energy's producers – on the other hand.[82]

The McMahon legislation, introduced December 20, 1945, and publicly endorsed by the president in February 1946, sought to achieve this tenuous balance by establishing a full-time five-member civilian commission appointed (with the advice and consent of the Senate) and serving at the pleasure of the president and establishing a joint committee in Congress with an unusually broad mandate for

80 Byron Miller, October 10, 1945 "Draft for Holifield Dissenting Report," "S. 1717 Prior to introduction" folder, Box 1, Miller Papers.
81 John Snyder to the President, November 14, 1945, n.d. [approximately December 13, 1945], note from "Edward" [Condon], "S. 1717 Prior to Introduction" folder, Box 1, Miller Papers. Condon, director of the Bureau of Standards, also served as science adviser to the Special Committee on Atomic Energy.
82 For Miller's liaison with the scientists, see Hewlett and Anderson, *New World*, p. 422.

intervention into executive affairs. Political control was further strengthened by an amendment – accepting Budget Director Smith's suggestion – that established a general manager appointed by the president with the advice and consent of the Senate.[83]

Some congressmen were openly distrustful of this power-sharing arrangement. McMahon had a hard time convincing some of his conservative colleagues – particularly Senator Johnson, who introduced legislation of his own calling for a commission made up of congressmen – that congressional interests could be served by strengthening executive controls. But the following exchange illustrates how the tacit understanding worked between a congressional entrepreneur and an administrator eager to counter a new centrifugal force. The issue of control was broached by Senator Johnson, who stated that he would go far beyond the McMahon bill's language and establish more congressional control. Harold Smith replied,

I don't think they are two separate types of control but a single problem. And I will argue that unless you have adequate Executive Control, you don't have adequate Congressional Control....I maintain that in fields such as this the Congress, unless it provides a sound administrative organization, has very few resources; and that, on the other hand, the administration's hands are tied.[84]

At this point, McMahon intervened to demonstrate Smith's point, asking the budget director if he thought it would be practical for Congress to set quotas for the production of fissionable materials. After Smith answered with a predictable "no," McMahon continued: "I am frank to say that the more congressional supervision, the more congressional opportunities to correct mistakes in this situation, the more I am in sympathy with it, but I don't see, taking that [production quotas] for an example, how the Congress could fix a quota."[85]

Congress and those charged with coordinating national policy in the executive branch drew together because of the mutual need to subject independent expertise to broader political control. Administrators in the AEC were granted discretion because they translated the demands of experts into a language comprehensible to politicians, identified the political implications of the policies proposed by experts, and ensured that their scope and direction conformed to the existing political landscape. In return, a select group of elected officials obtained greater access to the policy-formation process within the agency – the right to be fully and currently informed – and a greater voice in that policymaking. If this compromise was to work, however, atomic energy had to be set in "a field apart." That outsiders found it virtually impossible to gain access to expertise in the years immediately following the war ensured this

83 SCAE, "Report on S. 1717," pp. 6–7, 10–12.
84 U.S. Congress, Senate, "A Bill to Provide Temporarily for the Development and Control of Atomic Energy," S. 1824, introduced February 9, 1946 (Sen. Johnson), 79th Cong., 2nd sess., 1946; SCAE, "Hearings on S. 1717," pp. 38–9.
85 SCAE, "Hearings on S. 1717," p. 40.

arrangement. Legislators left atomic energy up to the Joint Committee, while federal agency heads and White House staff steered clear of this "mysterious" area.

At the height of congressional reaction to the expansion of executive authority – the consequence of the explosion in bureaucracy during the New Deal and the war – the seeds of a new alliance were thus sown. As scientific experts surged into crucial agency positions, managers were increasingly forced into the role of translators or middlemen. Their influence depended less on their own expertise and increasingly on their ability to translate the expertise of others into politically acceptable policies – a role not unlike that of elected representatives. As we shall see, shared turf produced its share of squabbles between administrators and legislators – Senator Bourke Hickenlooper's charges of "incredible mismanagement," for instance. Nevertheless, as legislators probed into the workings of executive policymaking, and administrators measured political reactions to supposedly technical decisions, each branch of government learned how to work more comfortably with the other.

James Newman, having represented both sides, was one of the first to recognize the significance of this development. Commenting on the Atomic Energy Act of 1946, Newman wrote: "The act was passed in the midst of an overpowering nostalgia to return to things as they were, before the rude innovations of the New Deal and the disturbances of war. In passing it, Congress implicitly recognized that under the disintegrating force of the atomic bomb, the ancient institutional forms – honored and familiar though they were – had become obsolete."[86]

What about the scientists who were crucial to expanding the scope of debate in the first place? They had been successful in stopping May–Johnson and as a reward played a hand in drafting legislation they could support. But they paid a price for this victory.[87]

The public lobbying effort opened a deep fissure in the scientific establishment. With the introduction of the May–Johnson legislation scientist administrators like Bush and Oppenheimer were pressured to endorse a bill that was less than perfect in order to retain a voice in the administrative apparatus. Having been excluded from the negotiations, but not from the frustrations of army administration for the past two years, laboratory scientists greeted May–Johnson as a less than acceptable starting point for dealing with their demands. They organized effectively, found support, and influenced the Atomic Energy Act.

In the year following the entry of this "reluctant lobby" into the public fray, however, some of its members began to understand the dangers of broadening the scope of debate and the vagaries of congressional versus army oversight. Even though "atomic scientists" was a magic term, as John Simpson put it, the awe

86 Newman, "America's Most Radical Law," p. 443.
87 "A Victory and an Impending Crisis," editorial, *BAS* 2 (August 1946): 1, 32; Merle Miller, "The Atomic Scientists in Politics," *BAS* 2 (August 1946): 242, 252.

that those magicians inspired could easily turn to distrust. At its extreme, this distrust surfaced in the continuous accusations running from 1946 through the mid-1950s that scientists had betrayed their country by giving away the secret of the bomb. Highly sensitive to this charge, Congress ultimately imposed the very security restrictions that laboratory scientists had most objected to in the original May–Johnson legislation.[88]

In his study of congressional attitudes toward science and scientists, Harry Hall pinpoints some of the reasons for the sort of distrust reflected in these measures. A number of scientists were foreign-born and spoke with heavy accents. Politicians perceived that the application of science was a powerful independent force – one they feared they would be unable to control. Senator Tydings – thinking out loud – expressed his unease to the Special Committee: "I want to get rid of this feeling of frustration. I want some door open some place in the room where I can get out of here and get into an atmosphere where I am making progress."[89] Byron Miller suggested that scientists made politicians feel inferior. Perhaps the most important suspicion was that scientists were not practical men. Consequently, legislators preferred the advice of scientist administrators such as Bush, Conant, and Compton. These were men who had their feet solidly planted on the ground. They moved in realms familiar to Congress – public service, industry, and voluntary institutions.[90]

For their part, many scientists were appalled at the ignorance – and arrogance – they discovered on Capitol Hill. Scientists had taken the initiative in broadening the debate, but they found controversy was not limited to those with expert knowledge. Politicians proved even more likely than army administrators to treat any scientific knowledge even remotely related to atomic research as a secret that could be used for national security purposes, but locked up for other purposes. Even worse, legislators specialized in publicizing their views. Their ability to capture the media, the scientists soon learned, did not depend on the accuracy of their information.

The scientists' political involvement contributed to the demystification of their role and provided the basis for their being treated as just another interest group. Labeling experts "political" undercut one of their greatest political advantages. As Senator Eugene D. Millikin told John Simpson during the SCAE hearings: "I have got to discount [your testimony] somewhat, just as I have to discount the testimony of the military, because I don't think any of the special pleaders have taken into consideration enough things that must be taken into consideration." Organizing explicitly to influence politics captured the nation's

88 Simpson, "The Scientists as Public Educators," p. 245; Edward Shils, *The Torment of Secrecy* (Glencoe, Ill.: Free Press, 1956).
89 Harry S. Hall, *Congressional Attitudes Towards Science and Scientists: A Study of Legislative Reactions to Atomic Energy and the Potential Participation of Scientists* (New York: Arno Press, 1979), pp. 20–32, 49–58; quotation from SCAE, "Hearings on S. 179," p. 326.
90 Hall, *Congressional Attitudes*, pp. 36, 106–7, 258.

attention. But in doing so, laboratory scientists spent their most valuable capital – the general public perception that science is apolitical.[91]

Furthermore, during the public debate many scientists ventured beyond the area of their expertise, leaving them subject to challenge from politicians and more circumspect experts. The young Alvin Weinberg – who later would mature into as smooth a scientist administrator as any – found himself testifying about the causes of war. "Let me preface the following remarks again with the statement that I am not an expert. The person that should be talking about these things, I am afraid, should be a professor of international law," insisted Weinberg, before launching into a discussion that included Spain's struggle to wrest seapower from Britain.[92]

For scientists newly exposed to politics the struggle for legislation to control atomic energy was a sobering experience. Those who testified before congressional committees and staffed the speakers' bureau of the Federation of Atomic Scientists sought to educate and instruct. They did raise the level of debate, yet their indoctrination into the world of high-visibility politics gave pause as well. Painfully they discovered that perhaps the scientist administrators who shared their values, but sought to achieve them by limiting participation rather than expanding it, had their reasons.

Scientists saddled with administrative responsibility – even those who had been sympathetic to the laboratory scientists – recognized the perils of mass political organization. Even at the peak of the laboratory scientists' success, Robert Bacher, the only scientist on the original five-member Atomic Energy Commission, knew which way the wind was blowing. In August 1947 he scribbled on a Federation of American Scientists press release a note to fellow commissioner Bill Waymack: "I take a somewhat dim view of it." Waymack replied: "So the American and Russian positions differ but little, eh?" Two months later Bacher again commented on a Federation release. "This stuff might have sounded all right two years ago."[93]

The dispute between Edward Teller and Oppenheimer over developing the hydrogen bomb, Oppenheimer's highly publicized fall from grace, and McCarthyite attacks on intellectuals certainly all took their toll on the scientists' public image. William Borden, the man behind McMahon's religious-like conversion to cold warrior in the fifties, claimed that the Atomic Energy Commission's guilty verdict in the Oppenheimer loyalty case removed "physicists from the category of an anointed species." Jerome Wiesner, who was attacked by Senator Joseph R.

91 Smith, *A Peril*, p. 346; Guy Benveniste, *The Politics of Expertise* (Berkeley, Calif.: Glendessary Press, 1972), p. 63; quotation from SCAE, "Hearings on S. 179," p. 327.

92 Lee A. DuBridge, "Science and National Policy," *BAS* 1 (May 1, 1946): 12; SCAE, "Hearings on S. 179," p. 355.

93 Notations on "Federation of American Scientists Statement," August 1947, and "United Nations and Atomic Energy," September 10, 1947, "Federation of American Scientists" folder, Box 1, The Office Files of Robert Bacher, Records of the Atomic Energy Commission, Record Group 326, National Archives, Washington D.C.

McCarthy in 1954, recalled that McCarthyism raised suspicions about a previously unexamined sense of mission toward federal service. For both sides of the scientist–public divide, the Oppenheimer trial and McCarthyism meant the end of innocence.[94]

Disillusionment, however, did nothing to loosen the experts' and federal administrators' embrace. The attraction had developed slowly over the century and was consummated during the permanent crisis that followed World War II. The scientists' loyalty had been challenged, but a public more eager than ever for technological breakthroughs to ensure the nation's security did not question the scientists' expertise. The solution? Add loyalty oaths; don't get rid of the experts.

For scientists – even those, like Weinberg, who suffered directly at the hands of McCarthy – there was no turning back. Only the federal government could provide the resources, the organizational infrastructure, and the demand for their research. The solution? Exercise influence from the inside – as had Bush, Conant, and, even though he ultimately lost, Oppenheimer; don't forsake the prominstrative state.

94 Herken, *Counsels of War*, pp. 71–2; Gilpin, *American Scientists*, p. 133.

3

Forging an iron triangle: the politics of verisimilitude

By setting up a civilian agency to administer America's atomic program and the Joint Committee on Atomic Energy to oversee it, the Atomic Energy Act gave the look of regularized politics to what had been a highly exceptional arrangement. Several factors assured that regularization as it was understood before World War II was skin-deep at best. The premium on military priorities and corollary restrictions on access to information that the national security state placed on the program ensured that its development would not follow prewar patterns. Robert Oppenheimer and his colleagues on the General Advisory Committee (GAC) summed up the AEC's priorities best, informing the president in June 1952 that the GAC could not say where, when, and under what circumstances civilian power would be generated because the bulk of the AEC's work had involved military demands and the production of fissionable materials.[1]

Two additional factors distinguished the politics of commercial nuclear power from politics as usual. The first was the degree to which atomic power was government-initiated. To the extent that interest groups were involved with the issue, only the organized scientists contributed very much to the shape of the program. Government-funded research had made atomic power an issue. For the time being, it would take government initiative to develop it. It was a combination of scientists and administrators, working almost exclusively within the Atomic Energy Commission, that initially kept hopes alive for economically feasible commercial nuclear power, even though these advocates were often the first to admit that this would not occur overnight.[2]

Only the federal government with its virtual monopoly of institutionalized expertise and its financial resources could possibly realize this objective. Although the federal government had relied heavily on contractors such as General

1 GAC to the President, June 14, 1952, Minutes of GAC Meeting #31, June 13–14, 1952, Append. A, Atomic Energy Commission Files, Department of Energy Archives, Germantown, Md., p. 2. On military incentive for civilian power, see Clarke, "The Origins of Nuclear Power: A Case of Institutional Conflict," pp. 474–87.
2 On the lack of business interest, see Lowen, "Entering the Atomic Power Race: Science, Industry, and Government," pp. 459–69; and Clarke "Origins of Nuclear Power."

Electric, Du Pont, and Monsanto during the war, only the government had the capital to finance such a massive undertaking and only the government could have produced the organizational infrastructure required to pursue multiple paths toward its wartime objectives. Equally important, during the war, government was able to attract academic scientists who had previously shunned opportunities in industry. Crawford H. Greenewalt, a Du Pont chemical engineer who went on to head the company, urged that Du Pont get out of the atomic business after the war. That "suitable physics personnnel will be difficult if not impossible to get" was the first reason Greenwalt cited for his position.[3]

Research scientists' suspicion of industrial management did not end with the war. Alvin Weinberg, who headed the physics division at Clinton Laboratories, summarized some of the reasons for this in a confidential conversation. The scientists at Clinton were alarmed at the prospect of Carbide and Carbon's assuming management of the labs in early 1948, Weinberg reported. Weinberg, who stated candidly that this group of scientists was not the best in the world, but that they were not the worst either, felt they had "a certain dignity which is of great importance in their work.... Psychologically, these scientists do not want to work for an industrial organization. They want to feel that they work for the Atomic Energy Commission and for a living concept called Clinton Laboratories." Weinberg added that although it might only be hearsay, rumors that Carbide and Carbon management was even more narrow-minded and restrictive than Monsanto Chemical had given the scientists "the willies." A story making the rounds had "two ogres in Carbide and Carbon Vice-Presidents' clothing [saying] 'Wait until we get in there – we'll tame those scientists.'"[4]

The experience of Atomic Energy Commission reactor division director George Weil confirms Weinberg's attitudes toward industrial research. As part of Fermi's team, Weil was one of the first scientists recruited to build the Hanford reactors that produced plutonium for the Manhattan Project. Weil immediately noticed the difference in research management. Du Pont was "tightly organized with well-defined lines of command and communication, highly regimented, rigid in its demands for unqualified, unquestioned support from its employees." Following the war, Weil jumped at the opportunity to work for the Atomic Energy Commission because he was eager to "be involved in establishing policies during these early stages of nuclear power development." At least in the early years, the environment at the AEC was anything but hierarchical and bureaucratic. Weil describes the AEC's atmosphere as "informal." "Delegation of responsibility was the rule rather than the exception." Lilienthal put his finger on

3 Greenewalt, quoted in David A. Hounshell and John Kenly Smith, Jr., *Science and Corporate Strategy: Du Pont R&D, 1902–1980* (New York: Cambridge University Press, 1988), p. 342.
4 Weinberg spoke to William T. Golden by phone. Quoted in Golden to Strauss, January 7, 1948, Clinton Laboratories folder, Box 16, Lewis L. Strauss Papers [hereafter LLS], Herbert Hoover Presidential Library [hereafter HHPL], West Branch, Iowa, p. 2.

the reason for this fluid situation: The commission was trying "to ride a bicycle while, at the same time, building it."[5]

Neither industrial firms nor federal government could stem the flow of scientists back to universities following the war. The federal government, however, soon moved forcefully to create the expertise it required through fellowships and investment in the nation's research universities. Determined to exercise some influence over their research agendas, yet eager to guide the application of any results that such research might yield, a growing number of scientists stayed in the federal government, or consulted on a regular basis.

For scientists working in the Atomic Energy Commission, "the bottom line" was hardly a consideration. But there was a political bottom line. Momentary public enthusiasm about the "endless frontier" might sustain costly long-term development projects briefly, but over the long haul, the nuclear power program had to construct more conventional sources of support. The fledgling Atomic Energy Commission and its potentially powerful partner, the Joint Committee on Atomic Energy, sought to create demand for civilian power by enticing a third party to join their political alliance. The third leg of this triangle would be a strong economic interest group – the nuclear power industry. But in 1947, that industry was little more than a gleam in its one-day partner's eyes.[6]

IRON TRIANGLES IN THEORY

Since the Progressive Era, federal programs have depended on the tripartite support of an interest group, a congressional oversight committee (or subcommittee), and an administrative bureau. Political scientists and, more recently, historians have generalized about these three-way relationships, labeling them iron triangles, whirlpools, policy networks, subsystems, and triocracies. Arthur Maass's case study, *Muddy Waters*, is a classic example of an iron triangle, or whirlpool, as Maass later called it, in action. The economic interest group in this case study (the National Rivers and Harbors Congress) was so influential with the Army Corps of Engineers and with congressional oversight committees that, in Maass's words, "Senators and Representatives request[ed] that the pressure group give approval to their individual projects so that these projects may have a better chance of being approved in turn by the United States Congress of which they are members."[7]

---

5 Weil, "Nuclear Power: From Promises to Problems," pp. 28–9, 67, 74.

6 For a comparative account with Western Europe during this period, where, particularly after the Suez crisis, demand was much higher (and subsidies more easily obtained), see Bupp and Derian, *Failed Promise*, ch. 1.

7 Arthur Maass, *Muddy Waters: The Army Engineers and the Nation's Rivers* (New York: Da Capo Press, 1974), pp. 47–50, quotation from p. 47; for examples from the political science literature, see J. Leiper Freeman, *The Political Process* (New York: Random House, 1965), David Truman, *The Governmental Process*, 2nd ed. (New York: Knopf, 1971), ch. 9, Theodore J. Lowi, *The End of Liberalism: Ideology, Policy, and the Crisis of Public Authority* (New York: Norton, 1969), Lawrence

Historians have differed about the relative influence of competing interest groups in the creation of iron triangles, but there seems to be agreement about the general pattern that formation of iron triangles followed. Interest groups pressed for legislation. The committee with legislative jurisdiction approved (usually after considerable discussion and delay) an administered solution creating a new agency or reorganizing an established one while passing along unresolved conflicts to that agency. The agency sought solutions through the daily give-and-take of what Robert Wiebe has labeled "continuous management." Each member of the iron triangle participated out of self-interest. For administrators, interest groups were crucial sources of political support. Indirectly, experts benefited as well. If interest groups could establish firm control over a policy area, experts might be granted more freedom from public scrutiny.[8]

Roosevelt's response to the Great Depression began to modify the way such alliances were constructed without reducing the benefits involved for each participant. The executive branch of the federal government began to initiate legislation leading to administered solutions. Much of this legislation shored up the organizational abilities of established interest groups. Thus the National Industrial Recovery Administration (NIRA), starting with one of the best-organized sectors of society – industry – suspended antitrust legislation and encouraged business associations to articulate production goals. The act imposed discipline that the forces of competition had seemingly destroyed. Commodity supports in the Agricultural Adjustment Act imposed the same kind of discipline in the agricultural sector. The National Labor Relations Administration eliminated barriers to organization, such as the open shop that had thwarted labor's organizing efforts. In tackling one of the New Deal's greatest challenges – social welfare policy – the Social Security Board sought to organize the potentially volatile poor around income maintenance benefits while ultimately charting a course toward a more middle-class constituency via contributory social insurance. Most New Dealers implicitly accepted pluralist interest group theory. Some interest groups, however, had to be enhanced by federal initiatives and authority for the economy to function adequately.[9]

C. Dodd and Richard L. Schott, *Congress and the Administrative State* (New York: Wiley, 1979), Jeffrey M. Berry, *The Interest Group Society* (Boston: Little, Brown, 1984). Historian Louis Galambos describes these three-sided governing alliances as "triocracies" in *America at Middle Age.*

8 For instance, compare the role of the railroads as portrayed by Gabriel Kolko in *Railroads and Regulation, 1877–1916* (New York: Norton, 1965) to the role of shippers as portrayed by Robert Wiebe in *Businessmen and Reform* (Cambridge, Mass.: Harvard University Press, 1962). For a discussion of "continuous management," see Wiebe, *Search*, pp. 145, 153–4, 181–5. On the relationship between administrative agendas and political support, see Grant McConnell, *Private Power and American Democracy* (New York: Knopf, 1966); Selznick, *TVA and the Grass Roots*; Balogh, "Democratizing Expertise."

9 The roots of government-initiated efforts to shore up imperfectly organized groups can be traced back to the Progressive Era and were largely directed toward business. See, for instance, Hays, *Conservation*, and, in the 1920s, "The Commerce Secretariat and the Vision of an Associative State," *Journal of American History* 61 (1974): 116–40. On the New Deal, see Ellis Hawley, *The New Deal*

Much of the scholarly literature on iron triangles is critical. It compares this style of governing to a more idealized pluralist model and finds time and time again that the common good has been sacrificed for narrower achievements. An implicit conclusion that can be drawn from the literature, however, is that these types of alliances were essential if programs were to pass muster politically. To be successful in the twentieth century, legislation and administration demanded iron-triangle nurture.[10]

At its inception, the program to develop commercial nuclear power did not fit the iron-triangle model, even as adjusted for the contingencies of the New Deal. Interest groups were more of an afterthought than a crucial catalyst. A massive wartime effort inadvertently created a technology that might prove to be of great social and economic value. Owing to its sudden introduction and the uncertainty that clouded its potential uses, support for – or, for that matter, opposition to – its development remained poorly defined. Explaining one of the reasons why the McMahon bill precluded the industrial use of atomic energy at the very time it called for the development of peaceful uses, University of Chicago law professor Edward Levi stated, "[We] do not know enough about the commercial possibilities of atomic power to have intelligent legislation on that subject."[11]

The utility industry saw potential long-term benefits but was cautious in its appraisal. A few leaders, such as Walker Cisler of Detroit Edison or Philip Sporn of American Gas and Electric, did see opportunities to get in on the ground floor. Potential equipment manufacturers saw more immediate possibilities. Charles Thomas of Monsanto, for instance, was a forceful advocate. But many of the large manufacturers that had been involved with the Manhattan Project were comfortable with the cost-plus arrangements under which they had been operating during the War Department's administration or, like Du Pont, were eager to get out of the field completely. As *Fortune* magazine put it in 1946, "The McMahon Act seems to offer great opportunities to private industry, but private industry looks at the Act's administrators, the AEC, and coldly asks, 'For example, *what?*'" Industry as a whole, which stood to benefit from cheaper electric

and the Problem of Monopoly: A Study in Economic Ambivalence (Princeton, N.J.: Princeton University Press, 1966); Christopher L. Tomlins, *The State and the Unions: Labor Relations, Law, and the Organized Labor Movement in America, 1880–1960* (Cambridge: Cambridge University Press, 1985), part 2; Martha Derthick, *Policymaking for Social Security* (Washington, D.C.: Brookings, 1979), part 1; Balogh, "Securing Support," in Hawley and Critchlow, eds., *Federal Social Policy*, pp. 55–78.

10 Lowi, *End of Liberalism*, and Henry Kariel, *The Decline of American Pluralism* (Stanford, Calif.: Stanford University Press, 1961), are particularly forceful criticisms. Galambos, *Middle Age*, on the other hand, emphasizes the triocracy's discrete problem-solving potential.

11 Levy, quoted in U.S. Congress, Senate Special Committee on Atomic Energy, "Hearings Pursuant to S. Res. 179," 79th Cong., 1st sess., 1945, p. 110.

    James R. Temples is one of the few scholars who has used the subgovernment framework to examine the politics of nuclear power. Temples correctly notes the evolution from an insulated forum to the far more participatory politics of the late seventies. Temples (like most political scientists who have worked in this field) does not address the question of how such a relationship was established in the first place. See "The Politics of Nuclear Power," pp. 239–60.

rates, recognized this possibility but was concerned that the rapid introduction of a new technology might upset the economy. Meanwhile, powerful utility executives such as John Parker of Detroit Edison told Congress that "much of the operation cost of present systems lie outside the generating plants – for example, in the distributing systems and in the multifarious services to customers. There is no present indication that atomic energy will relieve that major element in cost." Coal interests were perhaps the only group that was not ambivalent. The "coal boys" saw no possible good coming their way and opposed the development of nuclear power from the start.[12]

The most forceful advocates of the development of civilian nuclear power in the 1940s were the very scientists who had been directly involved in its development and who had placed it on the agenda in the first place. Scientists as a group – particularly those who had not worked directly on building reactors – were ambivalent or concerned about pushing civilian development too quickly. They supported long-term development but feared that the fledgling civilian program might interfere with the struggle to establish international controls. Most of them had testified that international controls should come first. As the cold war deepened, however, and international controls became less likely, the general scientific community was more willing to support peaceful development. It is likely that many of these scientists, as McMahon and Lilienthal had, saw civilian development as the sole way to head off complete military domination of the Atomic Energy Commission. It is certain that they hoped something good could come of the powerfully destructive force they had helped to unleash. Whatever their motives, it took more than their support to sustain a program of the AEC's magnitude.

If the interest group leg of the incipient iron triangle did not meet government specifications, another leg – the congressional committee alliance – was hardly constructed out of tempered steel. In fact, in its first few years, the Joint Committee on Atomic Energy existed mainly on paper. Created in an effort to gain control over a massive program – the opposite sequence to traditional iron-triangle politics – the JCAE enjoyed some unusually broad statutory powers. Because it was a joint committee, it could speak for both houses. The Atomic Energy Act also stipulated that the AEC was to keep the JCAE fully and currently informed.

12 For *Fortune* and Parker quotations, see Clarke, "Origins of Nuclear Power," p. 477.

As late as 1953, the Commerce Department was concerned that the introduction of nuclear power might disrupt established business interests. See Secretary of Commerce Weeks to the Director of the Bureau of the Budget, September 28, 1953, "AEC Legislation–Bureau of the Budget, 1954" folder, Box 39, Records of the Joint Committee on Atomic Energy, 1947–77, Record Group 128, National Archives, Washington, D.C. [hereafter JCAE Files]. For a good survey of industrial reaction, see *Nucleonics* 1 (November 1947): 1–3. The initial reaction of the coal industry was to stress that it would take a long time to develop power from atomic energy. ("Atom's Harnessing Seen Long Way Off," *New York Times*, August 8, 1945, p. 7.) For an account of the coal industry's more recent reaction to nuclear development see John W. Johnson, "The Coal Boys' Attempt to Split the Atomic Lobby: A Tale of Two Technologies and Government Policy in the 1960s," paper presented to the annual meeting of the History of Science Society, October 1987.

On the other hand, the JCAE did not initially have the power to authorize funds – a way of reducing the risk of policy pronouncements or riders being inserted by the appropriations committees or on the floor. Moreover, the JCAE lacked expertise and familiarity with the issues. McMahon had anticipated this at the hearings on S. 1717 when he insisted that Congress would have to work through the executive branch in many instances.[13]

Until it could develop its own expertise and act upon its statutory authority, the JCAE depended on the AEC in matters of substance. It was through the AEC that the JCAE initially gained access to expert policy advice. The JCAE also shared the AEC's concern with developing a dependable constituency for the civilian power program. But its first priority was using the techniques it had available to it to establish some control over the AEC and its experts.

Commercial nuclear power development turned the traditional policy development sequence on its head. Rather than responding to the competing demands of interest groups, the administrative agency had to generate such demands. While cognizant of the need to promote demand itself, the newly created joint congressional committee first sought to rein in a highly technical government program that had been operating before the Joint Committee even existed. Administrators and elected officials both tried to enhance their political clout by establishing, after the fact, the triangular set of relationships that had in the past been prerequisites for the very existence of federal programs. To the extent that these men could establish that more familiar style of politics, they would also increase their leverage over the experts they sought to guide.

THE JCAE: LEGISLATIVE CONTROL AND PROTECTION

The Joint Committee on Atomic Energy was a paper tiger at its inception. It lacked the staff, expertise, and experience to take full advantage of its statutory right to be kept abreast of policy developments. Its lawmaking authority was of limited value in a field that was changing too rapidly for the laws to keep pace. Its potential as a public forum was blunted by the committee's own obsession with security. Bourke Hickenlooper, chair for the committee's first full session starting in January 1947, conceded that his committee did not become effective until mid-1947. A review of the JCAE's files suggests it was not until the early fifties that it acquired the access to information and expertise crucial to effective oversight.[14]

13 The Atomic Energy Act of 1946, 79th Cong., 2d sess., Public Law 585, sec. 15 (b); the JCAE picked up the power to authorize in an amendment to the act in 1954. Atomic Energy Act of 1954, 83d Cong., 2d sess., Public Law 703, sec. 261.
    On the powers of the JCAE, see Temples, "Politics of Nuclear Power," and Green and Rosenthal, *Government of the Atom*.
14 Thomas, *Atomic Energy and Congress*, pp. 57–60; quotation from U.S. Congress, *Congressional Record*, 80th Cong., 2d sess., vol. 94, p. 9062, cited in Thomas, *Atomic Energy*, p. 57. See also Green and Rosenthal, *Government of the Atom*, pp. 6–8.

As late as August 1949, the Joint Committee was still negotiating with the Atomic Energy Commission for access to the minutes of the commission's meetings. This was a privilege that the committee thought it had established two years before at the hearings to confirm the AEC's commissioners. At times JCAE staff had to rely on rumors to brief the committee. On the other hand, it showed little capacity to digest the mountains of data that the AEC did provide it. AEC commissioner William Waymack blamed the committee for failing to absorb or even look at this information. Briefings were triggered more by crises of the moment than by any mutual agreement on management information needs.[15]

AEC chair David Lilienthal also questioned the committee's ability to deal effectively with the information that was provided. He was frustrated by the committee's immersion in detail and attraction to the sensational. After one particularly irritating session Lilienthal confided to his diary, "Such self-assurance and superficiality and just orneryness." The highlight of a more successful meeting, in Lilienthal's opinion, was a demonstration of an atomic chain reaction. "Vandenberg jumped a foot and everyone was delighted.... Hickenlooper came all the way up to the office Saturday ... to be photographed while the thing was going off."[16]

Hickenlooper was partially to blame for the Joint Committee on Atomic Energy's slow start. Born in Blockton, Iowa, in 1896, Hickenlooper served as that state's senator for nearly a quarter of a century, from 1945 to 1969. Despite his seat on the prestigious Foreign Relations Committee, the *Cleveland Plain Dealer* in 1959 dubbed Hickenlooper "the most anonymous man in the Senate." The *New Republic* suggested that this stalwart Republican Iowan "ought to be holding a pitchfork in Grant Wood's 'American Gothic.'" Even Edward and Frederick Schapsmeier, authors of a rather admiring profile for *The Annals of Iowa*, concluded that "Hickenlooper was a follower not a leader."[17]

Hickenlooper, like all committee chairs, selected the legislative staff. His hiring criteria were one reason that the Joint Committee on Atomic Energy failed to acquire an expert staff. Hickenlooper chose key staffers with backgrounds in security, not science or engineering. This reflected Hickenlooper's obsession with "locking up" the secret of atomic energy and, as we shall see, also served political purposes.

15 McMahon to Lilienthal, August 10, 1949, memo labeled "For Committee Use Only," August 1949, "AEC General, 1947–1949" folder, Box 10, JCAE Files. U.S. Congress, Senate Section of the Joint Committee on Atomic Energy, *Hearings on the Confirmation of the Atomic Energy Commission and the General Manager*, 80th Cong., 1st sess., 1947 [hereafter Confirmation Hearings], pp. 23–7; W. W. Waymack, "Four Years Under Law," *BAS* 7 (February 1951): 54.
16 David Lilienthal, *The Journals of David Lilienthal*, vol. 2: *The Atomic Energy Years, 1945–1950* (New York: Harper & Row, 1964), pp. 205, 256.
17 *Cleveland Plain Dealer* and *The New Republic*, quoted in Edward L. and Frederick H. Schapsmeier, "A Strong Voice for Keeping America Strong: A Profile of Senator Bourke Hickenlooper," *Annals of Iowa* 47 (Spring 1984): 263–4.

Even had the chair of the JCAE sought more technical expertise, it would not have been readily available. Nuclear experts in 1947 were not easy to find. Seeking to evaluate the AEC's massive construction program undertaken in the early 1950s as part of the military buildup, the JCAE ran head-on into these problems. Few engineering firms had the experience required to speak as authoritative experts in the field. A response from the firm of Jackson and Moreland was typical: "Even if our opinions are sound, they will be challenged due to this lack of experience," the firm informed the Joint Committee. Those that did have the experience were already connected to the AEC, and were so overburdened with AEC work that they simply did not have the personnel to take on new tasks. The committee's silence on many substantive issues also confirmed its concern about being blamed for giving the Russians "the secret." Nor can its members' preoccupation with their responsibilities on other committees be discounted when explaining the JCAE's short attention span.[18]

It is not surprising that the JCAE had difficulty finding the expertise it needed to evaluate AEC activities. In the vacuum created by wrangling over the details of Bush's National Research Foundation (which emerged as the National Science Foundation in 1950), the Atomic Energy Commission, the Office of Naval Research, the Defense Department, the National Institutes of Health, and a host of other federal agencies competed for a handful of highly trained experts. In fact, one of the Joint Committee's greatest concerns was how the shortage of scientific and engineering personnel would be addressed by the Atomic Energy Commission. As one JCAE staffer put it in an internal memo, "It is difficult for the JCAE to get into the issue, but I feel that the whole atomic energy expansion problem is in jeopardy unless the nation's scientific manpower policies are reversed."[19]

To address these problems and to realize the potential authority that its statutory foundation promised, the Joint Committee felt its way toward the "carrot and the stick" strategy that would eventually pay rich dividends. It began with the familiar. Seeking to counterbalance the AEC's advantage in information and expertise, the JCAE pressed the agency on matters that legislators felt they knew a thing or two about. Side by side in the JCAE files with reports on the magnitude of the AEC's operation and the skills of its work force – 6,500 scientists engaged by AEC contractors; 10 percent of the Ph.D. physicists, and 3 percent of the Ph.D. chemists in the country – are responses to the JCAE's probes into the agency's less esoteric operations. How were the railroad yards at Oak Ridge run? How large was the lumber stockpile at Hanford? Matters that the legislators

18 F. G. Coburn to Admiral Strauss, February 25, 1952; Strauss to Borden, February 27, 1952; Strauss to McMahon, June 24, 1952, "Committee Advisory Panel" folder, Box 154, JCAE Files.
19 On shortage of experts in executive agencies, see Newman, "Era of Expertise," pp. 42–3. For an example of the AEC General Advisory Committee's chronic concern about trained personnel, see GAC to AEC, February 17, 1952, folder 1, Box 1272, AEC, Department of Energy [hereafter DOE], p. 5 On JCAE, Walter Hamilton to William Borden, February 26, 1952, "Manpower Policy– AEC" folder, Box 58, JCAE Files.

could easily grasp – especially management, access to information such as the size of the nation's stockpile of atomic bombs, and security – pervaded the public exchanges.[20]

As the confirmation hearings held in 1947 signaled, a number of factors motivated the committee's forays into matters far afield from the AEC's central objectives. These excursions were, in part, a partisan effort to embarrass a Democratic administration. All Truman appointments were subjected to some tough questioning, given the Republican majorities in the Congress in 1947. David Lilienthal, the chair of the Tennessee Valley Authority and symbol of New Deal ideology, engendered a particularly strong reaction from the committee at a time when the New Deal and Truman's effort to build on it were under attack. Lilienthal was repeatedly questioned about his willingness to make the benefits of nuclear power available to the private sector. He assured the senators that this was his goal. Added to Lilienthal's problems was Senator Kenneth McKeller, from the TVA's home state of Tennessee. McKeller had been hounding Lilienthal for some time. When the JCAE invited McKeller to attend its confirmation hearings, the senator's merciless line of questioning eventually provoked Lilienthal's famous lecture to his cross-examiner on the essentials of democracy, concluding with "This I deeply believe."[21]

Along with the partisan scramble for advantage and McKeller's personal vituperativeness, the Joint Committee on Atomic Energy launched the first formal round of its quest for substantive control over the Atomic Energy Commission. The issue was broached when Senator Millikin asked whether members of the Military Liaison Committee had been in attendance at commission meetings, a question that led to a wider-ranging discussion of JCAE access to these meetings. The following day, Senator Vandenberg – who undoubtedly felt that he would soon find other ways to obtain this kind of sensitive information – took a cautious approach. Vandenberg stated that "I do not want our Committee to have any access to security secrets of that character [number of weapons]." It was Lilienthal's strongest supporter and Democratic compatriot, Brien McMahon, who argued, "As a Senator ... and as a member of this committee, I deny the right of this executive commission to keep secret anything from me that I may need to know to discharge my responsibility. And that might mean some very secret information." Lilienthal sidestepped the question of what specific information the commission would or would not reveal, but essentially agreed to Joint Committee

20 For example, see Wilson to Borden, July 8, 1949, and other memos in "AEC–General" folders, Box 10, JCAE Files. Green and Rosenthal, *Government of the Atom*, pp. 5–12, and Thomas, *Atomic Energy and Congress*, pp. 23–140, portray the JCAE as playing an essentially passive role in its early years. "Passive–aggressive" is probably a better description. Both authors agree that by the early fifties, the JCAE had become a protector of the agency as well as a more important force in substantive policymaking.

21 U.S. Congress, Joint Committee on Atomic Energy, *Hearings on the Confirmation of the Atomic Energy Commission and the General Manager*, 80th Cong., 1st sess., 1947, pp. 1–34, 81–98, 805–34; see pp. 131–2 for Lilienthal's rousing defense of civil liberties and attack on innuendo.

access to commission meetings – quite a concession for an executive agency to make to an oversight committee chaired by the political opposition.[22]

Bourke Hickenlooper did not know much about reactors, but he was sure he could tell the difference between a well-managed program and "incredible mismanagement." He charged the AEC with the latter in May 1949. Democrats had regained their majority in both houses of Congress, and the Democrat McMahon replaced Hickenlooper as Joint Committee on Atomic Energy chair. No doubt, Hickenlooper was both grandstanding for his upcoming reelection in 1950 and using fairly standard tactics for the political minority when he attacked the Democratic administration. But he was also continuing a pattern by now familiar to members of the Joint Committee. Hickenlooper was trying to use what little political leverage he had to offset the AEC's vast advantage in resources and access to expertise. These were the kinds of techniques the committee's members used to gain access to policymaking.

When the AEC demanded an opportunity to respond to Hickenlooper's charges, McMahon held hearings. The AEC concentrated on its performance in the past two years. Comparing the results of its programs to the condition that the program was in when it took over from the army in 1947, the AEC argued, was the best way objectively to measure the quality of its management. The AEC should be measured by the numbers and on the basis of its technical achievements. Hickenlooper quickly beat a retreat to more familiar turf, stating, "From the standpoint of actual production, the atomic energy program has gone forward due to the zeal and the loyalty of the scientific and technical personnel in charge of the various projects. The point of my objection is not to the activities of these people." Hickenlooper insisted it was not the scientists but, rather, Lilienthal's "administrative policies" that were objectionable.[23]

The JCAE also pounded away on the issues of security and secrecy – issues that required little technical expertise yet provided leverage over a defensive AEC. Even as the Special Committee on Atomic Energy sought ways to internationalize the atom, events abroad and at home dashed hopes for a thaw in relations between East and West. In March 1946, Churchill declared that communist fifth columns advanced in front of "an iron curtain." Soon after, the Soviet Union rejected the Baruch plan for international control of the atom. At home, Truman in March 1947 committed America to supporting freedom-loving peoples around the world. For some senators sensitive to the deepening cold war, Lilienthal's confirmation hearings were an opportunity to repudiate earlier criticism from organized scientists. Hickenlooper, for instance, reminded Lilienthal, "I remember very well…that last year the Senate special committee was subjected to criticism that was most caustic indeed, from various scientific groups

22 Ibid., pp. 13–15, 23–7; Vandenberg and McMahon quotations from p. 26.
23 Lilienthal to McMahon, May 25, 1949, U.S. Congress, Joint Committee on Atomic Energy, *Hearings on Investigation into the United States Atomic Energy Project*, 81st Cong., 1st sess., 1949, pp. 7–8; quotation from p. 14.

and others in this country, because first, we were trying to preserve some secrecy."[24]

Lilienthal, eager to show his vigilance in an area in which he was perceived to be soft, gave the committee what it wanted. He called the Smyth Report "the principal breach of security since the beginning of the atomic energy project." By the time Lilienthal was done, Senator Vandenberg was moved to comment, "You talk a little longer about this thing, and I will quit worrying about you and security."[25]

Before long, the JCAE had to choose between creating more technical experts and the unquestioning loyalty it sought from all personnel working in fields related to atomic energy. In January 1948, Lilienthal informed Hickenlooper that the AEC planned to expand the fellowship program administered for the commission by the National Research Council of the National Academy of Sciences. Though pleased about federal intervention to produce more experts to meet the nation's crisis, Hickenlooper was concerned about security clearances. In July, Hickenlooper asked Lilienthal about the commission's procedures for clearances, noting that failure to conduct background investigations might result in the government's subsidizing a "potential subversive" who was not eligible for employment on the atomic energy program. Lilienthal responded that a review of the applicant's character was conducted if the individual would be working where there was access to classified information; where there was no access to restricted data, no clearance was obtained. This policy was designed to attract the greatest number of applicants possible, Lilienthal emphasized.[26]

In January 1949, Hickenlooper took issue with the commission's policy. Acknowledging that it was an administrative decision, Hickenlooper nevertheless warned Lilienthal that "when the facts are known, the expenditure of Government funds for the education of a Communist is indefensible and could easily subject our atomic energy program to justified criticism." Commissioner Robert Bacher politely registered Hickenlooper's dissent, but stood by the commission's policy.[27]

What had been a tug-of-war between the agency and the chair of its oversight committee over a matter of administrative discretion, took on larger proportions when the program came in for public criticism on just the issue Hickenlooper had raised. On May 10, right-wing radio commentator Fulton Lewis revealed that one AEC fellowship had gone to an avowed Communist. "Probably the onset of a bit of rough water," Lilienthal noted in his journal. Responding to the

24 Ibid., p. 33. On cold war background, Bernard A. Weisberger, *Cold War Cold Peace: The United States and Russia Since 1945* (New York: American Heritge Press, 1985), pp. 54, 56, 61.
25 "Confirmation Hearings," p. 32; Vandenberg quotation from p. 33.
26 Lilienthal to Hickenlooper, January 29, 1948, Hickenlooper to Lilienthal, July 30, 1948, and Lilienthal to Hickenlooper, October 30, 1948, "AEC Fellowship" folder, Box 28, JCAE Files; quotation from Hickenlooper to Lilienthal, July 30, 1948.
27 Hickenlooper to Lilienthal, January 12, 1949, Bacher to Hickenlooper, January 31, 1949, "AEC Fellowship" folder, Box 28, JCAE Files.

resulting public furor and to debate in both houses of Congress, JCAE chairman McMahon sought to diffuse the matter and contain it within his committee's purview by immediately conducting hearings.[28]

Lilienthal refused to back down at the hearings. He emphasized that the fellows did not have anything to do with making bombs. Speaking on behalf of the commission, its scientific advisers, and the National Research Council, Lilienthal stated that the "introduction of security procedures into nonsecret fields ... would establish a precedent of grave and far-reaching consequence to our scientific and educational system."[29]

Lilienthal had every reason to assume that his position would be backed up by the National Research Council. The previous year, Detlov Bronk, chair of the NRC, had written to the AEC informing them that it was the hope of the fellowship boards that applicants not be barred from participating in the fellowship program just because they failed to obtain a security clearance. But in his testimony, Bronk referred to that expression as "past history." When Hickenlooper asked if there would be any difficulty on the part of the NRC in recommending people who would be subjected to AEC-administered tests to determine whether the applicants were "potentially subversive in their views," Bronk answered that he could not see any problems.[30]

It was now too late for informal accommodation. Senator Joseph O'Mahoney, chair of the Subcommittee of the Appropriations Committee that was scheduled to begin work on the bill containing the AEC appropriation, stated that he was in favor of security examinations for all fellowship recipients. He introduced an amendment to the fellowship appropriation requiring that the National Research Council deny funds to members of potentially subversive organizations.[31]

Wielding the "stick" on issues with which legislators were not only familiar, but comfortable, was one of the ways the Joint Committee on Atomic Energy jostled for control. Angling for leverage, the technically overmatched committee muddled through housekeeping and security measures in order to apply political

---

28 Lewis alluded to the fellowship problem in a May 9 broadcast and kept up a running commentary on the matter through May 19. See U.S. Congress, Senate, Subcommittee of the Committee on Appropriations, *Hearings on the Independent Offices Appropriation Bill for 1950,* 81st Cong., 1st sess., 1949, pp. 622–33; Lilienthal, *Atomic Energy,* p. 528; U.S. Congress, Joint Committee on Atomic Energy, *Hearings on the Atomic Energy Fellowship Program,* 81st Cong., 1st sess., 1949 [hereafter JCAE Fellowship Hearings], p. 1.

29 JCAE Fellowship Hearings, pp. 1–10; quotation from p. 8.

30 Bronk to Carroll Wilson, July 27, 1948, "AEC Fellowship" folder, Box 28, JCAE Files; JCAE Fellowship Hearings, pp. 63–4. George Kistiakowsky has suggested that James Conant may have believed Bronk to be involved in a more significant "double-cross" a year later when Conant's nomination for president of the National Academy of Science was ambushed by dissenting chemists. Conant withdrew his name and was replaced by Bronk, who served as the prestigious body's head for twelve years. Kistiakowsky, cited in James Hershberg, "Over My Dead Body," pp. 40–41.

31 U.S. Congress, Senate Appropriations Subcommittee, *Appropriations Hearings,* pp. 254, 274; Independent Offices Appropriation Act for 1950, Public Law 266, 1949, sec. 102–A. In its final form, the amendment also prohibited fellowship funds from going to persons believed to be disloyal on the basis of an FBI investigation.

pressure on the AEC for access to expertise and a larger role in central policy planning. The spate of espionage cases and deepening cold war made security, and then weapons development, logical issues for the committee to pursue. But its concern with access, security, and management was also the product of the committee's need to apply existing skills to a field where it had not yet established its own expertise.

The JCAE did not always have to wield the stick: it had carrots to offer as well. The flap over fellowships, for instance, demonstrated the potential advantages of a committee–agency alliance, a crucial link in iron-triangle politics. There was no question that Hickenlooper grandstanded, particularly because he ultimately coupled the fellowship affair with a number of other security and management issues when he accused the AEC of "incredible mismanagement." But it is equally clear that his January letter had sent the commission a message: The commission's position was not defensible given the current climate in the Congress. Just as the Congress was probably the best barometer for an agency seeking to read the public's mood, the AEC could use the JCAE as a weather vane when it was crucial to know which way the wind was blowing in Congress.

Besides an early-warning system, Hickenlooper offered the Commission a compromise that would have settled the matter quietly, leaving the AEC considerable administrative flexibility. Lilienthal took a courageous stand on the issue: for him it was a matter of principle, and he framed his testimony in the broadest possible terms. Everybody who received any federal funds for education was at risk. But with Bronk and the National Research Council backtracking and the JCAE expressing its unanimous disapproval, the chair had little beyond principle with which to work. Lilienthal's stand attracted broader attention, virtually ensuring the kinds of restrictions ultimately imposed by the Appropriations Committee.[32]

Lilienthal's own journal demonstrates how the JCAE alliance had functioned on similar matters in the past. Describing a phone call to Hickenlooper toward the end of 1947, Lilienthal recounts Hickenlooper's advice. "Perhaps the committee should later next week issue a statement on the whole subject to 'take the edge off J. Parnell Thomas if he blows up,'" Hickenlooper suggested. It would reassure the public. The committee and the commission could draft the statement jointly. "This is a remarkable attitude," an encouraged Lilienthal confided, "and bodes good for the future."[33]

Gordon Dean, who succeeded David Lilienthal as chair of the Atomic Energy Commission in 1950, recognized the advantages that a congressional ally could provide and strengthened ties between the JCAE and the commission. Brien

---

32 Lilienthal, *Atomic Energy*, pp. 528–9; JCAE Fellowship Hearings; Senate Appropriations Subcommittee, *Appropriations Hearings*, p. 276. Even the most liberal members of the committee backed away from Lilienthal's position: See, for instance, McMahon cross-examination, JCAE Fellowship Hearings, p. 59, Holifield, p. 100; Carleton Shugg to McMahon, December 14, 1949, JCAE Files.

33 Lilienthal, *Atomic Energy*, p. 263.

McMahon's former law partner and his handpicked candidate for AEC chair, Dean understandably pursued a more conciliatory approach to administration than had Lilienthal. Dean lectured his AEC colleagues in August 1950, "If we have suffered from any one thing it has been controversy....Sometimes I think we have encouraged it by insisting we are always right." Noting that this self-righteous attitude irritated people, Dean suggested that AEC staff be more willing to admit "that we have occasionally made mistakes – measured at least by the hindsight of Monday-morning quarterbacks. I think also that perhaps we have carried too many chips on our shoulders. On this I may be quite wrong, but I do believe there is such a thing as becoming punch-drunk from minor league fights only to lose out in the championship battles." Dean kept the Joint Committee on Atomic Energy fully informed about the Atomic Energy Commission's actions. As late as August 1949, the AEC and JCAE were still haggling over the question of access to commission meetings – an issue raised at the confirmation hearings more than two years before. By September 1951, McMahon was writing to Dean to thank him for the "cooperative spirit" on this and other matters regarding the AEC.[34]

It was shared turf that ultimately brought the AEC and JCAE together. At times it seemed that the Joint Committee on Atomic Energy spent more time monitoring the actions of potential rivals than it spent overseeing the AEC. A February 1950 memo from Chief of Staff William Borden to McMahon, for instance, warned that Senator Burnet Rhett Maybank of the Subcommittee on Small Business had proposed an "investigation in AEC subcontracts." Borden had also heard through the grapevine that the House Appropriations Committee was going to cut off funding for the Oak Ridge and Richland community management programs unless administration was transferred from the AEC to the General Services Administration. "The only way to get around the potential challenge to our jurisdiction," Borden advised, "is to appoint a subcommittee and tell the House Appropriations Committee it's already being studied, or amend the McMahon Act."[35]

As the Joint Committee on Atomic Energy and the Atomic Energy Commission moved closer to each other, the Joint Committee began to perceive issues through the more specialized lens of atomic development – as opposed to the varied perspectives its members had expressed as elected officials. The JCAE promoted a centralized, national program. Like all legislators, JCAE members

---

34 Transcript of Remarks of Gordon Dean to the Commissioners and Washington Staff of the AEC, August 9, 1950, AEC, DOE. Dean seems to have touched every athletic base in this talk. On actual improvements in the relationship see McMahon to Lilienthal, August 10, 1949, "AEC–General, 1947–1949" folder, McMahon to Dean, September 4, 1951, "AEC–General, 1951" folder, Box 10, JCAE Files.

35 Borden to McMahon, February 14, 1950, "AEC–General, 1950" folder, Box 10, JCAE Files. The JCAE chose the first option suggested by Borden, conducting hearings in executive session in April and June 1950. See U.S. Congress, Joint Committee on Atomic Energy, *Executive Hearings on Community Policy*, 81st Cong., 2d sess., 1950.

used their committee assignment to provide pork for their own districts. They didn't always choose the national perspective. The Joint Committee, for instance, pushed hard for smaller reactors in response to pressure from groups like the Rural Electric Cooperative Association, and the American Public Power Association, despite the fact that these smaller reactors were the least likely to produce commercially competitive power. For the most part, however, the JCAE was an important bulwark in Congress against such splintering tendencies.

States' rights was potentially the most powerful splintering threat. In practice, the states were remarkably deferential for the first decade of nuclear development. Occasional issues did crop up, however. For instance, a January 1952 Supreme Court ruling determined that under the wording of the Atomic Energy Act the commission's contractors were exempt from state and local taxes. States were outraged by the decision and called for legislation to ensure that private contractors would not be able to evade taxation. The JCAE, however, dragged its feet, tacitly supporting the tax breaks in opposition to the states. Ultimately the committee succumbed to mounting pressure on some of its members from state governors. It eliminated the tax breaks by amendment in July 1953. Nevertheless, the Joint Committee remained an important line of defense by deflecting local and congressional demands for AEC facilities. The JCAE helped ensure that site selection was left largely to criteria developed by the General Advisory Committee and Division of Reactor Development of the AEC.[36]

Access to the General Advisory Committee's expertise also helped committee members say yes to requests from less informed politicians for facilities that JCAE members themselves did not particularly care to have in their own backyards. The competition for the site of the National Reactor Testing Station was a good example of this pork rind politics. While the governor of Montana was petitioning the JCAE to place this project in his state, J. Robert Oppenheimer, chair of the AEC's General Advisory Committee, briefed committee members at a closed hearing on the need for an isolated location. The facility could be dangerous, Oppenheimer testified. "I was astonished," Oppenheimer continued, "to know that many people were wishing for this proving ground in their state." McMahon jumped in, bragging, "At least I said I didn't want it in Connecticut."[37]

36 See, for instance, James Grahl to James Ramey, May 14, 1958, and Grahl to Ramey, June 2, 1958, "Civilian Power Program, Part I" folder, Box 122, JCAE Files; and Clyde Ellis to Admiral Strauss, February 3, 1955, AEC 777/4, AEC DOE. On the JCAE's protection of its national interests, see Carson v. Roane Anderson Company 72 Sup. Ct., January 7, 1952; S. Walker to Files, March 19, 1952, Borden to McMahon, March 24, 1952, "AEC Legislation, Taxation Sec. 9-B" folder, Box 38, JCAE Files; "State Taxation of Atomic Energy Commission Activities," S. Rept. 694, 83rd Cong., 1st sess., July 28, 1953. For examples of localized requests for reactor sites, see R. F. Kitchingman, Manager, Great Falls Chamber of Commerce to McMahon, April 4, 1949, Governor John Banner to McMahon, May 18, 1949, and Lilienthal to McMahon, March 9, 1949, "Reactor Site General, 12/48–6/49" folder, Box 631, JCAE Files.
37 Minutes of an Executive Meeting of the Joint Committee on Atomic Energy, April 6, 1949, p. 18, "J. Robert Oppenheimer" folder, Box 10, Office Files of David E. Lilienthal, Records of the Chairman, Records of the Atomic Energy Commission, Record Group 326, National Archives, Washington, D.C.

It was in defending the program's budget, however, that the AEC and the JCAE established a long-term mutually advantageous relationship. The JCAE regularly intervened in an effort to protect the AEC's budget in the late forties and early fifties. Just as the AEC argued that it deserved enhanced discretion on matters ranging from reactor development to reactor safety because of the high level of expertise required, in the congressional forum, and particularly on budget matters, the JCAE mimicked this position.[38]

The Joint Committee on Atomic Energy argued that it had the experience to assist both appropriations committees with these kinds of decisions. In one of its earliest attempts to pass an amendment that would have given the JCAE the power to authorize funds, its leadership argued that because secrecy prevented the normal amount of public scrutiny in the area of atomic energy, the Joint Committee was the only "entity throughout the entire nation with sufficient knowledge to 'pinch hit' for public opinion." The statement argued that in most areas, "a kind of Congressional teamwork prevails, with the legislative committees serving as program experts in their respective fields." The JCAE's inability to authorize spending prevented these specialized skills from being exercised in the field of atomic energy. Lacking the statutory ability to authorize expenditures, the JCAE sought an informal voice on the appropriations committees, pointing to the JCAE's expertise to justify this role. This effort met with some success in the Senate, but in the House, the JCAE's strategy failed. Still, the JCAE persisted, intervening on behalf of the AEC in later stages of the budget debate. Its most successful defense came at the height of the atomic buildup in 1952, when last-minute bipartisan appeals by JCAE members helped stave off a 50 percent cut in expenditures proposed by the House conferees.[39]

The JCAE's campaign to obtain the power to authorize funds underscores just how fine a line distinguished mutual support from self-aggrandizement as oversight committee and administrative agency explored the potential as well as the limits of their relationship. The Appropriations Committee fellowship rider was the worst case: a committee less sympathetic to its needs than the JCAE could easily intervene in the program's future if the AEC sought too much independence from its congressional oversight committee. If there had to be congressional riders, better that they originate in the JCAE. Nevertheless, the AEC resisted Joint Committee efforts to acquire authorization power. The AEC sought to prolong the administrative discretion granted following the war.

38 Thomas, *Atomic Energy and Congress*, ch. 4; Philip Mullenbach, *Civilian Nuclear Power: Economic Issues and Policy Formation* (New York: Twentieth Century Fund, 1963), pp. 130–8.
39 Green and Rosenthal, *Government of the Atom*, pp. 83–7, 169–75; Thomas, *Atomic Energy and Congress*, ch. 4. Quotations from Memorandum Prepared by the Chairman and Vice Chairman of the Joint Committee on Atomic Energy Regarding Amendment to the McMahon Act, June 29, 1949, "AEC Legislation (General)" folder, Box 36, JCAE Files, pp. 1, 2. On JCAE intervention into budget debates, see Thomas, *Atomic Energy and Congress*, pp. 127–31.

Having cut the commission down to size by the kinds of direct assaults levied by Hickenlooper, and having demonstrated its potential value to the commission, especially in fighting for its budget, the Joint Committee on Atomic Energy pushed for more explicit controls and was now ready to play a more assertive role in developing a nuclear program. In 1954, the Congress formally acknowledged this when it ignored AEC objections and granted the JCAE power to authorize funds as part of the sweeping amendments to the act passed in that year. In 1957, the JCAE built upon this formal statutory authority when it began "informally" to sign off on specific projects to be built with funds from the reactor demonstration project – the commission's major subsidy for private construction of reactors. The committee did not hesitate to point out the crucial role its members had played in assuring adequate funding. As JCAE chair Carl Durham put it to AEC chair Lewis Strauss in August 1957, "I am sure you recognize that the members of the Joint Committee took an active part on the floor and in the House and Senate appropriations committees in obtaining and restoring necessary AEC funds."[40]

It was an aggressive policy agenda that ultimately powered the Joint Committee's newfound assertiveness. The JCAE may have owed its existence to the distinction that McMahon and some of the scientists drew between civilian and military uses, and to the potential peaceful uses that McMahon in particular emphasized at the Special Committee on Atomic Energy's hearings, but civilian development was not uppermost in McMahon's mind by 1950. Responding to the Soviet Union's detonation of an atomic bomb in the fall of 1949, the main thrust of the committee's oversight shifted, from what the AEC had been mismanaging to the new management project it faced in the race for the hydrogen bomb. McMahon, under the influence of his new chief of staff, William Borden, began pressing the AEC to increase its weapons material production.[41]

By 1950, McMahon, who had built a career championing the potential peaceful uses of atomic energy, spearheaded the quest for a hydrogen bomb. McMahon embraced the cold war buildup with the same zeal that earlier he had devoted to undermining military control of atomic energy. The senator who had carved out a place for the civilian atomic perspective in the first place now sought to articulate that vision. Ironically, in doing so, McMahon sounded very much like the congressional military and foreign policy competitors that the freshman senator had beaten to the punch on atomic energy in 1945.[42]

---

40 Green and Rosenthal, *Government of the Atom*, pp. 171–2; Thomas, *Atomic Energy and Congress*, pp. 132–40. On 1957 "informal" authorization see HHPL pp. xv–7–9. For Durham quote, see Durham to Strauss, August 24, 1957, Enclosure A of AEC 777/51, September 6, 1957, Box 1248/18, PDRP folder, AEC, DOE.
41 On Borden, see Gregg Herken, *Counsels*, part 1.
42 On McMahon's embrace of the weapons buildup, see Hewlett and Duncan, *Atomic Shield*, pp. 179–81, 371–3, 392–5.

BUILDING A CONSTITUENCY: THE DREAM OF MASS PARTICIPATION

Developing civilian atomic power and institutionalizing the kind of organization and sense of urgency that already existed for the military side of development were the AEC's distinctive mission. Simultaneously developing both civilian and military applications proved to be the greatest challenge for the AEC's first chair. David Lilienthal tried to steer a course that kept the issue before the public without fanning unrealistic hopes or inciting the coal industry to even fiercer opposition. Faced with increasing pressure from the military and the Joint Committee on Atomic Energy, Lilienthal made a crucial political decision. He opted to broaden the foundation of the AEC's support. If public interest and support had helped outflank the military during the AEC's embryonic stage, Lilienthal must have reasoned, it was a good bet to see the commission through the next stage of its development.

Lilienthal had plenty of opportunities to consider this high-risk strategy during the ordeal of his confirmation hearings. On March 15, 1947, he confided to his diary that atomic energy must be saved from the "routine government operation, bogged down, subject to the small potatoes of Congressional piddling." The way to prevent it, Lilienthal continued, "is an appeal to the public imagination, a series of such appeals, through every medium of communication. If my energy will hold out, I believe this can be done."[43]

Lilienthal's personality and experience as head of the TVA made him the ideal man to implement this strategy. Political scientist Erwin Hargrove portrays Lilienthal – a boxing coach in college – as a fighter who believed that an organization's effectiveness depended on public support and participation. Rhetoric had been his chief instrument at the TVA, and he used it once again in an effort to build a broader constituency for the AEC. As he wrote in his diary on the eve of the AEC's birth, the thing "that interests me the most is the problem of informing the American people of what lies ahead for them as a result of the entry of this new source of energy and means of learning about the world." The first fruits of Lilienthal's public relations effort on behalf of the peaceful applications of atomic energy grew out of his discussions with Edward R. Murrow, of CBS. Murrow ran an hour-long program entitled "The Sunny Side of the Atom" in July 1947. "A noble beginning," Lilienthal remarked."[44]

Lilienthal knew that his rhetoric required concrete examples if it was to mobilize the masses. At the TVA, he had pointed to lower electricity rates. Lilienthal was particularly fond of demonstrations. He described to a friend the techniques he had used for dramatizing the tangible results that the TVA could bring to the average farmer.

---

43 Lilienthal, *Atomic Energy*, pp. 160–1.
44 Erwin C. Hargrove, "David Lilienthal and the Tennessee Valley Authority," in Doig and Hargrove, *Leadership and Innovation*; Lilienthal, *Atomic Energy*, pp. 126, 217.

I used to make speeches before county crowds with a lot of farm machinery gadgets (grinders for feed, brooders, etc.) set up on a big table in front of me, and would work these into the talk, indicating how much some particular farmer somewhere had added to his net income when he had these machines.... Well, it was undignified as hell, like an Indian root doctor, but those farmers listened to every darn word, and came up afterward and handled the gadgets.[45]

While touring Manhattan Project facilities with his fellow commissioners-elect, Lilienthal met another master salesperson – Ernest Lawrence. Lawrence had moved well beyond the days of radioactive cocktails, and now presided over a burgeoning research empire at Berkeley funded in large part by the AEC. Lawrence stressed that atomic energy could produce electricity in a very short time if the existing Hanford reactor was modified to achieve this end. According to Lilienthal, Lawrence was fearful that longer delay in order "to hit the jack pot" (as General Electric and others had suggested) was a mistake and "might result in a reaction and skepticism against the whole built-up beneficial-uses story.... He hammered this hard," Lilienthal continued.[46]

Unfortunately for these public promoters, interest in the issue was declining in general. To reactivate his base of support, Lilienthal needed the kind of tangible results Lawrence had in mind, or at least the likelihood that such results were around the corner. With them, and the artful use of his powers of communication, Lilienthal saw the opportunity to reestablish the winning combination of his TVA days. Lilienthal turned to the AEC's statutory advisory committee of experts – the General Advisory Committee – for help.

The nine-member GAC was appointed by the president to advise the commission on scientific and technical matters. Required by the Atomic Energy Act of 1946, this committee was the legacy of Vannevar Bush's early proposals for a part-time commission staffed largely by experts. The first GAC was chaired by J. Robert Oppenheimer and included such luminaries as James Conant, Enrico Fermi, and Isidor Rabi. In the words of the AEC's Second Semiannual Report, the committee "rendered invaluable assistance ... in reviewing and advising on its major programs, furnishing technical advice on many specific problems and recommending additional programs and policies on its own initiative." Oppenheimer summed up its influence more accurately, if less formally: "In the early days we knew more collectively about the past of the atomic energy undertaking and its present state, technically and to some extent organizationally or some parts of it, than the Commission did." More impartial observers agreed with this assessment: in the early days the General Advisory Committee had an inordinate amount of influence.[47]

45 Hargrove, "David E. Lilienthal," pp. 45–6; Lilienthal, *The Journals of David E. Lilienthal*, vol. 1: *The TVA Years, 1939–1945* (New York: Harper & Row, 1964), p. 80, cited by Hargrove.
46 Lilienthal, *Atomic Energy*, p. 108. See also Lowen, "Entering the Atomic Power Race," p. 464.
47 U.S. Atomic Energy Commission, *Second Semiannual Report of the Atomic Energy Commission*, 80th Cong., 1st sess., S. Doc. 96, 1947, p. 22; U.S. Atomic Energy Commission, *In the Matter of J.*

The General Advisory Committee often discussed the political dimension of proposals: at times members seemed to be as concerned with the political consequences of their actions as with the technical side of things. The GAC, for instance, was just as concerned as Lilienthal about the need for concrete demonstrations of atomic energy's benefits; but it was equally concerned that expectations – even among informed scientists – were beginning to outstrip the reality of the AEC's power reactor development. At its first meeting, several members suggested that the committee should concentrate on the immediate construction of a reactor for the production of electric power. The committee returned to the issue at its next meeting – in February. Oppenheimer pushed for reactors in order to (1) advance the international aspects of atomic energy through demonstrations of its peaceful utilization; (2) affect public opinion in a similar fashion in the United States; and (3) ensure sufficient fissionable material so as to eliminate questions about allocation (between weapons or peaceful uses). He also pushed for a demonstration in a year or two; expanded power production should follow by the fifth year of the program. Enrico Fermi, who had consistently been among the most reluctant to develop commercial power, differed in his assessment of the political situation, arguing that weapons would be more important than Oppenheimer allowed, given the international situation. Nevertheless, he did agree that power production was a "worthwhile psychological factor," suggesting three years as a reasonable timetable.[48]

As the deepening cold war made it more apparent that demand for weapons material would remain high in the foreseeable future, the General Advisory Committee tried to temper the expectations that had developed outside the Atomic Energy Commission (and perhaps among the commissioners as well). At a meeting in March 1947, Oppenheimer warned his colleagues that it was their obligation to educate fellow scientists about the AEC's program. Lee DuBridge, president of the California Institute of Technology, proposed issuing a statement warning that the application of atomic energy for power was probably many years off. As the reason for this delay, DuBridge cited military demands, limited supplies of raw materials, and the time it would take to perfect breeder reactors. Fermi estimated it would be fifty years before atomic power production would even come close to existing levels of consumption.[49]

The General Advisory Committee was pessimistic because of the fuel situation and the steps required to alleviate this shortage. There was concern about the supply of uranium in its most frequent natural state – 238U. There was even greater concern, however, about the availability of "enriched" fuel. Weapons required this "enriched" form of uranium – 235U (which had to be separated out

Robert Oppenheimer: Texts of Principal Documents and Letters of Personnel Security Board, General Manager, and Commissioners (Washington, D.C.: 1954), p. 67. On the authority of the GAC in its early years see Hewlett and Duncan, *Atomic Shield*, p. 46; Gilpin, *American Scientists*, p. 12.

48 GAC Minutes #1, January 3–4, 1947, pp. 3–4; GAC Minutes #2, February 2–3, 1947, AEC, DOE.

49 GAC Minutes #3, March 28–30, 1947, AEC, DOE, p. 5.

from the 238U that occurred in nature at a rate of 99.3 percent compared to 235U's rate of 0.7 percent) or 239Pu (plutonium) – an entirely man-made substance. Reactors also required enriched uranium. Creating 235U and plutonium in quantity had been one of the major wartime challenges – the object of much of the massive plant at Oak Ridge and Hanford. In 1947 producing sufficient enriched fuel simply for weapons production was perhaps the AEC's most formidable technical task. Because the prospects of increasing production in the immediate future were slight, the surplus available for peaceful uses depended on the level of international tensions. In 1947, few could argue with Fermi's pessimistic outlook.[50]

That is where "breeding" entered the picture. If reactors powered by plutonium (which was a product of the reaction between a small 235U core and a larger 238U "blanket" that absorbed excess neutrons) could be built, the conversion process would actually produce more energy than it consumed. Reactors that could convert the relatively abundant 238U into plutonium were called "breeding reactors": the plutonium fuel, which had far more potential energy than that consumed by the production of plutonium, could then be inserted in the reactor's core. A small amount of plutonium could power the energy needs of the nation if this highly toxic, weapons-grade material could be "tamed" to produce heat rather than an explosion. The alternative to breeding – and the path eventually taken in America – was to make the process for separating 235U from the more frequently occurring 238U more efficient. Technologically, the prospects for both breeding and more efficient enrichment looked bleak, at least in the short run.[51]

At its next meeting the GAC agreed to draft a statement on the "realistic evaluation of the possibilities of atomic power" for public release. Oppenheimer informed the commissioners that he was alarmed by his conviction that atomic power was at least a decade away particularly when juxtaposed with the popular belief that commercial use was much closer at hand. That perception was not irrational "because people have said that a lot of good can come of the atom." This, Oppenheimer continued, has created "a rather bad discrepancy between expectation and probable reality."[52]

By the end of July – about a month after Murrow's "Sunny Side of the Atom" broadcast, the AEC's expert advisers demanded action from the commission. Oppenheimer raised the issue of a public statement about reactor development. Lilienthal wanted to know what its purpose would be. It would be aimed at three groups, Oppenheimer responded: scientists, who are currently discussing such matters in an unrealistic fashion; men of public affairs (Lilienthal and his col-

50 On fuel conversion, see David Rittenhouse Inglis, *Nuclear Energy – Its Physics and Its Social Challenge* (Reading, Mass.: Addison-Wesley, 1973), pp. 75–81, 99–100.
51 On breeding, see William Lanouette, "Dream Machine: Why the Dangerous and Maybe Unworkable Breeder Reactor Lives On," *The Atlantic* (April 1985), p. 37.
52 GAC Minutes #4, May 30–June 1, 1947, AEC, DOE, quotation from p. 17.

leagues, perhaps?); and industrialists. Conant added that "there is little chance in getting the general public to understand these things when the scientists themselves do not." DuBridge seconded this thought. He emphasized that there was a widely held belief among scientists that there was no technical reason for the delay. A logical conclusion might be that the commission had failed in this field.[53]

The commission balked. Lilienthal argued that a single statement would not solve the problem. He also warned that a public statement estimating that atomic power would not be available for a long time would deter young people from entering the field, further exacerbating the shortage of experts. The debate resulted in a compromise: the General Advisory Committee would draft a statement for discussion with the commission.[54]

The General Advisory Committee's proposed statement certainly was not cheery. The atomic advisers began by pointing out that new fields of discoveries and inventions might make present views obsolete, but nevertheless, knowledge had to be distinguished from vague hopes. The statement went on to report that because there was a shortage of natural uranium and enriched uranium 235, the most promising approach for the future was to develop reactors that could generate power at high temperatures while creating more fuel than they consumed (breeder reactors). Even with an intensive development program, it would take ten years to develop the right technology. After that, "Decades will elapse before stocks of nuclear fuel can have accumulated which will supplement in a significant way the present power resources of the industrial nations of the world."[55]

Both the General Advisory Committee and the commission were acutely aware that commercial uses for atomic energy had forged the political support for their mandate as a civilian agency. Commissioner Sumner Pike was distressed by the delay. What had happened to the GAC's estimate of development in three years? Committee members responded that three years was an estimate of a demonstration only. Later in the meeting Pike noted that the commission had been on trial from the start. Its already hostile external environment had worsened, according to Pike. "To say now that atomic power is very far distant, and that we now have in atomic energy only a military tool with some byproducts is not going to help the situation." Conant pointed out the unpleasant alternative. Should the optimistic impression continue without any actual result, scientists would conclude that either the commission was incompetent or was completely military-minded. Commissioner Lewis Strauss felt that the statement was so pessimistic that it would hurt the commission's operations by making funding more difficult. He proposed that the statement be delayed until the October meeting.[56]

In private, reactions were even stronger. Lilienthal began his diary entry with

53 GAC Minutes #5, July 28–9, 1947, AEC, DOE, pp. 3–4, quotation from p. 3.
54 Ibid., p. 4.
55 Draft Note on Atomic Power from the General Advisory Committee, July 29, 1947, Append. to GAC Minutes #5, AEC, DOE, pp. 1–2.
56 GAC Minutes #5, AEC, DOE, p. 7.

these words: "Had quite a blow today." Lilienthal felt that the GAC statement discouraged hope for atomic power for decades and was phrased in a way that questioned whether it would ever have an impact on energy supplies. "This pessimism didn't come from nobodies," Lilienthal noted.[57]

Lilienthal had gained an appreciation for the scientists' political sensibilities. The experts and the commission had the same long-term objectives in mind – protecting the integrity of civilian control and, more significantly, keeping the political basis for that control alive. "Actually, a properly worded statement deflating the atomic power *overoptimism* would be definitely in the interest of the Commission," Lilienthal confided to his diary. "[T]his was their [GAC advisers] purpose in proposing it.... It is curious how political reasons motivate scientists and savants," Lilienthal observed, "though they would be surprised if you pointed it out to them, and probably deny it." Those administrative scientists pursuing a strategy of political control from insulated forums undoubtedly would have denied that they were political. They were well aware that their political effectiveness depended not on the size of their constituency, but rather on discretion granted to them in deference to the "apolitical" nature of technical decisions. That "apolitical" aura, in turn, required a public consensus among the experts on technical matters.[58]

Oppenheimer also had some strong views about the interaction of expert advice and politics. In a letter to Commissioner Bacher (who had not been able to attend the GAC/AEC meeting), Oppenheimer expressed doubts about the Commission. "I was very much impressed and very much appalled at the way in which the Commissioners responded to the draft statement on power," Oppenheimer wrote to Bacher. "It was not that they were concerned about its publication. That was natural, and in my opinion sound. It was that no one of the Commissioners appears to have the least idea as to whether this was a true or false statement, whether it was a wise or a foolish one; nor did any of the commissioners evidence any sign of familiarity with the issues which we tried to raise."[59]

Over the next four months the GAC drafted a number of revised power statements, seeking a compromise between the technical and political realities of power development. The drafts and ensuing discussions reflected a slightly more optimistic tone, particularly in regard to the availability and price of natural uranium. Nevertheless, the draft finally approved by the commission concluded that even under the most favorable circumstances, the committee did not see how it would be possible to replace any considerable portion of the current power supply with nuclear fuel in less than twenty years.[60]

57 Lilienthal, *Atomic Energy*, p. 229. 58 Ibid.
59 Oppenheimer to Bacher, August 6, 1947, Bacher Files, p. 5.
60 GAC Minutes #6, October 3–5, 1947, p. 15, GAC Minute #7, November 21–3, 1947, AEC, DOE, pp. 8, 12, 16, 18; Draft Note on Power, October 23, 1947, November 17, 1947, and November 23, 1947, Energy History Collection, Department of Energy Archives, Germantown, Md.

Until the General Advisory Committee's statement the commission had been vague about the precise arrival date for atomic power. The AEC's Second Semiannual Report, issued in July 1947, for instance, stated that "it is speculative, at this stage of development, to estimate how long it will be until power production units feasible for large scale commercial use will be available." Coming on the heels of the GAC's statement, the next semiannual report was far more pessimistic. It broke the bad news by quoting a speech given by Lilienthal on October 6, 1947: " 'The first commercially practical atomic power plant is not just around the corner, not around two corners.'" The development of atomic power was from eight to ten years away, the report continued. Even worse, that estimate did not consider the time it would take to commercialize atomic power fully. It would take eight to ten years simply to " 'overcome the technical difficulties and have a useful practical demonstration plant in operation.'" The *Report* went on to quote the GAC's estimate of twenty years before any considerable portion of the energy supply would be provided from atomic sources.[61]

Robbed of the concrete benefits he so urgently needed to sell atomic energy, Lilienthal turned to a less tangible but potentially more powerful alternative: he launched a campaign to promote public involvement in the development of atomic energy. Lilienthal gave a number of speeches encouraging public participation, linking it to long-standing American traditions. "Atomic Energy Is *Your* Business," delivered to a community gathering at Crawfordsville, Indiana, and broadcast on the CBS network, was the best publicized.[62]

Lilienthal raced to add muscle to the statutory skeleton of civilian control that the entrepreneurial McMahon, Smith, and Newman embedded in the 1946 legislation. Lilienthal gambled that mass public involvement was just what the doctor ordered to protect the commission's political life, which was threatened by elite models of decision making – both technical and military. Taking issue with the notion that the subject of atomic energy was too "technical" for public discussion, or that "national security" required that the public be kept in the dark, Lilienthal argued that atomic energy was the people's business. The whole question of how atomic energy would be applied was more important than any of the technical details, Lilienthal told his national audience. "These are not scientific matters. What this vast problem requires is not technical judgment, but rather the human experience and good sense of the natural leaders of opinion within the thousands of communities of the country. There has never been any good substitute for the all-around common sense of an informed lay public."[63]

Lilienthal turned to a tactic inherited from the Tennessee Valley Authority –

61 AEC, *Second Semiannual Report*, p. 4; see also Lilienthal to W. H. Dukak, Merrill Lynch, February 3, 1947, "Copies of Letters Sent: Nov. 1946–Dec. 1949" folder, Lilienthal Office Files; U.S. Atomic Energy Commission, *Third Semiannual Report*, 80th Cong., 2d sess., S. Doc. 118, 1948, pp. 13–14.
62 "Atomic Energy Is Your Business," September 22, 1947, "Atomic Energy Radiation, 1950s" folder, The Papers of Abel Wolman, The Johns Hopkins University, Baltimore, Md.
63 Ibid., pp. 1, 7, 8, 9, quotation from p. 8.

he urged grass-roots community involvement. He sought to build a lasting constituency for the AEC and to bring the countervailing force of public opinion to bear against the would-be usurpers of the public's right to decide for itself. Lilienthal tried to create a permanent power base this way. As he urged his listeners, "Don't make this a crusade, a 'drive,' full of ardor and zing for a couple of weeks and then it's all over. This is not a 'campaign.' This subject will be with you for a long time, so take it in your stride, as part of your community responsibilities."[64]

Lilienthal was not sanguine about the prospects for success. Nobody recognized better than the former head of the TVA that to achieve this objective, he needed concrete benefits. No lasting program could be built upon the abstract principle of participation without some meatier inducements for the public to get involved. There were glimmers of drama in the medical uses of radioisotopes. But for Lilienthal these also served as painful reminders of what might have been had civilian nuclear power been ready to go. As Lilienthal confided to his journal several months after the Crawfordsville speech, "The cobalt radium story certainly went over big; best thing on peacetime applications yet, for unlike the power story it is here and now, and it is something that touches millions of people directly."[65]

By New Year's Day of 1949, Lilienthal confessed that the Atomic Energy Commission had failed to gain public support. In a memorandum to his fellow commissioners and the general manager, Lilienthal reiterated how difficult it had been to improve public understanding. He cited secrecy, the "impenetrable complexity" and mystery of the subject matter, a deteriorating international situation, and unfriendly military supervision as some of the culprits. Despite his best efforts and those of the commissioners, "the results," Lilienthal stated bluntly, "have been disappointing and short of any reasonable goal." He went on to issue the following warning to his colleagues: "If my antennae about public opinion are working at all well (and they have been fairly sensitive in the past) we are approaching a situation – in say 3 to 6 months – in which our initial large credit balance with the public may be gone."[66]

BUILDING A LASTING CONSTITUENCY

The reactor development program was beset by management problems in addition to the better publicized problems of fuel supply. For months, the GAC and the AEC debated just where reactors should fit into the larger research plan.

---

64 Ibid., p. 12.
65 Lilienthal, *Atomic Energy*, p. 325. Even in this area, an AEC hungry for specific examples of peacetime benefits may have claimed more than it delivered. John Teeter of the National Cancer Society complained that the AEC's press releases were great compared to the dollars devoted to cancer research. (Teeter to Francis Henderson, May 4, 1949, Misc. Correspondence Files, Box 1, Records of the Office of the Chairman.)
66 Lilienthal to the Commissioners and General Manager, January 1, 1949, AEC, DOE, pp. 1–2.

Meeting with the GAC at the end of March 1947, the commission and its staff fixed overall priorities: revitalizing weapons activities would be the first concern; steps to ensure a continuing supply of fuel for these weapons came next. Oppenheimer testified several years later that the GAC had concluded "without debate" but "with some melancholy ... that the principal job of the Commission was to provide atomic weapons and good atomic weapons and many atomic weapons."[67]

Debate about priorities within the reactor development program continued for the rest of 1947 and well into 1948. Efforts to coordinate reactor development were severely hampered because the commission had trouble attracting, then keeping, a research director. Its first director, James Fisk, a top-flight physicist from Bell Laboratories, resigned in the middle of 1948 to return to the private sector. The need for a coordinated approach to reactor development was particularly pressing because scientists had developed a number of different kinds of reactors while working for the Manhattan Project. The General Advisory Committee, given the relatively low priority of reactor development, could not possibly coordinate reactor efforts on a day-to-day basis. As the AEC official historians put it, "The Commission had not inherited a research program but a collection of laboratories, all uncertain of the future and each pursuing an independent course."[68]

Despite poor coordination, military preemption, and frustrating news about atomic energy's likely competitiveness and availability, there was strong pressure from scientists within the Atomic Energy Commission to develop economically competitive atomic energy to generate electricity. In the AEC's first five years it was the scientists who were the major advocates of building non-weapons-related reactors. This interest grew directly out of their wartime work. At the Clinton Labs, Eugene Wigner and Alvin Weinberg developed the concept for a closed-fuel-cycle plant that would never require refueling.[69] Wigner and Weinberg also developed during the war the concept for a reactor that would subject materials to intense neutron bombardment in order to test them for atomic durability. At the Argonne Laboratory, Walter Zinn was at work on the concept for a fast neutron-breeder reactor.[70]

Each group lobbied for the concept it had been working on in the past. In an effort to coordinate its reactor development program, the Atomic Energy Commission established a Committee on Reactor Development. Comprising sci-

67 Hewlett and Duncan, *Atomic Shield*, pp. 44–6. Oppenheimer quote from Herken, *Counsels*, p. 60.
68 Debate over reactor development can be followed in the GAC Minutes of 1947, AEC, DOE. For a good summary of reactors developed during and immediately following the war, see Oliver Townsend, "The Atomic Power Program in the United States," in The American Assembly, ed., *Atoms for Power* (New York: Columbia University Press, 1957), pp. 36–50. On research director, see CM 212, October 27, 1948, and CM 215, November 10, 1948, AEC, DOE. On Fisk, see Weil, "Nuclear Power," pp. 71, 83. For quotation, see Hewlett and Duncan, *Atomic Shield*, p. 29.
69 Weil, "Nuclear Power," p. 79.
70 Ibid., pp. 79, 82.

entists who advocated many of these different approaches, the group made little headway. As Weil reported, "it became apparent that, because of the strong personal attachments of the committee members to their own projects, this approach was of little, if any, help to the AEC."[71]

Like the decision at the various labs to begin reactor work once the bomb had been completed, pressure for development came from within the scientific community, not from interest groups pressuring the AEC or the Joint Committee on Atomic Energy for products or services. The national laboratories and the General Advisory Committee itself were crucial sources of demand for research-oriented reactors. The GAC did approve in principle Wigner and Weinberg's "high flux" reactor that would be used to test materials for use in future reactor development. The GAC also backed Walter Zinn's experimental breeder reactor at Argonne. None of these reactors would solve the practical engineering problems of large-scale power production, nor would they demonstrate to industry and the utilities that power could be produced economically. They would be valuable, however, in testing some of the physicists' theoretical assumptions, a step essential to orderly long-term reactor development. Needless to say, the General Advisory Committee, which was dominated by theoretical physicists and research scientists, was predisposed toward such efforts.[72]

When it came to short-term projects that emphasized demonstration and that did not advance theoretical knowledge, demand had to come from more powerful constituencies. Farrington Daniels learned this lesson the hard way. Daniels had served under Arthur Compton during the war at the Metallurgical Laboratory located at the University of Chicago, and replaced Compton after the war. On the recommendation of the Interim Committee's Scientific Panel, General Groves assigned the "Daniels Group" to Clinton Laboratory in Oak Ridge, Tennessee, to assist the Manhattan Project's design and construction consultant – Monsanto Chemical. They sought a shortcut to demonstrating atomic power.[73]

At the Atomic Energy Commission, Daniels led the charge for practical, readily demonstrable atomic power, even if the electricity produced would not yet be economically competitive with that produced by fossil fuels. Daniels had just the reactor to achieve this objective: it was his own "Daniels pile." Daniels never claimed that the reactor would advance existing theoretical knowledge about reactor design. His primary goal was to demonstrate the feasibility of power production from the atom. Because it had been developed under the management of Monsanto, and because it was not meant to advance theoretical knowledge, the reactor also was an important symbol. It represented the industrial

71 Ibid., p. 81.
72 GAC Minutes #3, AEC, DOE, pp. 16, 19; Richard G. Hewlett, "Beginnings of Development in Nuclear Technology," *Technology and Culture* 17: 475–7. The commissioner who was most involved the details of reactor development was the only scientist on the commission, Robert Bacher. For an example of Bacher's intervention, see discussion of AEC 152/1, November 10, 1948, CM 215, AEC, DOE, p. 107.
73 Weil, "Nuclear Power," 76. Hewlett and Anderson, *New World*, pp. 633–4.

engineer's approach to reactor development, demonstrating the practical applications of atomic power without time-consuming fundamental studies.[74]

In March 1947, the General Advisory Committee withdrew its support for the Daniels project. Although Daniels apparently had convinced the commission's advisory board, he had not persuaded the AEC's research division. Former chief of the AEC's Reactor Development branch George Weil subsequently explained the reasoning behind the staff's rejection. Ironically – considering that Daniels had headed the Metallurgy Lab – the Daniels pile used a metal that was very susceptible to cracking or breaking up into small pieces. Weil discussed this problem with Daniels, who responded that even if the reactor became inoperable after one day, it would be worth the cost – estimated to be more than $30 million – to have demonstrated that nuclear fission could generate electricity.[75]

It was more than technical problems, however, that did in the Daniels pile: there was little demand from the private sector for nuclear-generated electricity that would be far more costly than conventionally generated electricity. The demise of the Daniels pile and the rise of the nuclear navy taught a crucial lesson about the importance of demand when it came to projects that did not advance theoretical knowledge. Whereas the private sector failed to show any interest, demand from the military appeared to have no limits. Nobody appreciated this dichotomy better, and earlier, than Captain Hyman G. Rickover, the architect of the nuclear navy.

Rickover was born in Poland at the turn of the century and arrived in America in 1904. With the aid of his congressman and some cramming for the entrance exams, he entered the Naval Academy in June 1918. At the academy Rickover was excluded from social events because he was Jewish; he became a classic "grind." He moved from ship to ship during the interwar years after graduating from the academy, managed to get an MS in engineering from Columbia, and trained for submarine command at New London, Connecticut. His real break came when he was assigned to the Bureau of Engineering on the eve of World War II. It was here that Rickover began to develop a management style that would soon be his hallmark. Its essential components, as one of his biographers put it, were "a perfectionism and zeal that bordered on the fanatic, and a disregard for rank and status." During the war, Rickover also established relationships with key employees at General Electric and Westinghouse that would serve him well in his quest for a nuclear navy. Following the war, Captain Rickover was assigned to Oak Ridge to assess the possibilities for nuclear-powered naval propulsion.[76]

---

74 GAC Minutes #6, AEC, DOE, pp. 2, 8–12, for discussion of practical demonstration versus longer-range development; discussion of Daniels pile relies on Hewlett and Duncan, *Atomic Shield*, pp. 70–1, 105–6, 120.
75 Weil, "Nuclear Power," pp. 77–8.
76 Eugene Lewis, "Admiral Hyman Rickover: Technological Entrepreneurship in the U.S. Navy," in Doig and Hargrove, *Leadership and Innovation*, pp. 97–103; quotation from p. 100.

Surveying the scene at Oak Ridge, Rickover soon dismissed any possibility of economically competitive nuclear power. He was even more pessimistic about its prospects than the General Advisory Committee. "[N]uclear power could not today compete [with coal] even if the fissionable material and its reprocessing were made available at no cost to the nuclear plant operator," Rickover wrote to Vice Admiral E. W. Mills in June 1947. After arguing that naval propulsion was the only promising field for immediate development, Rickover enumerated some of the problems with the AEC's approach to reactor development from the perspective of an engineer obsessed with immediate development. He noted that the research effort was decentralized with little coordination. It lacked engineers to do what Rickover considered to be "largely an engineering development job rather than a scientific research problem." Any naval propulsion project would also face a shortage of experts, but Rickover argued that massive mobilization could adequately address that problem. Rickover anticipated the short-term future of reactor development in one omniscient sentence:

Since there is no economic or other reason which would impel the electric power industry to invest in the development of atomic power, and since A.E.C. has other immediate primary concerns, it would appear that if we are to have atomic power plants in naval vessels the inspiration, the program and the drive must come from the navy itself.[77]

Rickover almost single-handedly provided that inspiration, program, and, most of all, drive, leading the navy into the nuclear age. Although the General Advisory Committee recognized that naval propulsion would erode even further the slim base of civilian applications that in part justified civilian control, it was hard to ignore the overwhelming demand, particularly compared to the private sector's interest in reactors. By November 1947, the GAC had redirected established programs toward the specific objective of naval propulsion. This took some of the heat off the GAC and gave the Oak Ridge scientists a top-priority project on which to work. The minutes of the November GAC meeting reflected the committee's satisfaction at solving several problems in one stroke. Noting that too little attention had been paid to the practical engineering aspects of reactor development, the GAC expressed its pleasure with the trend

toward the consideration of a reactor in which engineering reliability may outweigh, at the beginning, material economy and in which there be such experimental facilities incorporated as are needed for engineering development rather than research in physics. Power for naval vessels seems to be a desirable interim objective.[78]

The military presented the Atomic Energy Commission with a never-ending list of nuclear-powered projects and fought hard to fund them. Industry, constrained by the profit motive, was more ambivalent about the potential of nuclear

---

77 Rickover to Vice Admiral E. W. Mills, June 4, 1947, Clinton Labs folder, Box 16, LLS, HHPL, quotations from pp. 4–5.
78 GAC Minutes #3, p. 20, GAC Minutes #7, p. 21, AEC, DOE.

power. The military's influence over the development of power reactors would only grow in the next few years. As Hewlett and Duncan have so ably detailed in *Nuclear Navy*, this influenced the technology upon which commercial nuclear power ultimately was built. The naval nuclear power program led directly to the Shippingport reactor – the first full-scale demonstration of electricity produced from nuclear power.[79]

Lilienthal's failure to establish a broad-based constituency as a countervailing force and the absence of any constituency for commercial nuclear power in the private sector left the commission dependent on military requests. The military was the AEC's only source of external demand for services and products. By 1950, a more active Joint Committee on Atomic Energy pressed to keep military production the AEC's top priority. This and the Korean War made Lilienthal's successor, Gordon Dean, even more accommodating. By the early fifties, relations among the AEC, the JCAE, and the military were thus easing into a traditional iron-triangle pattern. The military subsumed the role generally played by economic interest groups as a source of demand in the triad. Because cost was only a marginal consideration for the military, its demand for nuclear power was uninhibited if not voracious.[80]

If anything approximating this kind of demand was to emerge from the would-be manufacturers of civilian reactors and the utilities that might one day purchase them, the Atomic Energy Commission and the Joint Committee on Atomic Energy would have to create and promote it. Potential interest on the part of manufacturers and utilities was dampened by the abundant supply of cheap energy in America. Even the AEC was stymied by this economic fact of life. When the AEC's reactor development group met in April 1948, for instance, to discuss Ernest Lawrence's proposal to use the plutonium-producing reactors at Hanford to produce electricity, the group rejected the idea. There was too much

79 Hewlett and Duncan, *Nuclear Navy*, ch. 8. The Shippingport project grew out of an abandoned project to build a nuclear-powered aircraft carrier. Commissioner Thomas Murray joined forces with the Joint Committee to push for the civilian project, restoring construction funds in 1953. Hewlett and Duncan point out that although the power this reactor sold to the Duquesne Lighting Co. was not close to being economically competitive with power from fossil fuels, the project was extremely important to the development of light-water reactor technology, establishing it as practical.
   The broadened scope of competition between the United States and the Soviet Union – from military to economic as well – also provided an important source of demand for power-producing reactors. Stimulated by Eisenhower's "Atoms for Peace" address in 1953 and benefiting from both American and European subsidies, the export of reactor technology proved far more important than the domestic market in the late fifties and early sixties (Bupp and Derian, *Failed Promise*, ch. 1).

80 The best statement of how military demand replaced other sources of demand is the GAC Letter to the President, June 14, 1952, Append. A, GAC Minutes #31, June 13–14, 1952, AEC, DOE. For a summary of the AEC's priorities at the end of 1948, see "Reactor Development Program," AEC 152/2, November 24, 1948, and "A Proposed Reactor Development Program," AEC 152, October 20, 1948, AEC, DOE. For Dean's emphasis on military production see "Remarks Before the New York State Chamber of Commerce," June 7, 1951, "AEC General, 1951" folder, Box 10, JCAE Files.

actual and potential hydropower in the region to possibly consider nuclear generation. Hydropower could produce the same electricity at a fraction of the cost, the development group decided.[81]

Almost from its inception, the Atomic Energy Commission actively nurtured a fledgling nuclear industry. Failing to make the development of civilian nuclear power "the people's business," Lilienthal courted the nuclear industry and utilities as a more reliable and potentially better organized constituency. As early as March 1947, Lilienthal met with the Atomic Energy Committee of the Association of Edison Illuminating Companies. This group, consisting of prominent utility company executives, urged the AEC to increase industrial participation in the development of atomic energy.[82]

The AEC formalized its relations with industry in October 1947 by establishing an Industrial Advisory Committee charged with recommending ways to encourage industrial participation. The advisory committee emphasized the need for engineering skills to balance the more theoretical interests of the physicists. Once the physicists determined the characteristics of a project, the committee recommended, it should be turned over to engineers, ensuring that the design would be practical and that it would be executed in a reasonable time period. As with Rickover and the "Daniels Group," the industrial engineering approach guided the industrial objectives for and assessments of nuclear power. Theoretical physicists Weinberg, Wigner, and Zinn, on the other hand, pursued quite a different set of problems, in line with research at the cutting edge of their discipline's agenda. The heuristic framework imprinted by professional discipline left its mark on the way reactor projects, and their timetables, were defined.[83]

Different as these approaches were, they shared one thing in common: The inspiration, the program, and the drive, as Rickover had put it, came from the federal government. In its report to the AEC in December 1948, the advisory committee – which represented some of the nation's leading industrialists – warned that AEC contracts to firms employing thousands of technicians was no substitute for reactor development by competitive private interests. This was consistent with the strident free-enterprise rhetoric that American business trumpeted in the fifties. But the report carried a less publicized, though equally powerful, message for postwar business–federal relations. It stressed that government "must first provide the catalytic forces that will set more of the normal processes of industry to work." In a nutshell, the committee wanted far greater

81 J. B. Fisk to Carroll Wilson, April 16, 1948, IM 48–11/2, AEC, DOE.
82 Marks to Lilienthal, April 3, 1947, "General Counsel" folder, Box 6, Lilienthal Office Files; Trowbridge to the Commissioners, March 19, 1947, "Secretariat" folder, Box 17, Lilienthal Office Files.
83 Wilson to the Commissioners, July 7, 1947, Box 1, Office Files of Robert Bacher, Records of the Atomic Energy Commission, Record Group 326, National Archives, Washington, D.C.; Wilson to the Commissioners, September 17, 1947, "Robert Bacher" folder, Lilienthal Office Files; Harter to Parker, August 8, 1948, "Correspondence–Industrial Advisory Group" folder, Box 7, Lilienthal Office Files.

access by private industrial firms pursuing their own agendas, but acknowledged that only the government could stimulate such interest in the private sector.[84]

In January 1949, Lilienthal moved to implement some of the Industrial Advisory Committee's recommendations. He drafted a memorandum to AEC general manager Carroll Wilson. Calling for a freer exchange between industry and the AEC, Lilienthal cautioned that the AEC had defined industrial participation in far too narrow terms, concentrating primarily on securing more industrial contractors for the commission's own immediate applications. Lilienthal endorsed a policy that would let industry "take its own look at the program to see what's in it for them." But information would not be enough. Lilienthal recommended "that each committee be provided with a 'home....' While the general information aspects are obviously a responsibility of the information office," Lilienthal continued, "the program divisions such as Reactor Development for power, would seem to me to provide more comfortable homes for the committees." Perhaps this cozy relationship would provide the "catalytic" force that industry demanded.[85]

It was Philip Sporn who had suggested that Lilienthal find a "home" for industry. Sporn was president of American Gas and Service Corporation and a powerful figure in the utilities world. He had bombarded the Atomic Energy Commission with a steady stream of letters that urged a greater voice by the utilities in the development of atomic power, including a base in the AEC. Soon after Lilienthal's January 1949 memo the AEC established the Ad Hoc Advisory Committee on Cooperation Between the Electric Power Industry and the Commission, naming Sporn chair. It came as no surprise that the committee suggested a permanent committee representing utilities in order to provide the sound economic judgment of utility experts. Following up on this suggestion, the AEC's director of reactor development wrote, "Individual representation of each industry that is significantly affected, while creating the possibility of pressure groups, would appear to have an over-riding advantage of making most realistic the Commission's relationships with the actual problems of an industry."[86]

Unfortunately for the prospects of private development, a stolidly passive industrial rank and file sat out this mating dance between the AEC and a handful

---

84 Report to the U.S. Atomic Energy Commission by the Industrial Advisory Group, December 15, 1948, "Correspondence–Industrial Advisory Group" folder, Box 7, Lilienthal Office Files.
85 Lilienthal to Wilson, January 27, 1949, "Correspondence–Advisory Committee on Cooperation Between Electric Power Industry and AEC" folder, Box 1, Lilienthal Office Files, pp. 2–3. Lilienthal's proposal for disseminating more information came in response to a report issued by the AEC's Industrial Advisory Group on December 15, 1948, urging the commission to declassify far larger amounts of information, particularly information that would be useful for engineering and design. (AEC 184/2, July 6, 1949, p. 1, and Append. A, Folder 23, Box 1248, AEC, DOE.)
86 See, for instance, Sporn to Lilienthal, June 6, 1948, August 26, 1948, November 30, Box 1, Lilienthal Office Files; Sporn to Lilienthal, January 15, 1949, AEC 184, "Establishment of an Ad Hoc Advisory Committee," July 25, 1949, AEC 184/3, AEC, DOE; for the committee's recommendations, see "Report of the Ad Hoc Committee," March 29, 1951, AEC 184/22, AEC, DOE; for comments on setting up pressure group see "Report of the Ad Hoc Committee," April 10, 1951, AEC/23, AEC, DOE.

of aggressive industrial leaders. Although highly visible leaders such as Sporn, and Charles Thomas of Monsanto, sought to seduce the AEC, each time the commission responded, its enthusiasm was unrequited by the vast majority of manufacturers and utilities. Even Sporn and Thomas were unwilling to make any firm commitments. It was precisely for this reason that industry leaders had urged catalytic action by the AEC; conversely, AEC staff acknowledged the need and soon began actively to create private pressure for nuclear power. By 1952, the AEC contemplated "an actual full partnership of the AEC and private industry."[87]

The Joint Committee on Atomic Energy also recognized the need for a federal catalyst. McMahon and the committee had been a major catalyst in the massive nuclear weapons buildup of the early fifties. By 1953, the committee was a far more powerful and confident member of the nuclear team. The Republicans had reclaimed the presidency in 1952 and at least nominally recaptured the Congress. This elevated Republican Sterling Cole to Joint Committee chair. Republican staff felt even more strongly than their Democratic counterparts that the federal government had an active obligation to create an industry capable of commercializing nuclear power. The JCAE was well aware that committee and commission initiative were essential in creating the third leg of an iron triangle for civilian nuclear power. As JCAE staffer Walter Hamilton put it: "There is no atomic industry today, only a Federal monopoly. The relative absence of public pressure to change the law is testimony to that fact. The aim is to create something, not correct something."[88]

Federal agencies and legislative committees now took the lead in creating pressure groups. This reversed the traditional sequence of policy development. From the Progressive Era through World War II, it had been pressure groups that had normally lobbied congressional committees or their subcommittees for legislation establishing an administrative bureau for needed support – whether technical, informational, financial, or regulatory. They triumphed through the inherent threat of mobilizing broad-based or politically influential support. The AEC went beyond even the New Deal agencies that sought to shore up poorly organized constituencies. It sought to create from scratch the very constituency that would "pressure" it. Federally initiated policy agendas and constituencies were two vital new attributes of the prominestrative state.

The most telling critique of the older-style iron-triangle politics was that it tended to sacrifice broader objectives to narrower but skillfully articulated demands. Generally, however, you did not need a score card to tell the players or the teams they played for. After World War II, finding the catalyst for programs increasingly meant looking inside the federal government. It was also the place to start when

---

87 CM 738, August 27, 1952, AEC, DOE.
88 Borden to Trapnell, April 28, 1953, "AEC Legislation, Atomic Power Bill, HR 4687" folder, Box 38, JCAE Files; Walter Hamilton, "Memo of Views on Atomic Power Legislation," February 23, 1954, "AEC Staff Memorandums, 1954" folder, Box 39, JCAE Files.

considering the roots of an increasing number of new pressure groups. With the advent of state-initiated pressure groups, is it any wonder that for many government programs, public demand was nowhere to be found?[89]

A powerful set of alliances that certainly appeared to be an iron triangle was beginning to emerge by the time Eisenhower assumed the presidency. Over the course of the decade, the JCAE would overcome its initial disadvantage and become an aggressive overseer of agency policy, as well as protector of a shared perspective. Economic interest groups were created where none had existed before. With government aid, reactors that produced electricity for public consumption were actually built. Yet, as late as 1963, with all the elements of iron-triangle politics in place, something was still missing. That missing element was demand for the service the iron triangle was providing. Although the institutional players involved in iron-triangle politics could be constructed post hoc, the market for the services these three amigos provided proved to be less accommodating to the politics of verisimilitude.

89 The pattern found in the development of civilian nuclear power supports, and in fact may be a leading case for, an emerging reinterpretation by political scientists of the iron-triangle literature. The best summary of this revisionist school is found in Thomas L. Gais, Mark A. Peterson, and Jack Walker, "Interest Groups, Iron Triangles and Representative Government," pp. 161–6. Walker has paid close attention to changes in the historical developmental process in "Interests, Political Parties and Policy Formation in American Democracy," in Hawley and Critchlow, *Federal Social Policy*. Hugh Heclo has generalized about the effects of the changing pattern in interest group, agency, and committee relations in "Issue Networks and the Executive Establishment" in *The New American Political System*, ed. Anthony Ming (Washington, D.C.: American Enterprise Institute, 1978), pp. 87–124. An excellent account of the agricultural "policy network" that was born during a period dominated by iron triangles, yet evolved toward the "issue network" model after World War II, is Mark Hansen's *Gaining Access: Congress and the Farm Lobby, 1919–1981* (Chicago: University of Chicago Press, 1991). In "The Ever-Decreasing Grandstand: Constraint and Change of an Agricultural Policy Network, 1949–1980" (paper prepared for the American Political Science Association, Chicago, 1987), Hansen shows that despite ultimately succumbing to interests of Agricultural, Committee legislators who perceived an advantage in courting more urban constituencies, the original iron triangle demonstrated tremendous resilience in the face of postwar demographic shifts that in theory should have opened the field to consumer-oriented issue networks far earlier.

4

Triangulating demand: the Atomic Energy
Commission's first decade of commercialization

The Atomic Energy Commission, the Joint Committee on Atomic Energy, and
the nascent nuclear industry embraced military demand as the primary stimulus
to reactor development in the late forties and early fifties. This dwarfed demand
for civilian applications. Alvin Weinberg, research director of Oak Ridge National
Laboratory and a pioneer of power reactors, was one of the first observers to
acknowledge this dichotomy in demand. "There was a much deeper reason for
the post-war lull in reactor development" than the postwar scientific demobilization commonly cited, Weinberg told a University of Virginia audience in 1950.
"[W]hen the time came to ask for the many millions needed to build reactors,
there were few who rose vigorously to say the country needs nuclear power
reactors – reactors which are expensive, perhaps dangerous, certainly far more
complicated than coal – and needs them badly enough to pay the millions
required to finance them."[1]

The military was the exception to this rule. "[I]t was precisely the demands
of the military which have put vigor and push into the terribly difficult and
expensive job of extracting useful power from uranium fission," Weinberg continued. "The disadvantages of nuclear power – the radioactivity, the expense, the
fuel reprocessing which make private power companies remarkably disinterested
for the time being are for certain military purposes outweighed by the advantages – compactness and independence of oxygen supply." Weinberg cited the
nuclear-powered submarine and airplane as prime examples of how military
demand had advanced power reactor technology. He also noted that the nation's
other reactor projects were related to another military application – the production of more weapons-grade fuel. "[O]ur engineering technology has progressed
in certain areas to such a degree that the only user of the technology is the
military," Weinberg concluded. He cited the example of recent developments in
the aircraft industry – developments stimulated by military demand. "There is
today," Weinberg told his audience, "a very remarkable, though not too well

1 Alvin Weinberg, "An Address," July 12, 1950; "Alvin Weinberg" file, Box 119, LLS, HHPL,
  p. 8.

95

appreciated, resemblance between the atomic energy 'industry' and the aircraft industry. Both depend almost entirely on government contracts, and both are concerned in greatest degree with military devices of one sort or another."[2]

The nuclear community inched toward an independent commercial base. The only clean break from nuclear power's military past occurred on the rhetorical level. Ironically, defense officials, image-makers at the AEC, and industrial firms contracting with the AEC often cited the public's continuing association of atomic and military matters as evidence that the public was ill-informed. The military complained that Americans associated atomic energy with something magical, something not entirely subject to human control. Too many people, complained one Pentagon spokesperson, regarded atomic energy as "black magic." Writers on this topic, he continued, were projecting their own concern. A massive publicity campaign sought to set such misconceptions straight.[3]

Government officials urged the public to take a more realistic and scientific approach. As General Groves put it in 1949, "Much that has been written about atomic energy has inspired fear and confusion.... This is not a healthy state of affairs. Atomic energy must be explained." Displays like the AEC's "Man and the Atom" exhibit in New York's Central Park did that and more. Besides seeing a "chain reaction" set off sixty mousetraps, visitors received free copies of *Dagwood Splits the Atom*. Nobody could compete with the master popularizer Walt Disney when it came to civilianizing the atom. In a film shown on television and in schools in 1957, Disney taught Americans about "our friend the atom." The plot revolved around a genie released from a bottle. Over the course of the film, the menacing giant was reduced to an obedient servant by scientists.[4]

Little substance supported the public relations: civilian nuclear power promoters struggled to find a strategy that at least partially freed them from the military monopoly on demand. This was no easy task, since everybody conceded that economically competitive nuclear-generated electricity was years if not decades away. Weinberg, for instance, stated that unless the need for nuclear weapons disappeared,

nuclear energy production for civilian use will be tied to military use in much the same way that water power is tied to flood control and navigation.... A nuclear TVA in which the prime justification for the power plant is plutonium production; i.e., a military justfication in the national interest, but in which the by-product – power – is sold to the public, appears to me to be a very natural, an almost inevitable long-term pattern.[5]

2 Ibid., pp. 8–11.
3 Boyer, *Bomb's Light*; ch. 24; quotation from p. 297. See also Weart, *Nuclear Fear*, chs. 7–8; Smith, "The Physicist's New Clothes." For an assessment on how the military history continues to affect public attitudes today, see Hohenemser, Kasperson, and Kates, "The Distrust of Nuclear Power," pp. 25–34.
4 Groves, quoted in Boyer, *Bomb's Light*, p. 296; p. 297 for "Man and Atom." Weart, *Nuclear Fear*, p. 169, for Disney.
5 Alvin Weinberg, July 12, 1950, LLS, HHPL, p. 12.

The largest and most knowledgeable firms were conspicuously absent among interests pressing for legislative changes that would permit greater private development of nuclear power. In closed-door hearings in 1953, Congressman Chet Holifield pressed AEC chair Gordon Dean on this point. "It seems strange to me that this drive for immediate changing of the [Atomic Energy] Act is occurring from people like Charles Thomas of Monsanto ... and not from these big companies that really know what this thing is about." Dean explained that firms like General Electric and Westinghouse were hesitant about change because they were already major contractors with the AEC. He speculated that the experienced operators might feel a bit diffident about requesting changes given their contractor status. It is likely that more than diffidence was involved. These firms were quite comfortable with their existing contractual arrangements. Greater private participation could only mean more competition. Knowledgeable contractors were also hesitant, according to Dean, because "they know much more about the difficulties than people who have never built a reactor."[6] These firms, of course, were protected against the likelihood of cost overruns by cost-plus contracts. Financial liability for excess costs in the case of civilian power, on the other hand, is still being hotly debated today.

Aggressive industrialists continued to urge that the Atomic Energy Commission organize demand for nuclear power in the private sector. Following Weinberg's prediction, however, they relied on military demand for weapons-grade plutonium to make their proposals economically viable. In June 1950, while the Atomic Energy Commission and the JCAE were absorbed in the military buildup, Charles Thomas, who had been involved with the "Daniels pile" at Oak Ridge, submitted a proposal that relied on American industry to design, construct, and operate with its own capital an atomic plant that would produce both power and plutonium. The proposal stated that the necessary basic technology had already been developed. Thomas called upon the AEC to make it available and to enter into a long-term contract for the purchase of plutonium.[7]

In response, the AEC initiated its Industrial Participation Program. It invited industrial study groups to develop proposals for the production of power and plutonium. Industry's response at first was sluggish. *Business Week* reported that in 1950, the private sector was not interested in atomic power: "AEC people have had to beat on desks in order to find a company willing to take on projects." The Atomic Energy Commission had to put together two of the four groups (including Thomas's team) that were eventually accepted.[8] The AEC also began to advertise industry's "home" within the commission. In May 1952 it hired William Davidson to head up its new Office of Industrial Development. Davidson's

6 U.S. Congress, Joint Committee on Atomic Energy, Executive Hearings Transcript, May 26, 1953, no. 3516, National Archives, Washington D.C. [hereafter Executive Hearings], p. 51.
7 Thomas to Pike, June 20, 1950, AEC 331; Volpe to the Commissioners, August, 1, 1987, AEC 331/2, AEC, DOE.
8 *Business Week*, July 28, 1951, pp. 107–8; quotation from p. 105.

job, according to *Business Week*, was "to sell business on the idea that atomic energy is 'a venture in which they should be willing to risk their own capital.'"[9]

Up to 1953 over 90 percent of reactor development funds had been poured into direct military applications.[10] Nobody disputed the notion that the federal government should pay for producing plutonium for its bombs, or for reactors that might power its submarines. As the Monsanto proposal demonstrates, private interests shut out of AEC contracts sought to link civilian power with production of weapons-grade material. All four Industrial Participation groups made it clear that they were willing to invest, if "the Commission [would] guarantee to purchase weapon material at a sufficient price" for companies to produce electric power "at an attractive unit cost." A report from the Director of Reactor Development stated bluntly, "the companies indicated that they cannot risk the construction of a nuclear plant without some form of subsidy or guarantee against loss."[11]

A "NEW LOOK" FOR NUCLEAR POWER

The Atomic Energy Commission moved slowly because it lacked agreement on a broader approach to commercial development.[12] A controversy pitted Democratic advocates of public development of demonstration plants against Republican defenders of the free enterprise system within the Atomic Energy Commission and introduced blatant partisan politics into atomic politics outside the commission. Most scholars have focused on these familiar political fault lines to explain nuclear power's retarded development.[13] Until the public–private dilemma could be resolved, these interpretations argue, American nuclear policy could not move forward.

Yet virtually everybody involved in developing nuclear power in the 1950s – Republicans and Democrats – recognized that federal subsidies and leadership were required. Liberal Democrat Chet Holifield was fond of pointing out that billions of federal dollars had already subsidized reactor development up to this point. He wanted to see a power reactor built as quickly as possible, but did "not want to see the Congress be a party to bringing in an oil well that has been drilled 10,000 feet deep with the people's money and let someone come in and

9 *Business Week*, December 20, 1952, pp. 140–3; quotation from p. 140.
10 Richard G. Hewlett and Jack M. Holl, *Atoms for Peace and War, 1953–1961* (Berkeley: University of California Press, 1989), p. 23.
11 AEC 152/33, December 9, 1952, folder 1, Box 1306, AEC, DOE, p. 19.
12 The staff, for instance, urged the commission to take early action on a proposal submitted by Dow Chemical and Detroit Edison, "in order to retain that group's enthusiasm." Vacillation at the top, however, continued into 1953. On retaining Dow's enthusiasm, see CM 685, April 17, 1952, AEC, DOE. On the commission's inability to frame a specific response to Dow, see Dean to Smyth, Murray and Zuckert, January 14, 1953, folder 6 (Misc.), Box 2203, Smyth Files, AEC, DOE.
13 Hewlett and Holl's *Atoms for Peace and War* is a good example of this.

drill the other 200 feet and take the oil."[14] Millions of dollars in federal subsidies continued to flow, even after the commission agreed to develop nuclear power "privately." As Alvin Weinberg put it in 1953, the question of public versus private power was simply not germane to the problem of starting a civilian nuclear industry.[15]

There was a far more powerful, and far less analyzed, reason why the civilian reactor development program continued to falter. Demand for nuclear power was shifting from narrow military applications to applications driven by a far more inclusive definition of national security. The Atomic Energy Commission, the Joint Committee on Atomic Energy, and the emerging nuclear industry scrambled to recast their developmental goals to fit this broader definition of national security. When Weinberg addressed his Virginia audience in 1950, he already recognized that the cold war was the driving force behind nuclear power. "[I]t is taking a cold war to give motivation to the development of nuclear reactors for power in much the same way that it took a hot war to give motivation and point to the development of the original nuclear bomb," Weinberg observed.[16] What he could not anticipate in 1950, was the extent to which Eisenhower's "New Look" policy would mobilize nonmilitary resources to win that cold war. Always sensitive to the political environment in which he worked, Weinberg was one of the first to recognize the possibilities for shifting the nature of the demand for nuclear power. Weaning that industry of military demand and finding other sources of demand, Weinberg argued in 1953, was crucial. Acknowledging that this represented a change in his thinking since 1950, Weinberg now felt that an industry independent of the military could be carved out. This would not be easy because immense military demand – starting with the Manhattan Project – had distorted the evolution of nuclear power. Because of the bomb, America had entered the nuclear race at the wrong end. The nuclear industry would have to find applications on a far smaller scale than the bomb, the nuclear-powered submarine, or the nuclear-powered aircraft, Weinberg underscored. Those applications would have to meet stiffer market tests than those exclusively military applications had faced.[17]

Perhaps enlightened members of the nuclear community were thinking more creatively about other sources of demand because military demand had begun to falter. By the end of 1952, the Atomic Energy Commission, which had sought to promote a nuclear industry by linking civilian power to weapons material production, recognized that its own massive investments in the weapons complex at Savannah River, South Carolina, and Hanford, Washington, had paid off. By the time private groups such as Monsanto built dual-purpose reactors there was

14 JCAE, Executive Hearing #3516, May 26, 1953, p. 54.
15 Alvin Weinberg, "How Shall We Establish a Nuclear Power Industry in the United States?" c. March 31, 1953 (draft article for the *Bulletin of the Atomic Scientists*, p. 4).
16 Alvin Weinberg, July 12, 1950, LLS, HHPL, p. 1.
17 Weinberg, "How Shall We Establish?" p. 14.

no guarantee that the federal government – having built its own weapons production plants – would need the additional privately produced weapons material. Just as the first group of private entrepreneurs readied its proposals for civilian power subsidized by guaranteed federal purchase of weapons-grade by-products, the AEC shifted its objective to privately constructed reactors that would produce economically competitive electrical power without reliance on a market for by-product weapon materials.[18]

Civilian power was pushed by declining military demand and pulled by the possibility of contributing more than "massive retaliation" to Eisenhower's "New Look." National security was redefined under the Eisenhower administration to include such nonmilitary considerations as domestic economic strength. Secretary of State John Foster Dulles, for instance, warned, "If economic stability goes down the drain, everything goes down the drain."[19] Combined with Eisenhower's conviction that military spending was out of control and commitment to balanced budgets, this meant sharp cutbacks in the direct military application of nuclear power. One of the first items to go was a proposal for a nuclear-powered aircraft carrier. The Defense Department canceled its requirements for this reactor in the spring of 1953.[20] Since the military had been the AEC's only external client, its cutbacks sent shock waves through the commission.

Two often overlooked components of America's New Look national security policy intertwined, providing a new basis for the development of nuclear power. Secretary of State Dulles was committed to alliances. He even listed them ahead of nuclear deterrent capability in his 1954 *Foreign Affairs* article that outlined the New Look. Charged by critics with "pactomania," Dulles felt there was little room for neutrality in a world divided by the cold war. In a world without neutrals, winning over previously unaligned third world countries was essential.[21]

The New Look also promoted a new technique for persuading would-be allies and demoralizing foes: psychological warfare. Psychological warfare was vaguely defined, and in Eisenhower's words could be anything "from the singing of a beautiful hymn up to the most extraordinary kind of physical sabotage." Patriots anxious to promote the nation's security soon joined promoters eager to create civilian demand for nuclear power in a chorus of support for the atom's potential to harmonize America's international interests.[22]

Beginning early in 1953, the two-word refrain sung by this chorus in order to justify commercial development of a technology that still elicited little en-thusiasm from those who were ostensibly to benefit most from it – utilities and electricity consumers – was "international prestige." Virtually all proposals for reactor development domestically and abroad stressed the need to enhance

18 AEC 152/33, DOE, p. 4.
19 Gaddis, *Strategies of Containment*, pp. 133–4; quotation from p. 134.
20 Hewlett and Holl, *Atoms for Peace and War*, p. 28.
21 Gaddis, *Strategies*, pp. 152–4.
22 Ibid., pp. 154–7; p. 155 for quotation.

America's prestige by being the first to rely on nuclear power as a regular source of electricity. America could strengthen her alliances by improving its psychological appeal to technologically underdeveloped nations.

Unfortunately, several sour notes shattered the harmony: prestige, it turned out, was harder to price than plutonium. Nor was it self-evident to the private sector that it alone should shoulder the price of prestige. Nuclear-produced power was far from economically competitive with power produced by conventional sources in America. George Weil, for instance, was encouraged that, according to some estimates, nuclear power might be produced at two to four times the cost of conventional fuel. If direct military applications did not justify the additional costs, indirect benefits to the nation's security would, nuclear advocates argued. But, as Weil also pointed out, the only way to establish the true cost of nuclear power was to build demonstration plants for the sole purpose of producing electricity. In a hypothetical analogy that closely resembled Weinberg's real-life aircraft industry comparison, Weil asked what would have happened if Henry Ford had been commanded to design a vehicle that could serve the U.S. Army in times of crisis and the civilian population the rest of the time? Assume further that the army had subsidized the construction effort, Weil speculated. Such vehicles might have been called "tankxies." Do you suppose that Mr. Ford would have been able to evaluate the economic feasibility of his plans for strictly civilian use, Weil asked? The situation for nuclear power was no different. If the government and industry wanted to develop economically viable nuclear power, they would have to invest in demonstration plants for the sole purpose of producing electricity. If the objective was economically competitive electricity – not just a race for kilowatts for international prestige at any cost – the technology would probably have to be far more sophisticated than the light-water reactor that soon powered nuclear submarines.[23]

Left unresolved was the multimillion-dollar question of who should pay for this development. Ironically, Weil's solution to this dilemma was as reminiscent of the twenties as the horseless-buggie analogy. Weil looked to private, nonprofit funds to break the deadlock.[24] Had nuclear power been developed in the twenties, one could certainly envision Herbert Hoover calling upon the "Associative State" to coordinate just such an effort. In the prominstrative state, however, it would be the federal government that led the way.

It did not take long for the AEC and Joint Committee on Atomic Energy to adapt their program to the broadened definition of national security. Commissioner Thomas Murray, who had been personally involved in the aircraft reactor project since its inception, and who was Rickover's strongest backer on the commission, read the handwriting on the wall. In April 1953, Murray proposed

23 George L. Weil, "Single Purpose Nuclear Power Plants," November 28, 1952, talk delivered to New York Section, New Jersey Division of American Institute of Electrical Engineers.
24 Ibid., p. 19.

that the aircraft carrier project be converted into a project to develop a nonmobile full-scale power reactor for the sole purpose of producing electricity.[25] This full-scale reactor, ultimately constructed with government funds at Shippingport, Pennsylvania, relied on light-water reactor technology already developed by Rickover and Westinghouse for submarine propulsion. Although the reactor could not establish nuclear power as economically competitive, it might demonstrate that nuclear power could be safely produced and supplied to the electric power grid. Improvements in the technology, the nuclear community argued publicly, would soon make it economically competitive as well. In more insulated forums, opinion was less sanguine. Commissioner Smyth summed up the origins and the potential of the project best, when he told the AEC's General Advisory Committee in May 1953, that "the decision was ... an attempt to salvage what they could of the program." Or as AEC chair Gordon Dean put it, "Since the military requirements had been knocked out, it was necessary to reevaluate the various mobile reactors in terms of the reactor development program."[26]

The nuclear community's first steps toward a new source of demand were rewarded at higher levels. In April 1953, Eisenhower approved a National Security Council (NSC) policy statement that proclaimed, "The early development of nuclear power by the United States is a prerequisite to maintaining our lead in the atomic field." At the same time, the NSC underscored that "such early development should be carried forward primarily through private, not government, financing."[27] Even at this early date, before much was known about Soviet peaceful applications, John Foster Dulles emphasized, and Eisenhower agreed that "it would look very bad if the United States lagged behind" other countries.[28] The NSC was more concerned, perhaps, with Canadian and British developments. Both countries had successfully operated atomic power plants.[29]

It was not long before direct competition with the Soviet Union for psychological advantage in the court of international world opinion spread to the peaceful application of nuclear power. After the Soviet Union exploded a thermonuclear device in August 1953, the Eisenhower administration recognized it was in for a prolonged struggle that would require all of America's resources. In December 1953, before the United Nations General Assembly, Eisenhower pledged that America would devote "its entire heart and mind to find the way by which the miraculous inventiveness of man shall not be dedicated to his death, but consecrated to his life."[30] NSC 5507/2 spelled out some of these objectives. The

25 Hewlett and Holl, *Atoms for Peace and War*, p. 28.
26 On the history of Shippingport, see Hewlett and Duncan, *Nuclear Navy*, ch. 8. For Smyth and Dean reactions, see GAC Minutes #35, May 14–16, 1953, AEC, DOE, pp. 40–1.
27 AEC 331/87, May 25, 1953, "IRA General Policy" folder, Box 1245, AEC, DOE, p. 1. For background, see AEC 655, June 8, 1953, AEC, DOE.
28 Dulles quoted in Lowen, "Entering the Atomic Power Race," p. 475.
29 George T. Mazuzan and J. Samuel Walker, "Developing Nuclear Power in an Age of Energy Abundance, 1946–1962," *Materials and Society* 7, nos. 3/4, (1983): 308.
30 Eisenhower quoted in Hewlett and Holl, *Atoms for Peace and War*, p. 209.

nuclear program would "maintain United States leadership in the development of the peaceful uses of atomic energy, particularly atomic power"; it would also "forestall successful exploitation of the peaceful uses of atomic energy."[31] The stakes were high, according to the National Security Council. "If the United States fails to exploit its atomic potential, politically and psychologically, the USSR could gain an important advantage in what is becoming a critical sector of the cold war struggle."[32]

The National Security Council's planning board expected some concrete results. Those familiar with nuclear technology, however, knew better. The Atomic Energy Commission's representatives on that board recognized that economic nuclear power was still at least a decade away. In fact, the AEC strongly opposed an international version of the full-scale power reactor on the grounds that such a project might face operating difficulties and certainly would not be economically competitive. The project, they argued, might do more harm than good.[33] This was not the last time that AEC officials balked at promoting overseas what they had already initiated at home for fear that problems might arise. Summing up the industrywide meetings in December 1957, reactor development director Kenneth Davis concluded: "While some prototypes should be built abroad, it is important that they also be built in the U.S. particularly... those which may involve any hazards."[34]

NSC Progress Reports also exposed the increasing urgency with which the psychological competition with the Soviet Union was pursued and the growing concern that America might not be able to deliver when it came to nuclear power. The NSC's Operations Coordinating Board (OCB) was pleased to report at the end of 1955 that a United States Information Agency poll reported substantial pluralities of Europeans felt that the United States had done more than any other country to develop peaceful uses and that the United States had shifted its emphasis from military to peaceful uses of the atom.[35] The memo also reported that the free world's scientific community was virtually unanimous: the United States had maintained its lead over the Soviet Union in peaceful applications. But the same progress report sent danger signals as well. As in America, there was a lag in visible accomplishments: this could lead to popular disillusionment. "Few laymen realize that any significant economic changes from the peacetime use of atomic power are still at least 10 to 15 years in the future." With pamphlets like "Atomic Power for Peace," distributed to 6.5 million people in thirty-four languages, such misunderstanding was not surprising.[36] The

---

31 National Security Council 5507/2, March 12, 1955; summary from National Security Council Progress Report, April 22, 1957 p. 5, Declassified NSC documents, public documents collection, Harvard University [hereafter NSC] (available on microfilm).
32 NSC 5507/1, March 1, 1955, NSC, p. 10.
33 Hewlett and Holl, *Atoms for Peace and War*, p. 199.
34 W. Kenneth Davis, "Conclusions from Industry Meetings," December 9, 1957, "Power General, Dec. 1957" file, Box 84, LLS, HHPL.
35 NSC Progress Report, December 21, 1955, NSC, p. 4.
36 Ibid., p. 3.

Operations Coordinating Board urged greater emphasis in the program's public statements "on the problems, as well as the promises of atomic energy."[37]

Eight months later, the OCB warned that "there are signs that the early emotional over-optimism on peaceful uses may turn into a corresponding emotional disillusionment."[38] At the same time, the OCB stated, "the Soviet Union has emerged as a challenger to U.S. leadership in Atoms-for-Peace programs."[39] According to the OCB, the Soviets had benefited by suddenly unveiling their power development program after years of secrecy and by setting specific targets – calling for 2.5 million kilowatts of nuclear power in their five-year plan.[40] By April 1957, a Joint State Department–AEC progress report struck a defensive tone.

The large programs for the production of nuclear power which have been announced by the United Kingdom and the Soviet Union have been used to give support to charges that the United States is lagging in power reactor technology. These charges, however, are based almost entirely upon comparisons of the number of kilowatts of electricity produced.

They overlooked American technological advances domestically and abroad, the report argued.[41] Even before Sputnik, the Eisenhower administration was hard pressed to explain why Ivan could produce nuclear kilowatts when Johnny could not.

Both the Atomic Energy Commission and the Joint Committee on Atomic Energy had played the "international prestige" card long before the NSC built that justification into national policy. The Russians might "beat us at developing the peaceful side of the atom," David Lilienthal warned as early as 1950.[42] In March 1953 the commission forwarded a policy statement that bluntly warned, "It would be a major setback to the position of this country in the world to allow its present leadership in nuclear power development to pass out of its hands."[43] Dean reiterated that point more bluntly behind closed doors in an executive session with the JCAE. Senator Millikin, who felt that the military buildup should not be sidetracked by the more indirect benefit of international prestige, asked AEC chair Gordon Dean what the consequences of another country's developing economically competitive power would be. "I think probably the prestige element would be rather large," Dean responded.[44] The foremost advocate of government-stimulated development for reasons of international prestige was Commissioner Thomas Murray. International prestige was one reason Murray had pressed hard for construction of the full-scale power reactor. In October

37 Ibid., p. 5.
38 NSC Progress Report, August 15, 1956, NSC, p. 2.
39 Ibid., p. 3.   40 Ibid., p. 4.
41 NSC Progress Report, April 22, 1957, pp. 5–6.
42 Lilienthal cited in Lowen, "Entering," p. 472.
43 NSC 145, March 6, 1953, NSC, p. 1.
44 JCAE, Executive Hearings, #3516, May 26, 1953, p. 6.

1953, he depicted a "nuclear power race" with high stakes. "[O]nce we become fully conscious of the possibility that power hungry countries will gravitate toward the USSR if it wins the nuclear power race, ... it will be quite clear that this power race is no Everest-climbing, kudos-providing contest," Murray warned.[45]

From the AEC's perspective, there was an even more pressing reason to move quickly on this front. As Dean wrote to Eisenhower on March 4, 1953, "The need for an early statement of policy is further accented by the approaching hearings of the Congressional Joint Committee on Atomic Energy."[46] The JCAE had shown relatively little interest in developing nuclear power in its first five years. Representative Chet Holifield, for instance, admitted in executive hearings that "this committee is as much to blame as anybody else because we had our eye on the military weapons ... and we never pushed the Commission on the civilian aspects of it."[47] With military demand on the wane, and Britain, Canada, and the Soviet Union apparently gaining on the United States, the Joint Committee on Atomic Energy began to push harder, eventually emerging as a powerful force for development.[48] It, too, relied on a broader definition of national security – one that encompassed international leadership in scientific and technological development – to justify greater government and private efforts toward developing nuclear power. That the AEC's and Joint Committee's efforts to create an economic interest group in the private sector had begun to pay dividends was another factor behind the congressional push. By the spring of 1952, more companies wanted to get in on the action. The AEC invited additional proposals, getting a response from seven groups.[49]

In December 1952, the Joint Committee on Atomic Energy issued a compendium of business opinions and information about nuclear power. The study warned that the United States must be ever vigilant in guarding against foreign competition so that it could never again "be truthfully said that the reactor of

---

45 Murray quoted in Mazuzan and Walker, "Developing," p. 309.

46 Dean to Eisenhower, March 4, 1953, in NSC 145, March 4, 1953, NSC.

47 JCAE, Executive Hearings #3518, June 11, 1953, p. 49. Although the JCAE was eager to see the Atomic Energy Commission more actively develop industrial ventures, it did not care to see this achieved at the cost of losing control over its carefully tended turf. Inherent in the very goal that both the JCAE and the AEC sought – commercialization – was an opening for other political players. In a May 1953 letter to AEC chairman Dean, Congressman Cole prodded the commission toward closer relations with industry; he also pointed out that several other federal agencies, including the Commerce Department, were planning studies of atomic power. An obviously upset Cole wanted to know the statutory basis for involvement by agencies other than the AEC. Dean assured Cole that these other agencies were merely assisting the AEC in studying problems that might be faced in nuclear power development. They were not mounting threatening efforts to control the field. In the proposed legislative amendments, the AEC (Dean assured Cole) would retain the principal responsibility for research and development. (Cole to Dean, May 18, 1953, Dean to Cole, June 25, 1953, "Atomic Power – Atomic Energy Commission" folder, Box 96, JCAE Files.)

48 The best account of the JCAE's growing assertiveness is Harold Green and Alan Rosenthal's *Government of the Atom.*

49 "Increased Industrial Participation in Reactor Development," AEC 655/1, June 23, 1953, AEC, DOE, pp. 2–3.

the most advanced design and performance operates anywhere but in the United States."[50] It had been JCAE pressure and the threat of hearings that forced the AEC to hammer out a policy statement on nuclear power in the first place.[51] At the hearings, held in early 1953, a broad spectrum of witnesses urged the Eisenhower administration and the JCAE to act in order to preserve America's international prestige and leadership through hastening its development of nuclear power.[52]

Although reactor development driven by the quest for international prestige was more sensitive to the issue of economic competitiveness than development undertaken for strictly military purposes, the nation's only full-scale project to produce nuclear power was still a "tankxie." The commission, the JCAE, and the Eisenhower administration supported it, not because it promised any new concepts that would lead to economically competitive power. Rather, it would provide some more engineering experience at the same time that it contributed to America's all-out battle for the hearts and minds of the world. Declassified AEC minutes make it clear that even the project's strongest supporter was aware that the AEC was building a "tankxie," not a Model T. Commissioner Murray, the minutes reported, "[a]cknowledging that the first reactor might not be competitive ... favored going ahead in the interest of furthering reactor technology and demonstrating large-scale nuclear power production."[53] The justification for this project, which eventually produced electricity in December 1957 at more than ten times the cost of conventionally fueled plants according to Rickover's own estimates, was America's race for international prestige.[54] The Atomic Energy Commission and the Joint Committee were willing to authorize up to $100 million in 1953 dollars to achieve that end.[55]

The decision was challenged by only a few AEC reactor physicists. They were working on a new generation of reactors that at least promised to produce economically competitive power. In a meeting with the Atomic Energy Commission and the Joint Committee, the General Advisory Committee kept its doubts to itself. Speaking for the GAC, its chair, Isidor Rabi, told the JCAE that the committee had been concerned that "Russia might get the jump on us in various ways in the international field." Rabi also told the Joint Committee that although the GAC had not participated in the selection of the technology for the AEC's full-scale power reactor, its members had no doubts about the project.[56]

50 U.S. Congress, Joint Committee on Atomic Energy, "Atomic Power and Private Enterprise," 83rd Cong. 1st sess., p. 4.
51 Cole to AEC, August 19 1952 in AEC 331/41, August 26, 1952, AEC, DOE.
52 JCAE, "Atomic power and Private Enterprise"; Mazuzan and Walker, "Developing," p. 308.
53 CM 885, July 9, 1953, AEC, DOE, p. 408.
54 Rickover's estimate cited in Hewlett and Holl, *Atoms for Peace and War*, p. 421.
55 GAC 35, p. 41, for original estimate of cost.
56 GAC #35, pp. 10–12.

Behind closed doors, however, General Advisory Committee members expressed their frustration at not being consulted. Eugene Wigner, one of the AEC's reactor pioneers, observed that building the reactor at Shippingport was better than doing nothing but that the GAC had been entirely left out of the decision process. Wigner was very doubtful about the decision, even though he personally favored the light-water system the full-scale reactor employed.[57] Several months later, at a July 1953 Commission meeting, the AEC's reactor development staff sought in vain to substitute economic viability for international prestige as the AEC's primary objective. As one staffer put it, "Only if the decision to build is based on economics of operation will there be sufficient incentive to design a really advanced reactor."[58] Weinberg, eager to demonstrate some practical value without sacrificing long-term prospects for economically competitive power, advocated applying existing technology to compact power sources while continuing government development of breeder reactors.[59] Because the Pressurized Water Reactor (PWR) – full-scale reactor – at Shippingport would do little to advance scientific knowledge or improve nuclear power's economic competitiveness, advocates of more sophisticated approaches dubbed the PWR, "Power Without Reason."[60]

Spearheaded by Murray with the blessing of Joint Committee on Atomic Energy chair Sterling Cole, the pressure to demonstrate that nuclear power could produce electricity, regardless of the cost or scientific knowledge gained, overwhelmed esoteric (and private) criticisms raised by a few theoretical scientists. Thomas Murray announced the full-scale reactor project to an electrical utility convention in October 1953. "For years," Murray concluded, "the splitting atom, packaged in weapons, has been our main shield against the Barbarian – now, in addition, it is to become a God-given instrument to do the constructive work of mankind." There was no mistaking the psychological motivation behind the nuclear community's decision. As *U.S. News and World Report* summed up, "An international race for supremacy has started. Britain, with one atomic-powered project, is in the race. Russia probably is starting. Now the U.S. is jumping in."[61]

The full-scale reactor project, while an important symbol, hardly constituted an aggressive development program. In May 1953 the Atomic Energy Commission issued a policy statement on nuclear development designed to stimulate the private sector and signal the commission's continuing commitment to create an industry, even in the absence of private demand. Acknowledging that a difficult development period lay ahead, the report was nonetheless optimistic that economically competitive nuclear power could be attained within a few years. It was

57 GAC 35, p. 34.   58 CM 885, p. 407.
59 Alvin Weinberg, "How Shall We Establish?" p. 26.
60 Lowen, "Entering," p. 476.
61 Murray quoted and *U.S. News* cited in Hewlett and Holl, *Atoms for Peace and War*, pp. 194–5.

imperative that "we create a favorable atmosphere which will hasten that day," the commission pledged.[62]

The private sector participants had reservations about making firm commitments, even assuming the proposals proved economically feasible. The AEC's Division of Reactor Development summarized these: "The problems of ownership, patent rights, hazards, and public liability are all to be faced before any full-scale effort can be undertaken."[63] With the exception of liability, these problems were addressed in the sweeping amendments to the Atomic Energy Act of 1946. The Cole–Hickenlooper bill, which in effect rewrote the earlier law, was passed in 1954. Its most important features, from the perspective of those promoting commercialization, were that it granted the commission the authority to license privately owned and operated reactors (although the fuel for these would continue to be leased from the AEC) and that it liberalized patent provisions.[64]

The Atomic Energy Commission, under increasing pressure from a Democratically controlled Joint Committee on Atomic Energy, announced its Power Demonstration Reactor Program (PDRP) in 1955. The program enlisted private resources to demonstrate the technical and economic feasibility of power reactors. It provided a number of incentives including fuel charge waivers, research at AEC laboratories, and fixed-sum research contracts with the industrial reactor group.[65]

Private response nonetheless was sluggish. By January 1957, when the AEC issued its third round of invitations, the PDRP had snagged only eleven proposals. The most advanced project was the Yankee Electric Co. for a reactor to be built in Rowe, Massachusetts. A contract had been signed; completion was estimated for 1960.[66]

More persuasive to industry than the AEC's incentives was the threat of public construction. In announcing its third round of invitations, the AEC issued a not so veiled threat. It stated that the commission would request funds to initiate

---

62 "AEC Preface to Statement on Policy on Nuclear Power Development," May 26, 1953, "AEC Legislation, Atomic Power Bill HR 4687" folder, Box 29, JCAE Files.

63 AEC 331/38, AEC, DOE, p. 6.

64 Atomic Energy Act of 1954, Public Law 703, 83rd Cong., 2d sess., 1954. For a comprehensive compendium of legislation, hearings, and other related documents, see U.S. Atomic Energy Commission, *Legislative History of the Atomic Energy Act of 1954*, vols. 1–3, Washington, D.C., 1955.

As noted earlier, the 1954 legislation also granted the JCAE its long sought after power to authorize funds for construction.

65 The PDRP was revised in three subsequent rounds of invitations. See "Power Demonstration Reactor Program," AEC 777/11, June 30, 1955; "Power Demonstration Reactor Program – 2nd Round, " AEC 777/14, September, 9, 1955; "Power Demonstration Reactor Program," AEC 777/19, April 13, 1956; "Proposed Third Round Power Reactor Program," AEC 777/27, December 17, 1956; "Proposed Modification to Third Invitation Under Power Demonstration Reactor Program," AEC 777/42, June 4, 1956, AEC, DOE. A good summary of the PDRP can be found in Robert Perry et al., *Development and Commercialization of the Light Water Reactor, 1946–1976* (Santa Monica, Calif.: RAND, 1977), pp. 10–24.

66 NSC Progress Report, April 22, 1957, NSC, n.p.

commercially viable projects on its own if industry did not respond to its invitation. Legislation introduced by Senator Albert Gore and Representative Holifield in 1956 had come very close to mandating public construction a year earlier. The 1956 Gore-Holifield bill, which authorized $400 million for AEC construction of demonstration reactors, would have created such a public presence had it passed in the House (it did pass the Senate). Utility trade journals, fearing the "socialization of the electrical power industry," monitored developments closely.[67] Though "Gore-Holifield" was defeated by the appeal to "free enterprise," the very firms that pressed for its defeat also pressured the federal government for additional incentives – particularly in the form of government indemnification.[68]

Utilities were caught between the rhetoric of free enterprise and the economic reality of nuclear power. AEC commissioner Libby put his finger on the problem, writing to Lewis Strauss in November 1957. "The industry representatives rather hedged on the matter of subsidies as a means of accelerating the program," Libby reported of his meeting chaired by Representative Melvin Price. "They talked around the subject and gave the impression that they thought they needed governmental help but still did not like the subsidy idea."[69] Several weeks later Kenneth Davis concluded that only direct governmental assistance would accelerate the program, "since neither the equipment manufacturers nor the utilities can pay the whole

---

67 For a discussion of the defensive motivation for private development, see Mullenbach, *Civilian Nuclear Power*, pp. 9–14. For a review of trade journals and quotation on "socialization," see Lowen, "Entering," p. 463. On Gore–Holifield, see Green and Rosenthal, *Government of the Atom*, pp. 15–16, 153–6, and Hewlett and Holl, *Atoms for Peace and War*, pp. 344–5.
68 Mazuzan and Walker, *Controlling the Atom*, chs. 4 and 7.
69 Libby to Strauss, November 25, 1957, "Power General, June 1957–Nov." file, Box 83, LLS, HHPL. At least in private, nuclear manufacturers had no qualms about expressing their need for subsidies. As the Allis-Chalmers representative told the nation's leading nuclear manufacturers and the chair of the Joint Committee on Atomic Energy at a closed-door session, "we are running out of utilities" to get support from. His plea for subsidies was seconded by top executives from General Electric, and Babcock and Wilcox ("Notes to industry meeting, November 22, 1957, "Power General, June 1957–Nov." file, pp. 4–5).
   One of the frankest discussions of the need for public subsidies to utilities took place at a meeting of leading manufacturers sponsored by the Atomic Industrial Forum on November 14, 1957. Chauncy Starr, vice-president of North American Aviation, forwarded these minutes to Strauss in order to convince the admiral that public subsidies were crucial if nuclear power was to be accelerated (Starr to Strauss, December 20, 1957, "Chauncey Starr, 57–70" file, Box 106, LLS, HHPL). The minutes report agreement that "the atomic power industry is today not economic and probably will not be for a period of at least five to ten years." Manufacturers also agreed that "it is not likely that atomic power development can be accelerated in the United States, during the present uneconomic period, without increased expenditure of public funds." Manufacturers agreed that subsidies must be directed toward utilities because they were the ones, ultimately, who made the decision to contract for nuclear power. However, "doubt was expressed that the government would be willing to grant a subsidy to the atomic industry without insisting on a substantial measure of control over the resultant program." In an unconscious expression of postwar large-scale capitalism's approach to "free enterprise," the manufacturers expressed the view that once public funding was provided, "if the government refrained from exercising technological and fiscal control over the program, *the free competitive enterprise system* [my emphasis] could work more effectively in helping the United States to find the best route to economic power." The manufacturers concluded that there was strong support for government subsidies, but no consensus as to how such arrangements might be administered.

cost."[70] At the industrywide meeting of architectural and engineering firms, reactor designer Walter Zinn asked, "How can we get prospective utilities to accept subsidies which are palatable to them?"[71] The industry was in a quandary. Everybody knew that nuclear power was not yet economically competitive. By the late 1950s, "industry was no more eager to invest in atomic power than it had been ten years earlier," according to Rebecca Lowen's review of utility trade journals.[72] Visible public subsidies were anathema to private utilities because they inevitably prompted demands for public control of power. Was it worth a financial loss to keep the government out of the power business? If so, how would this be explained to rate payers?[73]

There had long been speculation that the threat of public power was the only incentive that could move the nation's utilities to invest in nuclear power. As a reporter for *Fortune* put it in 1955, "One of their chief reasons for acting, perhaps the most important one, is referred to only guardedly. The utility industry is mortally afraid of public power."[74] More conclusive evidence lies in the files of Lewis Strauss. Private development's most forceful advocate, Strauss was privy to a number of confidential discussions within the industry. In early December 1957, his aide Robert Zehring reported back to Strauss that the president of a major nuclear manufacturing firm told him that the real reason the private utilities had agreed to invest their own capital in nuclear power was "fear of public power action by the new Congress and not because of any present economic pressure or fear of conventional fuel shortage."[75]

Zehring also forwarded a set of notes recapping the comments of utility presidents at a closed-door high-level meeting in preparation for the industrywide meetings on accelerating nuclear power. Zehring suggested that the notes be destroyed after Strauss read them; fortunately for scholars, Strauss ignored this advice. The utilities were hardly eager to embrace nuclear power, Zehring's notes reported. The utilities would put up more capital if the government insisted. But they would do so "in order to meet an indicated Governmental need [for international prestige, international competition, etc.] and not because [of] the country's power requirements....If the Government insists that the national interest requires these large scale plants...they should be built in the areas of the country where present power costs are high," Zehring reported. "The majority of the Committee appear to feel that they are behind the eight ball and will have to give some ground and invest more capital funds...in order to avoid complete Government ownership and operation of the atomic power industry."[76]

70 Davis, "Conclusions from Industry Meetings."
71 "Highlights of Meeting of December 5, 1957 (Architect-Engineer Firms)," attached to Zehring to Strauss, [misdated 12/2/57], "Power General, Dec. 1957" file, Box 84, LLS, HHPL, p. 2.
72 Lowen, "Entering," p. 477.   73 Ibid.
74 Quoted in Lee Clarke, "The Origins of Nuclear Power," p. 480.
75 Zehring to Strauss, December 5, 1957, "Robert W. Zehring, 53–57" folder, Box 124, LLS, HHPL.
76 Zehring to Strauss, November 26, 1957, "Power General, June 1957–Nov." folder, Box 83, LLS, HHPL, pp. 2–3.

FROM SELLING SHOES TO SELLING REACTORS

Lewis L. Strauss – a devout patriot and equally dedicated proponent of nuclear power – directed the chorus that sought to protect the nation's security by winning the race for international nuclear prestige. First as a presidential adviser, and then as the chair of the Atomic Energy Commission, Strauss defended the free enterprise system at the same time that he cajoled his friends in the private sector, urging them to invest in the nation's and their own future by committing funds to nuclear power. To many, Strauss was simply a tool of the private sector. What distinguished him, however, from corporate leaders – particularly among nuclear manufacturers – was his resistance to any form of federal assistance. Not only did Strauss want private utilities and nuclear manufacturers to "own the oil," to use Holifield's phrase, he insisted they pay for at least some of the drilling. When even Eisenhower felt that nuclear power "should be developed without too much concern about the role of private industry," and with manufacturers and eventually even some utilities clamoring for subsidies, Strauss faced a stern political test.[77] Undaunted, he became nuclear power's leading salesman. Far from being a tool of the industry, Strauss in his quest to develop nuclear power epitomized the AEC and Joint Committee's attempt to triangulate demand and create such an industry.

Strauss was born in 1896 in Charleston, West Virginia. In 1990 the family moved to Richmond, Virginia, where Lewis's father bought a partnership in his brother-in-law's firm, Fleischman, Morris, wholesaler of shoes and other goods to small retailers throughout the South. Valedictorian at his high school, Lewis received a scholarship to the University of Virginia starting in 1913, but a business recession forced him into the family business instead.[78]

For three years Strauss worked for the firm as a "drummer," the trade term for traveling salesman. He soon became a polished promoter, developing skills that would come in handy decades later when he sought to peddle nuclear power. There was a major difference, however. Shoes were an item everyone needed, and, as his biographer, Richard Pfau, puts it, Strauss "rarely left a store without an order." Power reactors, on the other hand, were hardly in demand in the fifties, and Strauss's nuclear "customers" were well aware of their economic drawbacks. Two character traits undoubtedly motivated Strauss to take on this most difficult sales job. First, he was extremely stubborn. Second, as Pfau notes, "Developing new territories was Strauss's favorite task. He loved to be the first salesman from Fleischman, Morris to visit a potential customer because this gave him the opportunity to prove himself."[79]

Strauss entered virgin territory in 1917 when he camped outside Herbert

---

77 Eisenhower quoted in Hewlett and Holl, *Atoms for Peace and War*, p. 240.
78 Richard Pfau, *No Sacrifice Too Great: The Life of Lewis L. Strauss* (Charlottesville: University of Virginia Press, 1984), pp. 3, 8.
79 Quotations from ibid., p. 8.

Hoover's door, eventually convincing the nation's newly appointed food administrator to take him on as an unpaid assistant. From office boy, Strauss rose to private secretary by the end of the war. In 1920, he joined the investment banking firm of Kuhn, Loeb. By the time Strauss was thirty, he was earning the lordly sum for those days of $75,000 a year. His wealth would only grow. Strauss, in the Hoover tradition, remained committed to public service. When war broke out again, Strauss served under James Forrestal, retiring in 1945 with the rank of admiral. Few men in 1946 shared Strauss's unique combination of wealth, experience in public service, and political connections. A lifelong Republican, he counted Robert Taft and Herbert Hoover among his closest friends, but he was also extremely close to Forrestal, Truman's first Secretary of Defense.[80]

Despite Strauss's partisan leanings, President Truman tapped "the Admiral" – as his friends called him – to serve on the first Atomic Energy Commission. Uncomfortable with the way the commission handled security matters, Strauss favored far more restrictive policies than his colleagues did. The ultimate break came in early 1950 in heated and prolonged debate over developing a hydrogen bomb. Siding with the far more hawkish Joint Committee on Atomic Energy against the majority of his fellow commissioners, as well as against Oppenheimer and the General Advisory Committee, Strauss carried his crusade to the White House. Presidential backing ultimately tipped the balance. With America's massive strategic buildup secured, Strauss resigned his AEC post several months later; but Dwight Eisenhower convinced him to return to atomic energy. Strauss was appointed chair of the Atomic Energy Commission in July 1953 and remained in that position for five years.[81]

Pfau calls Strauss the "father of nuclear power."[82] Based on "the Admiral's" prodigious efforts to sell commercial nuclear power to the general public; his crucial role in shifting the justification for nuclear power from strict military needs to international prestige; his never-ending attempts to badger private sector investors to invest in the economically questionable technology; and ultimately his leadership, which blunted efforts to launch public power – Strauss might better be called the "father of private nuclear power." No immaculate conception ensued, however. Commercial nuclear power in the 1950s had not yet found a mother.

Strauss shared his views on the future of nuclear power with the Virginia State Chamber of Commerce on April 14, 1950. Although the AEC's primary objectives were clearly military, its secondary objectives, Strauss confided, were all "benign." Some enthusiasts incorrectly predicted that "atomic energy eventually is going to supplant all other sources of power and that the realization of a golden age of cheap power is almost at hand," Strauss asserted. Others, however, claimed that practical nuclear power was at least a generation away. Strauss carved out

80 Ibid., pp. 12, 36, 84.
81 Ibid. chs. 8 and 9; Hewlett and Holl, *Atoms for Peace and War*, pp. 20, 30–1.
82 Pfau, *No Sacrifice*, p. 220.

a position far closer to the first of these two popularly accepted fallacies, as he called them. "It may be available far sooner than many expect," he proclaimed. As for its special hazards such as dangerous radiation, "in future years those problems will probably be regarded as no more difficult of solution than were the dangers to persons working around high-pressure, high-temperature steam not so very long ago." Even at a time when the demand for reactors was exclusively military, Strauss predicted exciting developments to come in civilian nuclear power.[83]

Returning as chair of the Atomic Energy Commission in July 1953, Strauss put commercial nuclear power at the top of his promotional agenda. At his swearing-in ceremony, "the Admiral" quoted Micah 4:3: "They shall beat their swords into plowshares, and their spears into pruning hooks." Strauss told the assembled dignitaries that "I hope that my return to the Commission will coincide also with an era of vigorous progress in the benign uses of this great natural force... for industrial power, for healing, and for widespread research." Little more than a year later, Strauss delivered a speech to the National Association of Science Writers, predicting "it is not too much to expect that our children will enjoy in their homes electrical energy too cheap to meter."[84]

Strauss carrried out the most effective portion of his promotion behind the scenes. He was the go-between for W. Kenneth Davis, the AEC's aggressive new director of reactor development, and the private sector on matters involving public relations. Davis urged Strauss to push utilities into "'shows'" and "ceremonies" that would publicize the peaceful atoms.[85] A card-carrying member of the National Science Writers' Association, Strauss flooded friendly editors with information about progress in reactor development. David Lawrence of *U.S. News & World Report* offered Strauss relatively unrestricted access to a national audience. *Newsweek*, on the other hand, was a constant source of irritation, infuriating Strauss by checking his facts.[86]

Strauss shifted the justification for nuclear reactors toward the race for international prestige and away from strict military applications. As Eisenhower's adviser on atomic energy, Strauss responded to the administration's budget-balancing by recommending deep cuts in some military applications. As he was

83 Address of Lewis L. Strauss Before the Virginia State Chamber of Commerce, April 14, 1950, "Power General, 1950–56" folder, Box 83, LLS, HHPL, pp. 1, 3, 5.
84 Strauss quoted in Pfau, *No Sacrifice*, p. 187 (speech delivered September 16, 1954).
85 Davis to Strauss, August 3, 1956, and Davis to A. L. Atwood, December 21, 1956, "Davis 1956" folder, Box 22, LLS, HHPL.
86 On press in general, see Lawrence and Holles files in LLS, HHPL. On *Newsweek*, see Strauss–Muir correspondence in "Power General, June 1957–Nov" file, Box 83, LLS, HHPL. Strauss had a keen eye for details that could help or hurt. He insisted on announcing any good news about reactor development personally so as to increase press coverage (see, for instance, LLS to Holles, December 10, 1957, "Holles, 1957" folder, Box 42, LLS, HHPL). On a whirlwind tour of nuclear facilities with Sir Edwin Plowden, Strauss apologized to his press aide for not getting any photographs of Shippingport into the newspapers. "[I]t is a hard plant to photograph effectively as about 4/5 of it is below ground level, in which respect it resembles an iceberg," Strauss groused. (Strauss to Holles, May 11, 1956, "Holles, 1956" folder, Box 42.)

one of the nation's strongest proponents of the nuclear buildup in the early fifties, as well as a supporter of military hardliner Robert Taft in the 1952 Republican presidential race, Strauss's credentials could hardly be challenged from the right. He successfully led the charge against the nuclear reactor slated for an aircraft carrier – the same reactor that was reincarnated as the full-scale reactor project at Shippingport.[87] Although Strauss was hardly enthusiastic about spending public dollars to demonstrate power, he did pioneer the ideological justification for such a project. Sputnik, launched by the Soviets in the fall of 1957, dramatically underscored the case for international technological competition. At a closed-door meeting between industry and the commission in December 1957, Strauss reiterated what had long since become his central argument for private development of nuclear power. As the notes put it, Strauss "emphasized the national need to advance nuclear technology to keep world leadership."[88]

On that point, there had been little disagreement. Where interests did diverge, however, was over who would foot the bill. As Strauss put it, "we cannot afford to let this go forward with only natural economic forces pressing it at this time."[89] Unlike many leaders in the private sector, however, Strauss did not mean by this public subsidies. He meant that the private sector should invest, even though the economic conditions were anything but favorable. Strauss spearheaded a crusade to convince private industry to invest in nuclear power while publicly insisting that America was not behind. It was not an easy dichotomy to sustain.

Strauss played hardball when it came to realizing his political objectives. His files are filled with traces of efforts to smear opponents – particularly where Strauss felt national security was concerned. Strauss's most heated battle was over the issue of the Atomic Energy Commission's weapons tests. These tests had raised fears in the minds of many Americans about the health effects of increasing levels of fallout. As the Admiral wrote to his embattled colleague W. F. Libby during the heat of the 1956 presidential campaign, "In view of Governor Stevenson's citation of Ralph Lapp as his great scientific authority on weapons tests … will you give consideration as to whether or not some scientists might not characterize him [Lapp] for the fraud he is." Strauss was equally tough when it came to reactor development. In the heat of the battle over public power, Strauss commanded his press aide to dig out Senator Albert Gore's past statements about British superiority in developing reactor technology. Strauss planned to charge that Gore had influenced the Japanese to buy British rather than American.[90] Strauss soon leveled charges of un-Americanism against his longtime

87 Pfau, *No Sacrifice*, p. 139.
88 "Highlights of Industry Meeting, December 3, 1957 (Utility Organizations – Public and Private)," attached to Zehring to LLs, [misdated 12/3/57], "Power General, Dec. 1957" folder, Box 84, LLS, HHPL, p. 9.
89 Ibid.
90 Strauss to Libby, October 10, 1956, "Libby 56" folder, LLS, HHPL; Strauss to Holles, December 21, 1956, "Holles 1956" folder, Box 42, LLS, HHPL.

antagonist Clinton Anderson. As Strauss wrote to one staffer, Anderson, by questioning the AEC's progress, was "encouraging other countries to assume that we are second best." Strauss instructed the staffer to try to line up Carl Durham, a conservative Democrat on the Joint Committee, to lead the public attack.[91]

Few people claimed that Lewis Strauss was a nice guy. Pfau, a relatively sympathetic biographer, who labeled his book *No Sacrifice Too Great*, concludes that although charming in private, "the public man, especially when he faced opposition, was irascible and unpleasant."[92] Despite his faults and vendettas, Strauss does deserve credit for adhering privately to the principles he espoused publicly – no government subsidies.

Strauss's closest ally among utility executives was Walker Cisler – the president of the Detroit Edison Company and partner with Dow Chemical in the AEC's industrial participation program. Cisler framed his proposals in the ebullient rhetoric of can-do free enterprise. Lobbying for amendments to the original Atomic Energy Act in the spring of 1953, for instance, Cisler argued that "traditional competitive free enterprise is ready and able to undertake the task" of developing atomic energy. "The closer is the approach to American competitive free enterprise... the more efficient will be the country's utilization of atomic energy potentials," Cisler continued.[93] "Business is built on the principle of taking risks in the hope of earning a profit. Government, on the other hand, isn't risk-taking by organization or by disposition, inasmuch as it does not strive to make a profit," the newly formed Atomic Industrial Forum echoed several days later.[94]

Strauss had trumpeted similar rhetoric for several years. Yet for him, it proved more than just rhetoric. The Admiral wanted a firm commitment from the private sector. As he wrote to Cisler on April 28, 1953, "One school of thought believes that regardless of the passage of necessary amendments very little private money will be forthcoming. On the other hand, there are those who assume what you and others have in mind, and specifically your group, is that the bulk of financing will be provided outside of public funds.... Certainly my assumption has been ... that all of the financing is to be private," Strauss emphasized.[95] Cisler wasted no time, responding by telegram. "Your understanding correct," he cabled. But in the next sentence, Cisler reiterated the need for federal R and D – a crucial government subsidy. What Cisler did not spell out was a subsidy so significant that its demise ultimately killed most of the initial industrial participation projects (although it only killed Cisler's economically) – federally guaranteed long-term plutonium buy-back arrangements. Nor did Cisler mention the question of federal indemnification against losses from nuclear accidents,

91 Strauss to Holles, March 9, 1957, "Holles 1957" folder, Box 42, LLS, HHPL.
92 Pfau, *No Sacrifice*, p. ix.
93 Cisler to Strauss, April 14 1953, "Cisler 48–53 folder," Box 16, LLS, HHPL, pp. 1–2.
94 "Background on Atomic Industrial Forum," April 16, 1953, "Cisler 48–53" folder, p. 3.
95 Strauss to Cisler, April 28, 1953, "Cisler 48–53" folder.

a government subsidy obtained by the industry in 1957 and still in effect today.[96]

Despite Cisler's assurances, Strauss's fears were well founded. The years following the amendments requested by Cisler and the rest of the nuclear industry were years of increasing frustration for Strauss. Those few projects like Cisler's that had committed private capital to construct power-producing reactors were moving slowly. In these cases, Strauss pushed his contacts in the private sector to speed up operations. Cisler, for one, did not take Strauss's requests lightly. As he wrote in a "Personal and Confidential" memo to Strauss in June 1956, Detroit Edison planned to bring in a new officer to expedite the project. The overall result, Cisler intimated, "would be more rapid progress which would be in keeping with your request to me at Geneva."[97]

The major problem, however, continued to be industry's reticence to invest at all. Strauss, no doubt, was anxious to speed up Detroit Edison's groundbreaking in order to thwart growing pressure for publicly financed and publicly distributed nuclear power. Throughout 1956, Strauss led the battle against the Gore–Holifield legislation that would have financed this public approach. He continued to use *U.S. News & World Report* to disseminate his views on the evils of public power. The first step was to staunch the torrent of public criticism targeted at America's unenviable position in the kilowatt race.[98] When the *New York Times* criticized "the Lag in Atomic Power" in a March 1956 editorial, Strauss struck back, urging Kenneth Davis to issue a public response. In it, Davis argued that America's Shippingport reactor would produce amounts of power comparable to Britain's Calder Hall reactor by the middle of next year. Davis then went on to dismiss the "arbitrary" goal of measuring nuclear progress in terms of kilowatts produced.[99]

In conjunction with Kenneth Davis, Strauss sought to muscle utilities into committing funds toward reactor projects, arguing that this was the only way to blunt the thrust toward public power. Davis, from his post as reactor development director, spotted potential opportunities. Strauss then moved in for the high-level sales pitch. Although Strauss and Davis were relieved that the Gore-Holifield legislation failed, in 1956, the threat that such legislation might pass in the next session loomed large. As the *Wall Street Journal* reported on December 6, 1956, "Backers of the bill talk more confidently of success next year. They think the recent debut of Britain's new atom power plant helps dramatize foreign progress in the race to develop atomic energy for peacetime uses."[100]

In August 1956, Davis wrote to Strauss, reminding him that Pacific Gas and Electric was the largest utility in the United States. Davis suggested that Strauss

---

96 Cisler to Strauss, May 1, 1953, "Cisler 48–53" folder.
97 Cisler to Strauss, June 17, 1956, "Cisler 54–56" folder, Box 16, LLS, HHPL.
98 Strauss to Lawrence, May 24, 1956, "Power General, 50–56" folder, Box 83, LLS, HHPL.
99 Davis to Editor, *New York Times*, April 9, 1956, "Davis 56" folder, Box 22, LLS, HHPL.
100 *Wall Street Journal*, December 6, 1956, p. 1.

"urge that PG&E consider a large reactor in an area of high fuel costs."[101] In November 1956, Strauss sent Davis a list of some of the nation's largest utilities, and asked the reactor development director which ones had failed to participate in "our" program.[102] Strauss apparently launched a campaign to persuade some of the nonparticipants to change their minds. His point man seems to have been William Webster of Yankee Atomic Electric Company. Although there are no records of the Admiral's conversations, a telephone report from Webster has survived. Webster named five groups that were "certain to go ahead" and that all planned to make announcements. He also listed a number of other groups that might be persuaded. Webster felt he had started "2 or 3 fires that might fan this a little bit."[103] With the Joint Committee's annual hearings on the status of nuclear power coming up in February, the pace soon quickened. As Davis reported to Strauss about his conversations with the president of PG&E: "I stressed the importance of an announcement before February 19th, and really concrete action before summer."[104]

Although most of Strauss's efforts were directed toward persuading utilities to invest, interest even among manufacturers could not always be taken for granted. The reason PG&E cited for delaying its plans to announce a large reactor project was that General Electric, with whom the utility had been negotiating "very earnestly," suddenly walked out.[105] On February 12, 1957, Strauss called on the president of GE, Ralph Cordiner, to discuss the matter. In a memo to the files written the next day, Strauss confirmed that, indeed, GE had walked out, citing past overoptimism about the nuclear market and overexposure, given the $100 million that GE had already invested. According to his notes, Strauss then put on the hard sell. He asked Cordiner if the investment could not be considered "costed operations" – an accounting device that would make the expenses tax deductible. Cordiner acknowledged that perhaps $25 million might be considered "costed operations."[106]

Strauss was not bashful about spelling out the political implications of GE's withdrawal. "I said that if the General Electric Company persisted in the position he had stated that then no matter how much help the utilities wished to give to the principle of private ownership, I was sunk." Cordiner, according to Strauss, responded to this appeal: "'If we took on the Pacific Gas and Electric Company job and the Southern California Edison job would that help you in your presentations to the Congress?'" Strauss replied, "[P]rovided I got the word quickly enough." After receiving assurances that Rickover would not be involved in the

101 Davis to Strauss, August 3, 1956, "Davis 56" folder.
102 Strauss to Davis, November 19, 1956, "Davis 56" folder.
103 Virginia Walker to Strauss, notes on William Webster call, November 19, 1956, "Power General, 50–56" folder, Box 83, LLS, HHPL.
104 Davis to Strauss, January 25, 1957, "Davis 57" folder, Box 22, LLS, HHPL.
105 Ibid.
106 Strauss, "Memorandum for the Files," February 13, 1957, "Memo for the Record" folder, LLS, HHPL. I would like to thank William Lanouette for sending me a copy of this document.

project, Cordiner, according to Strauss, said he hoped to make an offer on both plants by the end of the week, but that he would not commit his company beyond these two plants.[107]

Although the industry managed to stave off public power and at the same time pass the Price-Anderson indemnification legislation in 1957, Strauss continued to push for private investment. He was fighting a two-front war. In public, Strauss led the charge against the "atoms-lag" critics. Strauss insisted that America led the race for nuclear power, in part because it had relied on the private sector. In private, however, Strauss lambasted the nuclear manufacturers and would-be nuclear utilities for not doing enough. A November 1957 memo from staffer Robert Zehring captures this dilemma. Zehring tracked utility investment in nuclear power as a percentage of total investment for Strauss. Reporting on investment figures for the past five years, Zehring noted that "when one compares these huge spending figures [$1.2 billion for expanded generating plants in 1957] with the estimated present *annual* spending by the same group of only $50 million...for all phases of nuclear development...one realizes how far we have to go in persuading the private utilities to get into the nuclear power game at an accelerated rate."[108]

Considering the enormity of his task, Strauss had little ammunition. It was one thing to oppose public control of nuclear power as much of the industry did. It was another to oppose public guarantees for the private development of power. Yet because of his vision of free enterprise – one sharply out of line with corporate America's – Strauss continued to oppose the kind of subsidies that utilities and manufacturers sought before they were willing to invest. Thus, he had little to go on other than the threat of public power – an approach that all knew he personally despised – and his innate sales ability. Zehring hoped that this unlikely combination, and the embarrassing performance of the private sector, would pay off for the Admiral. "Perhaps if the private companies know that you know these figures," Zehring speculated, "and they keep in mind your own reputation for aggressive action as well as bearing in mind the public power threat looming up in the new Congress, they will commit some more funds to atomic power plant construction and research and development."[109]

If Strauss failed to produce much in the way of reactor starts, he did help cement a formidable interest group. Of course, the major objective of this third link in the nuclear triangle was precisely the kind of federal subsidies that Strauss opposed. Every power reactor built before 1963 received some kind of direct or indirect federal assistance. This "was perhaps the strongest evidence for the persistent lack of industry confidence in the commercial future of nuclear

---

107 Ibid.
108 Zehring to Strauss, November 19, 1957, "AEC Power General 1957 June–Nov." file, Box 83, LLS, HHPL.
109 Ibid.

power," concluded a RAND Corp. evaluation of the commercialization process.[110]

Strauss had directed his most forceful appeals toward the nation's leading utilities. The utilities, however, remained skeptical about the economics of nuclear power. By the mid-1960s, an increasing number of utility executives began to take the economic gamble. As difficult as this audience was to crack in the 1950s, there would be a tougher group to convince – the American electricity consumer. Reference to these unorganized customers was nonexistent at the industrywide meetings held at the end of 1957, with one ominous exception. Acknowledging that his company had spent little so far on nuclear development, a Baltimore Gas and Electric executive told his colleagues that he feared the public's reaction "when they find out that present nuclear plants are so high-cost and that [the] public is really paying the bill for high-cost juice."[111] Though prescient, the fear was premature. It took several more decades for a significant number of customers to recognize that nuclear juice often left a bitter aftertaste.

110 Perry, *Development*, p. 17.
111 "Highlights of Industry Meeting, December 3, 1957 (Utility Organizations)," p. 3.

5

The centrifugal push of expertise: reactor safety, 1947–1960

Just as the most basic political relationships between agency, Congress, and interest groups had to be constructed after the AEC's inception, health and safety regulation – usually sharply demarcated from development – was also constructed within the AEC. Contemporaries were quick to point out the unusual nature of this arrangement. But few challenged self-regulation in the early years. National security is the factor most commonly cited. One need look no farther than today's Department of Energy and the degree to which it continues to regulate its own weapons production plants to recognize how influential national security can be in distorting traditional patterns of oversight and review. Expertise and the way it was organized, however, were also central to the political evolution of safety review as it related to the development of commercial nuclear power.

Because nuclear power was not far removed from the laboratory, only a handful of experts understood the complexities of nuclear power: those who designed reactors were also the leading authorities on their safety. The AEC's initial approach to safety was modeled on the Manhattan Project's techniques. There, reactor designers were nothing less than jacks of all trades. As Enrico Fermi's assistant noted, Fermi "eagerly participated in constructing the 'piles,' in making measurements, in repairing instruments, etc." Fermi was also in charge of safety. He was prophetic about the nature of reactor problems. After a lengthy brainstorming session with his assistants and Du Pont one day, Fermi was asked which of the many problems discussed might prove most troublesome. "The ones which we haven't yet discussed," Fermi answered.[1]

Because nuclear power was developed by highly skilled professionals, the professional model of "self-review" governed its development. Just as the American Medical Association once dominated the state bodies that regulated and enforced state regulatory and certification bodies, a small group of scientists and engineers dominated safety review of nuclear development. The proliferation of expertise in this field and the multiplicity of disciplinary perspectives that ultimately addressed safety questions eventually dramatically changed this pattern. But in its early years, independent safety review was a fragile and insulated art at best.

---

1 For Fermi quotations, see Weil, "Nuclear Power," pp. 25, 26; on Fermi and safety, see p. 13.

Organizational influences were equally powerful. From an organizational perspective, the Atomic Energy Commission's developmental and safety objectives were directly at odds. Developers sought to disseminate reactor technology. They hoped to interest private and public utilities and ultimately they hoped to serve the everyday needs of electricity consumers. The nuclear community's greatest challenge when it came to development was to create broad-based demand. Safety objectives presented the opposite organizational challenge: The objective was to centralize authority for safety determinations, to ensure that conflicting standards did not jeopardize public safety, or retard development. Originally, virtually all of the jurisdiction over safety issues rested with the Atomic Energy Commission. Maintaining that centralized control was crucial.

Four characteristics conspired against centralized and integrated safety regulation from the very start, although their cumulative effect was not felt until well into the 1960s. The first of these – specialization – created organizational imperatives for considering problems in isolation. In the case of safety determinations, this encouraged perspectives that often were antithetical to development interests in the nuclear community. This chapter explores that phenomenon as it occurred within the Atomic Energy Commission. Chapter 6 examines another characteristic that worked against centralized control: the need to subject many of the AEC's complex problems to interdisciplinary scrutiny. Participation by experts trained in a variety of disciplines diversified the increasingly specialized units, further increasing the variety of ways in which problems were defined and addressed. Chapter 6 also explores two other characteristics – easy entry by issue networks, and competing jurisdictional claims by states and local government – which were stimulated by the political environment and culture in which this specialization and interdisciplinary jostling took place.

THE ORGANIZATIONAL ARTICULATION OF EXPERTISE:
THE REACTOR SAFEGUARD COMMITTEE

As with reactor development, the General Advisory Committee initially formulated the Atomic Energy Commission's safety policy. Reflecting its member's wartime experience, the GAC did not initially draw many boundaries between development and safety-related issues either in its own deliberations or in its assignments to AEC staff. As we have already seen, the General Advisory Committee's major challenge had been in finding ways to speed development, which had proceeded at a frustratingly slow pace. Mounting concerns about safety further slowed that pace and forced the GAC to recommend a more specialized approach to resolve safety questions.

AEC chair David Lilienthal raised questions about safety at the General Advisory Committee's first meeting in January 1947. The commission wanted the GAC to review a safety report prepared by the Manhattan Project about operations at Oak Ridge. Enrico Fermi, confirming a tradition that assumed that the men who developed reactors were best qualified to determine the safety of their equipment, suggested that the matter be referred back to Clinton Laboratories at Oak Ridge. The General Advisory Committee was quick to embrace specialization where its members felt they lacked sufficient expertise themselves. At the January meeting, the committee discussed the possibility of establishing advisory committees on geology and mining, medicine and biology, and the social sciences. But reactor safety was so integrally related to reactor development, in the opinion of GAC, that there was no discussion of a specialized approach to reactor safety.[2]

When one commissioner raised the issue of safety again at the GAC's May meeting – asking about the possibility of a reactor blowing up – the General Advisory Committee recognized that safety was going to be a recurring problem. The GAC was comfortable with a fluid arrangement where developers also addressed safety issues, but the commission required more independent assurances. Enrico Fermi was one of the first to recognize that the days of undifferentiated development and safety review were numbered. Not only was the Atomic Energy Commission simultaneously building and riding a bicycle, it had to demonstrate that cyclists could enforce their own traffic regulations. In executive session, Fermi emphasized that where safety was concerned, each problem was different; many of the problems would require a broader array of experts than the General Advisory Committee included. To deal comprehensively with these problems, Fermi continued, a standing committee composed of a chemical engineer, a reactor expert, and a nuclear explosion expert was required. Meeting with the commission, Oppenheimer informed its members that general answers to safety questions would never be adequate. Each reactor had to be examined with reference to its surroundings, construction, and planned operation. The GAC recommended a standing committee (à la Fermi) that would address the commission's safety questions. "We don't think a group that meets around a table once every two months can give you a satisfactory answer," Oppenheimer told the commission. Organizational and professional specialization was the obvious solution.[3]

Oppenheimer drafted a letter, approved by the commission, that implemented this approach. Called the Reactor Safeguard Committee, this group of experts from a variety of fields met for the first time in November 1947. Unlike the General Advisory Committee, whose authority rested on a statutory requirement, the RSC reported to the director of research and served exclusively at the pleasure of the commission.[4]

2 Lilienthal to Members of the GAC, January 3, 1947, Box 1, Bacher Files; GAC Minutes #1, January 3–4, 1947, AEC, DOE, pp. 5–6.
3 GAC Minutes #4, May 30–June 1, 1947, and Transcript of Discussion with the AEC, May 31, 1947, p. 2, appended to Minutes, Summary Report of the Reactor Safeguard Committee, August 11, 1949, Info Memo 206, AEC, DOE (hereafter IM 206). It appears that the commission's concerns about safety at the GAC meeting were related to a reactor proposed by General Electric near Schenectady, New York. ("Items of Interest to the GAC," July 24, 1947, Energy History File, AEC, DOE, p. 4.)
4 Richard Hewlett, "The Evolving Role of the Advisory Committee on Reactor Safeguards," March 11, 1974, *Controlling the Atom* File, Nuclear Regulatory Commission, Washington, D.C. (hereafter *CTA* File), p. 1; IM 206, p. 3.

The Reactor Safeguard Committee made up in stature what it lacked in statute. It was chaired by Edward Teller, a physicist at the University of Chicago's Institute of Nuclear Studies. Teller's wartime work at Los Alamos made him a leading authority on the explosive capacity of nuclear reactions. Best known as "the father" of the hydrogen bomb, Teller also displayed a keen interest in the question of reactor safety. Teller more than any other atomic scientist appreciated the fact that successful commercialization depended on rigorous safety standards. Teller insisted that a committee specializing in safety matters be established.[5]

Reflecting the diverse disciplines the GAC sought for the Reactor Safeguard Committee were members John Wheeler, a reactor specialist from Princeton, and Joseph Kennedy, chair of Washington University's Department of Chemistry. Invitations were also extended to and accepted by Benjamin Holzman, a meteorologist in the air force, Manson Benedict, a chemical engineer who had worked on the Manhattan Project's gaseous-diffusion fuel-processing plant, and Abel Wolman, a sanitary engineer at The Johns Hopkins University.[6]

Letters to the charter members of the RSC requested they serve the commission as "disinterested experts." Should they accept, the invitations continued, they would have "no legal responsibilities in the event of a disaster." The letters also were quite clear on an issue that was to be a continuing source of controversy: Members would review safety in detail on a case-by-case basis. By defining the RSC'S organizational mission as exclusively safety-related at the same time that it left the RSC relatively free from direct AEC control, the General Advisory Committee formalized the distinction between development and control inherent in the AEC's mission from the start.[7]

The Reactor Safeguard Committee, however, was not entirely insulated from pressure to develop reactors quickly. George Weil, chief of the reactor development branch, disclosed that one of his first assignments on joining the Atomic Energy Commission was to represent the AEC on the Reactor Safeguard Committee and "assure that the committee's reports would not be worded in such a fashion as to usurp the Commission's prerogative to make the final decision" regarding reactor siting. Weil intervened only once to explicitly change the committee's language – and that with Teller's blessing. Committee members were extremely sensitive and sympathetic to the goal of developing nuclear power quickly.[8]

Nevertheless, as the sole body formally concerned with safety issues, the

---

5 IM 206, pp. 2–3; Alvin M. Weinberg, "The Maturity and Future of Nuclear Energy," *American Scientist* 64 (January–February, 1976): 19. Teller himself claimed that the power of the explosion at Hiroshima frightened him. He realized after working with the three reactors at Hanford that a single accident could wreck the hopes of a peaceful atom (Edward Teller, *Energy From Heaven* [San Francisco: Freeman, 1979], pp. 159–60). Recent accounts of the H-bomb emphasize that Teller was not the sole father. See, for instance, Daniel Hirsch and William G. Matthews, "The H-Bomb: Who Really Gave Away the Secret," *BAS* (January–February 1990): 22–30.

6 IM 206, pp. 3–4.

7 "Summary Report of the RSC," March 31, 1950, WASH-3, *CTA* File, p. 2.

8 Weil, "Nuclear Power," pp. 73, 75–6.

Reactor Safeguard Committee soon became the target of the General Advisory Committee's frustrations. It was not long before those less worried than the RSC about reactor safety dubbed it the "Committee for Reactor Prevention." The GAC expressed its frustration explicitly in the spring of 1948. The issue was the proper location of the AEC's high-flux reactor, in which materials crucial to the development of power reactors would be tested. In a series of difficult decisions, the General Advisory Committee had determined to consolidate power reactor development at the Argonne Laboratory, under the direction of Walter Zinn. Zinn himself had questions about the safety of building the high-flux reactor so close to the Chicago area and was the first to suggest that some reactors be built at a remote proving ground. The RSC report on the question (February 1948) addressed these concerns in a formal fashion. The report concluded, "Although the possibilities of dangerous contamination of the air or surface water of the Chicago region with radioactive fission products from this pile can be made very small by proper design and careful operation, the Committee is not convinced that they can be so far reduced as to make the Du Page County site a proper location for this pile." The RSC's report destroyed a portion of the AEC's consolidation plan and virtually assured that a remote site would have to be selected for the high-flux reactor.[9]

The General Advisory Committee had hammered out the consolidation plan after months of debate and political infighting. Its members were not pleased that the RSC had torpedoed its efforts, particularly since many of the concerns raised did not seem to be shared unanimously by RSC members. Furthermore, the Safety Committee had not been able to quantify its concerns. Oppenheimer, for instance, "felt the Committee [GAC] received an impression of disparity among the experts and also had a feeling of concern that the conclusions were based too greatly, perhaps, on qualitative considerations." Hood Worthington, of the Du Pont Chemical Company, felt that the seriousness of the problems raised had been overemphasized by the RSC. In Worthington's opinion, that was to be expected from a committee zealous to ensure safety. It was virtually built into their mission. Nobel Laureate physicist Isidor Rabi, who in 1952 would replace Oppenheimer as head of the GAC, searched for a more constructive response to the problem. Already frustrated by the General Advisory Committee's inability to expedite the reactor program, Rabi proposed that the GAC play a more active role in finding solutions to the problems raised by the RSC. All agreed that the high-flux reactor was crucial and that the RSC's safety concerns could not simply be dismissed. If the RSC considered the proposed site unsafe, more safety features

9 For reactor prevention sobriquet, see Teller, *Energy*, p. 161; for a good discussion of the AEC's decision to consolidate reactor development, see Minutes of the Special Meeting of the GAC, December 29–30, 1947, pp. 1–12; Zinn to Fisk, July 23, 1948, p. 1; Weil to Fisk, May 26, 1948, pp. 1–2; "Report of the Atomic Energy Commission Reactor Safeguard Committee on the Proposed High Neutron Flux, Water Cooled Thermal Reactor," February 10, 1948, IM 48–48, AEC, DOE, pp. 7–8; Hewlett and Duncan, *Atomic Shield*, pp. 186–8, 208–9.

would have to be built in or the reactor removed to a less populated site.[10]

The General Advisory Committee's report stated that a strenuous effort must be made to design a structure that would contain potential radiation in order to meet objections raised by the RSC. The report did not challenge the RSC findings, conceding that the committee's arguments were sound. It did state, however, that the GAC was "greatly disturbed by the prospect of the delay in the reactor program that would be brought about if the high flux pile were not built in the near future, or if built, were placed at a remote site."[11]

By its October meeting, the General Advisory Committee had given up on the possibility of adequately containing the high-flux reactor at the Argonne site, and the Reactor Safeguard Committee had rejected proposals for a lower power reactor at Argonne. In view of the increased concern that the Reactor Safeguard Committee voiced about reactors built near population centers, the need for a remote test facility had become an inescapable conclusion. Accepting this conclusion did not mean that the GAC was pleased by it or even that its members necessarily agreed with the RSC's arguments. One GAC member expressed the view that the RSC was retarding development by emphasizing the special hazards of reactors without any adequate estimates of the probability of dangerous events occurring. Another member urged patience, reminding his colleagues that the national labs, already actively pursuing reactor development, would eventually become more sensitive to safety. Over time, this would put them in a stronger position vis-à-vis the Reactor Safeguard Committee. For the time being, the General Advisory Committee would just have to be aware that the RSC was a retarding influence and try to counteract it.[12]

The articulation of expertise in the RSC foreshadowed a number of patterns that would recur throughout the development of commercial power. Establishing the RSC formalized the safety review process. Safety had always been considered in building reactors, but it had been considered by the very people responsible for designing, operating, and promoting them. Before the RSC, safety had not been distinguished from the development process itself. Even though development and safety regulation were the responsibility of the same agency, formalizing safety review solidified organizational boundaries and ensured that safety would be a permanent topic of discussion between organizational units within the AEC. That replaced the less formal pattern of discussion between staff members working on a specific reactor. The RSC also changed the medium through which safety was discussed. Increasingly, the safety agenda was dominated by written reports, submitted on a case-by-case basis by the RSC.

10 GAC Minutes #9, April 23–5, 1948, AEC, DOE, pp. 17–19, 25.
11 "Report of the General Advisory Committee on Safeguards for the High Flux Reactor," April 24, 1948, appended to minutes of ibid., pp. 1, 3.
12 Zinn to Weil, August 13, 1948, Teller to Weil, September 13, 1948, in "Research Reactor for Du Page County Site," October 5, 1948, AEC 149, DOE; GAC Minutes #11, October 21–3, 1948, AEC, DOE, pp. 12–13.

Specialization proved a source of both strength and weakness for the RSC. Because the RSC probed safety in such depth, accumulating experience and knowledge as it matured, it emerged as the unassailable authority on technical safety matters. Expertise derived through specialization had its drawbacks, however. Other actors – even other expert bodies, as we have seen with the General Advisory Committee – discounted the RSC's views precisely because its mandate was so narrow. Limited or not, the Reactor Safeguard Committee's expertise could not be ignored. Because the RSC consisted of leading experts who were relatively independent of any organizational constraints, their advice shaped the debate. Even the General Advisory Committee's illustrious members who were impatient with the slow pace of development and who raised doubts about the RSC's decisions balked at overriding decisions hammered out by an independent group of safety specialists.

The Reactor Safeguard Committee's expertise was politically valuable because it operated in an area where there was little experience and where judgment was essential. On the crucial issue of reactor siting, the first methods used by the RSC, according to Teller, were rudimentary and depended on rule-of-thumb determinations as to the degree of isolation required. The RSC used a formula that excluded the general public from an area covered by the radius in miles equal to one-hundredth of the square root of the normal operating level of a reactor (measured in kilowatts). Even this rather arbitrary guideline was hardly a fixed standard. As the RSC was quick to point out, "It is not possible to set the limits of the hazard area by any formula because not only are the type and power of reactor significant but also the local meteorological, topographical, hydrological and seismological conditions lead to different evaluation of hazards in different directions."[13]

To deal with variables such as these the Reactor Safeguard Committee soon began to draw upon expert consultants and create its own subcommittees, extending even further the specialization process. The RSC also expanded its requirements for data to include detailed reports on site-specific conditions. At the heart of its data requirements was an effort to describe and evaluate a theoretical major catastrophe (soon to become known as a "maximum credible accident"). The amorphous state of the field led to wide-ranging discussion and not a little disagreement among the safety experts themselves. This, in turn, prompted efforts to develop an agreed-upon formula for deciding safety issues.[14]

Meanwhile, the General Advisory Committee, frustrated by what it considered

13 Teller, *Energy*, pp. 162–4; IM 206, pp. 17–18, quotation from p. 18; "Estimate of Radiation from Cloud Containing All of a Reactor's Fission Products," Append. A to IM 206.
14 IM 206, pp. 4–5, and "List of Items of Information Necessary to the Safety Evaluation of a Proposed Reactor Installation," Append. H for reporting requirements including evaluation of the possibility of a major catastrophe; Mazuzan and Walker, *Controlling the Atom*, p. 229; Okrent, *Nuclear Reactor Safety*, pp. 14, 32–4; RSC charter member Abel Wolman has discussed the nature of disagreement in the early RSC in an interview with the author (hereafter Wolman Interview), December 18, 1985.

to be an overemphasis on safety, responded in five ways that would recur among subsequent proponents of more rapid development. It never challenged the RSC's findings directly. Nevertheless, the GAC was cognizant of the divergent views among RSC members and used anything less than absolute RSC consensus as a pretext for questioning the Safety Committee's recommendations. The GAC also pushed the Reactor Safeguard Committee for a more quantitative, and less qualitative, formulation of its criteria; the GAC was the first in a long series of development advocates to lobby for standardized evaluations. It also pioneered another mechanism that would be used again by other development advocates. It looked to a parallel source of expertise – in this case, the national laboratories – to develop safety capabilities in an organizational structure linked to practical developmental considerations. Finally, the GAC placed its faith in new technological solutions to the problem. It realized that containment was the ultimate solution to what it considered an overly conservative siting policy. In this conclusion the RSC agreed wholeheartedly. Teller, in fact, advocated building all reactors underground.[15]

To discern emerging patterns without commenting on the scope of the debate involved is like listening to Ravel's "Bolero" on the stereo with the volume turned down. Crucial to the process I have described is the degree to which debate was limited to forums housed within one agency: specifically, the national labs, the General Advisory Committee, and the newly formed Reactor Safeguard Committee. These forums were far removed from the public at large, and at the time they were insulated from congressional politics and, to a large degree, even the Atomic Energy Commission.

Would the debate remain insulated? At a time when nuclear power was struggling to compete economically, siting and safety considerations were two of the most important "controllable" economic variables. If safety considerations precluded sites close to population centers, or required expensive design adaptations, transmission or construction costs would offset any other possible economic advantages that nuclear power had to offer. Because siting and safety considerations were so crucial to the overall cost of nuclear power, Lilienthal recommended that they be left to the experts. As he wrote to Brien McMahon in November 1949, "Choice of location and selection of design vitally affect the cost of building and operating a reactor.... The resolution of these factors... is thus a question of broad public policy. Nonetheless, such a decision must begin with consideration of the technical problems involved in reactor design and operation. The Reactor Safeguard Committee has been formed to bring to such technical problems the skill and experience of experts from a wide range of scientific fields." Despite the broad-ranging policy implications of safety review, the AEC managed to keep the scope of the debate extremely limited for some time. For instance, the Joint Committee on Atomic Energy, hardly an outsider

15 See, for instance, C. Rogers McCullough to the Files, August 17, 1960, undated box labeled "ACRS," Wolman Papers, The Johns Hopkins University.

to the nuclear community, remained remarkably uninformed about the RSC as late as 1954.[16]

By the mid-fifties, the AEC had experienced more than its share of widely publicized disputes between distinguished experts. In 1950 a fierce debate raged inside the AEC over the question of developing a hydrogen bomb. It was soon carried all the way to the president's desk and in highly public fashion. The battle prompted James Conant to conclude that the nation had "not yet evolved a satisfactory procedure for evaluating differences of opinion among technical experts." Several years later AEC experts were embroiled in a prolonged public debate over fallout from its nuclear tests. By 1956, the debate over fallout and nuclear testing had attained such prominence that Adlai Stevenson made it a major issue in his campaign. Both these controversies tarnished the AEC's reputation. Particularly in the debate over fallout, the agency suffered a great loss of credibility as independent standard-setting authorities regularly reduced acceptable levels of radiation and as the AEC was forced to concede that fallout from tests was higher than originally reported.[17]

Remarkably, there is little evidence that the public was particularly concerned about reactor safety. Although intense debate characterized the safety discussion within the AEC's insulated forums, the professionals engaged in these debates were also committed to containing – as opposed to publicizing – their differences. There was nothing conspiratorial about this. It was the result of a unique combination of the limited numbers of experts who dealt with such matters, extreme centralization of decision making within the AEC – and a culture that encouraged professional solidarity vis-à-vis the public.

The General Advisory Committee's response to disruptions in its program created by the Reactor Safeguard Committee's safety review is instructive. The GAC's tack was carefully circumscribed. Despite fairly intense criticism within the General Advisory Committee, there was no hint of public criticism. Nor did the GAC ever seek to disband the Reactor Safeguard Committee. Rather, the GAC banked on mobilizing alternative bastions of expertise in the national labs – experts who might be more sympathetic to reactor development and less exclusively devoted to questions of safety. As the Reactor Safeguard Committee neared the completion of its report on reactor siting in early 1950, the GAC did express concern about the RSC's mission. The report itself was hardly shocking in light of the RSC's negative high-flux reactor determination. The report summarized the RSC's views on siting policy, stating that it was sensible to restrict

16 Quotation from Lilienthal to McMahon, November 25, 1949, Joint Atomic Energy Committee Correspondence, January 1947–50, Box 2, Records of the Office of the Chairman, Records of the Atomic Energy Commission, National Archives, Washington, D.C. As late as October 1954 the Joint Committee on Atomic Energy was still asking the AEC for a statement of the Reactor Safeguard Committee's functions (see Allardice to Nichols, October 25, 1954, "Advisory Committee on Reactor Safeguards, Vol. I" folder, Box 14, JCAE Files).
17 Conant cited in James Reston, "Secrecy and the Reporter," *Atlantic Monthly*, April 1950, p. 42. On the national fallout debate, see Chapter 6, note 18.

the location of high-power and potentially dangerous reactors to more isolated spots. But these restrictions were hardly immutable. It might be possible, the report continued in a more optimistic vein, to relax these restrictions in the future. The best hope for this lay in airtight containment shells.[18]

While the General Advisory Committee grumbled in private and sought alternative routes for safety review, some sought to disband the Reactor Safeguard Committee entirely. Commissioner Smyth – who had replaced Robert Bacher as the only AEC commissioner with a scientific background – was one such proponent. In the long run, the GAC's approach, while seeking to rein in the RSC, helped stave off more dramatic alternatives: it preserved the RSC as an institution. At the same time it sought to integrate safety review and reactor development through the national labs.[19]

Weil took advantage of more draconian proposals, such as Smyth's, to suggest concrete steps that implemented the GAC's vision of parallel tracks for expertise. He proposed that the national labs developing reactor plans assume primary responsibility for the safety of the reactors. The Reactor Safeguard Committee would continue to function, but it would serve as a board of review and criticism. Not surprisingly, the General Advisory Committee endorsed this plan. In a letter to Lilienthal in early 1950, Oppenheimer wrote that the GAC viewed the proposal as an important step in advancing reactor development. Saved by this compromise, the RSC remained at the center of reactor safety issues for at least another decade and a half.[20]

In January 1951 the commission established yet another track that it hoped one day might preempt to Reactor Safeguard Committee's safety reviews and replace them with reviewers that would weight developmental factors more heavily. The commission created the Industrial Committee on Reactor Location Problems (ICRLP). This committee was made up entirely of representatives from large industrial firms and one member employed by a large insurer. Its official mandate was to reexamine the hazards associated with the operation of facilities for the production of fissionable material. Substantial evidence suggests, however, that by creating the ICRLP, the commission sought to promote a sibling rivalry with the RSC.[21]

The Industrial Committee on Reactor Location Problems "reexamined" the Reactor Safeguard Committee's work. The RSC had raised a number of questions about the Hanford, Washington, production facilities, resulting in the committee's determination that public uses would have to be prohibited within an exclusion area of five miles. Because that exclusion zone took in portions of the fertile Wahluke Slope, the Department of Interior pressured the AEC and some members

18 WASH 3, pp. 1, 19–30; Hewlett, "The Evolving Role," pp. 1–2.
19 Atomic Energy Commission Meeting Minutes #352, November 1, 1950, AEC, DOE.
20 Hewlett, "The Evolving Role," p. 2; WASH 3, pp. 1, 2, 19; Oppenheimer to Lilienthal, February 1, 1950, Append. A to GAC Minutes #19, January 31–February 1, 1950, AEC, DOE.
21 CM Minutes #499, November 27, #1950, pp. 646–7; "Advisory Committee on Reactor Safeguards," July 2, 1953, AEC 661, DOE. p. 2.

of Congress to make more of the excluded area available for farming. Most recently, the RSC quietly had raised questions about radioactive contamination of the Columbia River.[22]

The ICRLP's mandate was prospective as well. Besides Hanford, the other item discussed at the committee's maiden meeting was the 240,000 acres of land the AEC hoped to acquire for the Savannah River plutonium production center. Commissioners Smyth and T. Keith Glennan, who had already visited that site, reported they had been impressed by the importance of the evaluation of hazards associated with the operation of production facilities. They warned the commission that one committee – the RSC – had the power to determine the exclusion area. Consequently, the RSC could determine just how much farmland would have to be sacrificed. A slight change in the formula used to arrive at the decision could free up a lot of land. As the commission minutes reported, "The visit by the two Commissioners to the Savannah site had shown that the chief communities of the vicinity...and the best farm land were on the periphery of the site recommended by DuPont, and that relatively slight differences in the final boundaries of the site would markedly affect the number of people to be moved." The intent in establishing the ICRLP, and its mandate to "reexamine hazards," was to establish a track parallel to the RSC's. This might blunt the Safety Committee's determinations and ameliorate their impact on farming interests on the perimeter of production facilities.[23]

Industry representatives who advised the Atomic Energy Commission on development were not bashful about expressing their views on the Reactor Safeguard Committee's siting policies. The Report on Industrial Participation in the Reactor Program (August 1952), which was an important catalyst toward some of the legislative changes made in 1954 and 1957, stated that "all of the groups have indicated, and some of them emphatically so, that their participation may be impossible if the area of land ownership or control surrounding the reactor is that required by the Reactor Safeguard Committee's exclusion formula."[24]

Meeting in executive session with the Joint Committee on Atomic Energy, in June 1953, Monsanto president Charles Thomas told the committee that

22 AEC 206, p. 37; on Wahluke Slope, see Hickenlooper to Lilienthal, August 9, 1948, Box 1, and Lilienthal to McMahon, August 17, 1949, Box 2, JCAE Correspondence, AEC Files, National Archives; on radioactivity in Columbia River, see Report on RSC Meeting #11, September 5, 1950, AEC 172/6, DOE.
23 CM Minutes #499, p. 647. At the committee's first meeting in December, Commissioner Glennan told the members that the AEC wanted assistance "bringing the industrial approach" to bear on the problem of nuclear hazards and reactor locations. The charter proposed by the AEC stated that it was appropriate for the committee to reexamine considerations that required the isolation of reactors. Technical and scientific aspects of reactor hazards, described as the domain of the RSC, should be weighed "against the non-technical aspects of reactor location which were not considered by the RSC," such as the "impact on adjacent communities" (Minutes of the First Meeting of the Industrial Committee on Reactor Location Problems, AEC 396/1, December 4, 1951, Nuclear Regulatory Commission Freedom of Information Request 85–646, pp. 1, 4).
24 "Industrial Participation in Reactor Program," August 25, 1952, AEC 331/38, DOE, pp. 5–6, 11, quotation from p. 10.

under the Atomic Energy Commission, you have what is known as a Security Committee. [Thomas later corrected himself and said he meant the RSC.] It is made up of men from many walks of life, who were called in as consultants. Frankly, I think that the teachings of that committee are really ludicrous. They state that you can not put an atomic power plant on any land unless you have something like a hundred square miles.... If we have to operate under that, we might as well get into something else, you see. It is just impossible.

Adding that he had been a consultant himself, Thomas told the committee that "the easiest thing to do is to make it as safe as possible.... But it is not realistic when you are coming in to try to make an industry out of it."[25]

Relaxed safety standards were one of the subsidies that the industry sought before it would invest in nuclear power. At a minimum, the industry wanted standards that were clear-cut and that would not depend on a body of academic experts. Private promoters wanted more than just clarity. They wanted significant relaxation in the standards currently being applied. John Grebe, director of the Nucleonics Department of Dow Chemical, hammered home this point in the privacy of Joint Committee executive hearings, arguing that development should be left to the private sector. When asked by Holifield to get down to specifics as to the private sector's "quid pro quo," Grebe told the committee that "obviously, we would have to immediately depend on reducing the enormous cost of exclusionary safety. We would have to cut that out, and we cannot afford [those] kind[s] of expenses." The private sector did not monopolize pressure to relax safety standards. A number of academic institutions expressed interest in developing reactors for research or demonstration purposes. Because many of these institutions wanted to build in heavily populated areas, the RSC's exclusion zone was an immediate stumbling block. In January 1955, for instance, Commissioner W. F. Libby wrote to Strauss informing him of Northwestern University's interest in building a reactor located in downtown Chicago, which could be used to heat the campus. Libby was extremely encouraging on the possibility of obtaining AEC subsidies for the project. Libby felt compelled to raise one problem: "in my personal opinion," Libby informed the Northwestern dean, the proposal's "principal hurdle might be the Reactor Safeguard Committee."[26]

In July 1953 the commission merged the Reactor Safeguard Committee and the Industrial Committee on Reactor Location Problems, forming the Advisory Committee on Reactor Safeguards (ACRS). Since the ICRLP's original mandate was to "re-examine" hazard evaluations, it is not surprising that in the words of the AEC report recommending consolidation, "the responsibilities of the two

25 "Public Power Policy: Monsanto–Union Electric Group," June 12, 1953, doc. #3521, pp. 25–6, JCAE, Executive Hearings.
26 JCAE, Executive Hearings, June 11, 1953, #3518, p. 87, for Dow Detroit proposals; p. 40 for Grebe quotation. For Libby quotation, see Libby to Strauss, January 25, 1955, "Power General 50–56" file, Box 83, LLS, HHPL.

committees were becoming very closely related." Consolidating the two committees tilted the newly formed ACRS toward industrial representation because several members who worked for industrial concerns had already been added to the RSC. The Advisory Committee on Reactor Safeguards' first chair, C. Rogers McCullough, symbolized this shift. McCullough had chaired the Industrial Committee on Reactor Location Problems. A long-time employee of the Monsanto Chemical Company, McCullough oversaw development of the Daniels pile for Monsanto after the war and continued to work for the company as he served on the ACRS. According to RSC member Abel Wolman, McCullough brought the high-flying physicists down to earth. In 1956, McCullough took a leave of absence from Monsanto to serve as a full-time employee of the AEC. When the ACRS became a statutory body in 1957, the AEC's general manager asked Strauss to call Charles Thomas, president of Monsanto, and request that Monsanto start paying McCullough again. As the general manager put it, "We are interested in McCullough being able to serve on [the ACRS] but to do so, he can't be an AEC employee. We are therefore interested in his rejoining their company in order to serve on the Committee." McCullough's multiple roles capture nicely the intimate relationship that existed between commission, its safety advisory committee, and the private sector.[27]

As in 1950, the AEC directed that the Reactor Safeguard Committee (now the Advisory Committee on Reactor Safeguards) eschew specifics and concentrate on more general safety-related matters. In spite of this mandate from the top, the ACRS continued to do what it had been asked not to – determine whether specific reactor proposals were safe. George T. Mazuzan and J. Samuel Walker, in their excellent study of the origins of the regulatory side of the AEC, point to one reason for this quiet act of defiance. Their explanation centers on the hazards summary report, which required the reactor designer to describe the reactor and site and to summarize all known potentially hazardous features. Because this central report was routed through the ACRS, the ACRS retained its case-by-case grasp on the safety review process.[28]

There are two additional reasons why the Advisory Committee on Reactor Safeguards remained at the center of the site approval process. The first was organizational inertia: once the Reactor Safeguard Committee established its pattern of review, it was difficult to stop. Perhaps the most important explanation, however, was the prestige of the ACRS's scientists and the experience that the committee had accumulated in dealing with the specialized problems of safety.

---

27 AEC 661, quotation from p. 3, see Append. B for proposed membership of ACRS. Wolman Interview, December 18, 1985. On McCullough leave, see Mazuzan and Walker, *Controlling the Atom*, p. 68. For general manager's quotation, see Virginia Walker to Strauss, August 24, 1957, phone transcript of General Fields' telephone conversation, "C. Rogers McCullough" file, Box 63, LLS, HHPL.
28 Mazuzan and Walker, *Controlling the Atom*, pp. 62–3; AEC 661, p. 3, and "Draft Bulletin GM-RDV 2 (revised), Append. A, AEC, DOE.

Merging the RSC and ICRLP was a commission strategy to dilute the overly "conservative" RSC. The addition of men with industrial backgrounds like McCullough's probably did bring the ACRS down to earth. There was an unintended consequence of the merger, however. The newly formed ACRS contained a more representative cross section of experts, making it all the more difficult to ease it out of the site selection process. Even while being paid by Monsanto, chemical engineers like McCullough, who concentrated exclusively on safety, were forced to hammer out a consensus among others whose mission was similarly narrow but whose disciplinary perspectives were quite different. ACRS members, serving an organization whose mandate was to ensure safety, began to raise questions that developers were not eager to hear, let alone answer. Developers were grateful, however, for the extraordinarily limited scope of the debate. At least such doubts were confined to forums unanimously staffed by friends of nuclear power. Even in the heat of battle, voices rarely were raised above a stage whisper.

The Atomic Energy Act of 1954, which encouraged private development of reactors and established the formal basis for regulating private operators, presented another challenge to the ACRS's practice of reviewing specific reactor proposals. Because the act encouraged private development, it soon increased the workload for the ACRS and threatened to backlog the review process. The ACRS asked for more staff support. General manager Kenneth Nichols responded to this request and addressed the new statute's regulatory requirements when he established a small permanent staff – the Hazard Evaluation Staff (Safety Staff) – that reported directly to the general manager. Like the ACRS, the Safety Staff would evaluate specific reactor proposals. But it appears that Nichols also hoped to take the ACRS out of case-by-case review and fill this function with permanent staff more accountable to AEC management. As the AEC had done twice before (1950 and 1953) Nichols suggested that the ACRS review safety standards and codes. This time the general manager had a safety staff directly accountable to him that could assume the burden of specific site review.[29]

Nichols realized that this was a delicate issue. A letter he wrote informed the JCAE – which by now, at least, knew the ACRS existed – that his new group would develop safety codes and standards. This, of course, was precisely the role that Nichols intended for the ACRS. More significantly, there was no mention of the new staff's authority to review reactor proposals and changes – the role Nichols hoped to ease the ACRS out of by transferring it to the Safety Staff.[30]

29 For increased workload interpretation, see Mazuzan and Walker, *Controlling the Atom*, p. 63, and Minutes of ACRS Meeting #19, October 14–15, 1956, p. 12, NRC-FOIA 85-646, where ACRS members agree that the Hazards Evaluations staff has relieved the committee of a large part of the preliminary review workload and provided advice on minor problems.
30 For a statement on proposed ACRS and Safety Staff responsibilities and letter to JCAE, see "AEC Reactor Hazard Evaluation," April 15, 1955, AEC 661/4, *CTA* File.

The commission approved the Safety Staff and also emphasized that hazards should be evaluated in accordance with a standard set of criteria. But on the matter of clarifying the relationship between the newly created staff and the ACRS, the commission punted. It noted that Nichols's proposal "does not modify the position of the AEC vis-à-vis the ACRS," not a terribly helpful pronouncement when the definition of that position was precisely the topic of debate. Again, C. Rogers McCullough's role captured the amorphous situation. McCullough took a leave of absence from Monsanto to serve as deputy director of the Safety Staff while continuing to chair the ACRS.[31]

As it had since its predecessor's inception, the Advisory Committee on Reactor Safeguards continued to review each hazards summary report and transmit its conclusions directly to the AEC. Once established, this relatively independent unit specializing in safety proved remarkably resistant to the efforts of both the commission, and the general manager, to limit its review of specific reactor proposals. In the meantime, one track for parallel review – the ICRLP – had come and gone, while a potential new one – a safety staff reporting to the general manager – had been established.

As plans for reactors moved closer to development, the Atomic Energy Commission intervened more actively in determining how safety questions would be resolved. Even though the commission's actions were not decisive, the Advisory Committee on Reactor Safeguards was aware of the commission's interest. In January 1956, when the commission sought to make the position of ACRS chair a full-time employee of the commission, ACRS members resisted. Harvey Brooks commented that the ACRS was subject to all sorts of pressure. The pressure, according to Brooks, came not only from private interests but even more from special interests within the AEC, "whose job is to push certain projects at all costs."[32]

As the number of organized actors participating in these decisions grew, the potential for controversy increased. Two factors helped to contain the situation. The ACRS had not actually rejected any sites since the time it raised questions about the high-flux reactor at Argonne. Also, although organizational missions were solidifying, virtually all safety expertise was still housed within the AEC. What's more, these experts fully appreciated the value of confining the discussion within the commission's walls. That soon changed with the ACRS's review of the first proposed commercial breeder reactor and the JCAE's intervention.

31 CM #1076, April 18, 1955, AEC, DOE, p. 3. When, later that year, the commission approved Nichols's proposal that the Safety Staff be transferred to the newly created Division of Civilian Application – a unit charged with coordinating development and safety – the commission again felt compelled to add language that would ensure the status quo for the ACRS. The commission added the statement to its directive, noting that "the ACRS is to be administratively independent of AEC offices and divisions" (CM #1135, October 5, 1955; "Reactor Evaluation Staff," September 26, 1955, AEC 804–3, *CTA* File; Mazuzan and Walker, *Controlling the Atom*, pp. 63–4). See Mazuzan and Walker, p. 68, for a discussion of McCullough's multiple responsibilities.
32 Brooks to J. Z. Holland, January 23, 1956, "ACRS, 1957–60" box, Wolman Papers.

CONTAINING THE DEBATE:
THE AFTERMATH OF THE PRDC REACTOR CONTROVERSY

In the mid-1950s, experts specializing in reactor safety were acutely aware of the need to maintain consensus in public at the same time that they were committed to vigorous debate in insulated forums. The consequences of the atomic scientists' public crusade were readily apparent. Effective at first, public lobbying ultimately threatened to undermine the scientists' most valuable political resource – an apolitical image. Scientists in the nuclear community sought to avoid repeating that outcome, consistently choosing the political techniques used by the the war's science administrators. They worked quietly within official channels at the same time that the specialized pursuit of safety questions made achieving such a consensus more difficult. When there were disagreements, the scientists sought practical compromise. Consensus, they recognized, was a crucial source of authority in this fragile new field.

Commissioner Thomas Murray shattered the routine mechanisms used to achieve insulated consensual decisions on June 29, 1956, when he released to the commission a letter from the Advisory Committee on Reactor Safeguards that raised doubts about the safety of a reactor proposed by the Power Reactor Development Company (PRDC). Murray released the confidential letter to embarrass Lewis Strauss, with whom he had been feuding over the issue of public versus private development of nuclear power and a host of other issues. Strauss considered the attack to be not only ideologically motivated but stimulated by partisan politics related to the upcoming presidential campaign. Whatever Murray's motivations, the letter certainly proved embarrassing for Strauss personally, and the commission as a whole. Strauss, it turned out, had already accepted an invitation from the PRDC to attend a groundbreaking ceremony for the reactor, which was to be constructed on the outskirts of Detroit. Because the Atomic Energy Commission had not yet approved a construction permit, Strauss's acceptance implied that the AEC chair might be less than objective when it came time to vote on the permit. Another longtime Strauss Democratic antagonist, Joint Committee on Atomic Energy chair Clinton Anderson, seized the opportunity to lambast Strauss publicly. Strauss's critics all pointed to the ACRS letter as evidence that there was more than a case of theoretical conflict of interest at stake: the public's safety might well be threatened if the ACRS concerns were swept under the rug.[33]

Strauss and his colleagues on the commission refused to back down. Presenting

33 The most comprehensive and balanced account of the "PRDC controversy" is Mazuzan and Walker's *Controlling the Atom*, ch. 5; see specifically pp. 136–40; see also John G. Fuller, *We Almost Lost Detroit* (New York: Reader's Digest Press, 1975), pp. 1–118; for a contemporary account, see *BAS* 13 (1957): 111, 310, 343. Robert Ken Woo, Jr., provides useful insight into labor's role in the controversy in "Walter P. Reuther and the United Auto Workers' Decision to Intervene in the Power Reactor Development Company Controversy, 1956–1961" (unpublished undergraduate honors thesis in History and Science, Harvard University, 1989).

the AEC regulatory staff's position to the commission on August 1, Harold Price (Division of Civilian Application director, to whom the Safety Staff reported) noted that there were differences of opinion about the likelihood of the reactor's problems being satisfactorily resolved. Price recommended that a construction permit be issued with the condition that the reactor's problems must be resolved before issuance of a final operation license. The commission voted to do that in a three-to-one vote; Murray voted no. The staff report reflected the profound optimism with which the AEC approached technical problems. It stated that "although there is some doubt that the safety problems can be solved in time to meet the schedule proposed by PRDC, the Commission believes that the safety problems associated with this reactor will prove to be of a kind which can be solved within a reasonable time." The commission also argued that it was only issuing a conditional construction permit. Another review was required before the reactor would actually be allowed to operate. Years later, President Eisenhower told AEC chair John McCone just what he thought of such reasoning: It was "an illusion, because the investment practically prejudges the further decision – unless the courts should rule against it." In fact a federal appeals court did rule against it, but the commission's decision ultimately was upheld by the Supreme Court in 1961.[34]

The PRDC controversy was the most publicized reactor safety controversy in the AEC's first two decades. There can be no doubt that partisan politics, the public–private power debate, and personal antagonisms between Strauss and Murray as well as Anderson all contributed to the public and strident course the controversy took. Yet these ingredients were present throughout the fifties and, for that matter, were hardly confined to the politics of reactor development. A crucial catalyst for the PRDC controversy, and an event that foreshadowed future trends in the politics of expertise, was the breakdown of expert public consensus on the safety of the proposed reactor.

The ACRS had struggled over its response to the PRDC proposal, extending its deliberations over the matter into the second and third day of its meetings in early June. No siting decisions were easy in this new field. But the PRDC reactor and its proposed site were particularly challenging for this body of experts. The PRDC wanted to build a "fast breeder" reactor. Breeders are reactors that produce more nuclear fuel than they consume. The prospect of this nuclear alchemy has captivated scientists ever since the world's first chain reaction. The

---

34 Mazuzan and Walker, *Controlling the Atom*, pp. 142–3. For the staff report, see AEC 331/114, July 30, 1956, AEC, DOE. For Eisenhower quotation, see A. J. Goodpaster, Memo of Conference with the President, December 17, 1960, White House Office, Staff Sec., Subject Agenda, "AEC VOl. III" file, Eisenhower Presidential Library, Abilene, Kans., p. 4. (I would like to thank John Yoo for bringing this and related documents to my attention.)

The commission's vote was challenged in administrative proceedings by a group of AFL-CIO unions, and ultimately in court. The Court of Appeals ruled against the AEC in June 1960; the Supreme Court overturned the lower court's decision a year later (Mazuzan and Walker, *Controlling the Atom*, ch. 6).

Argonne Lab's first director, Walter Zinn, pioneered this concept and completed the nation's first fast breeder reactor – the Experimental Breeder number 1 (EBR-1) – in 1951. Breeding was far more dangerous than the light-water reactors most commercial ventures ultimately relied on. The AEC recognized this, requiring that Zinn construct his fast breeder at the isolated Reactor Testing Station in Idaho. That decision, although it caused quite a bit of inconvenience, proved wise when in November 1955, more than half the EBR-1's fuel melted, blowing radioactive gas out of the building. Fortunately, nobody was injured. The accident did raise questions about the safety of future breeders. To make matters worse, from a safety standpoint, the PRDC planned to build its reactor approximately forty miles from Detroit.[35]

The fateful June letter, released by Murray, that ultimately was bandied about in public reflected the effort of the Advisory Committee on Reactor Safeguards to reach some sort of consensus. The first paragraph made the general problem clear: "The proposed PRDC reactor represents a greater step beyond the existing state of the art than any other reactor of comparable power level which has been proposed by an industrial group." The specific problem was that such a reactor had never been successfully tested. In fact, questions about EBR-1's stability under certain circumstances had still not been resolved. Although the AEC planned further tests, the ACRS concluded in its June 6 letter that unless these experiments were "amplified and accelerated," it was doubtful that sufficient experimental information would be available in time to assure safe operation of the PRDC reactor. More significantly, the letter stated that even such an accelerated program might not yield sufficient information to permit safe operation at the proposed site. "The Committee considers it important that bold steps be taken to advance the development of the fast breeder reactor concept," the letter concluded, "and commends the willingness of the Power Reactor Development Company to risk its capital and prestige in advancing the development of this reactor concept. But the Committee does not feel that the steps to be taken should be so bold as to risk the health and safety of the public."[36]

At a meeting on June 21 between the ACRS Subcommittee on the PRDC and AEC staff, subcommittee chair Harvey Brooks said, "There had been a wide divergence among Committee members" as to the probability that an experiment could provide assurance. Brooks pegged his own confidence level at 95 percent; some committee members, however, felt that there was little more than a 50 percent chance that an experiment would prove conclusive. The June 6 letter was a compromise the entire committee could support. The compromise, however, had barely contained the most divergent set of opinions that the AEC's safety experts had expressed to date. Like all ACRS correspondence with the commission, it was not available to the public (or even the Joint Committee on Atomic

35 Lanouette, "Dream Machine," pp. 36, 41–2, 43.
36 McCullough to Fields, June 6, 1956, attached to Minutes of Meeting of PRDC Fast Breeder Reactor Subcommittee with AEC Staff Members, June 21, 1956, NRC FOIA-85-646.

Energy). It was hardly built to withstand the intense political scrutiny that followed once the letter was leaked by Murray.[37]

Strauss won the battle on the PRDC reactor, but the Democratic leadership of the Joint Committee on Atomic Energy had the last word on the use of experts. Up to the PRDC controversy, the JCAE had played a relatively inconsequential role in safety matters. The PRDC case catapulted safety experts into the public spotlight and ensured a far more visible role for the ACRS. As a consequence, in 1957 the JCAE leadership successfully coupled a reactor indemnity insurance amendment crucial to the industry with another provision that established the ACRS on a statutory basis and required that it issue public reports based on its review. Although the provision did not go so far as Congressman Chet Holifield's proposal that would have bound the commission to the ACRS safety determination, it did provide statutory authority for the ACRS's right of review. Even though the ACRS sought to maintain a united public front, the inadvertent revelation of a debate that raged between the ACRS safety experts enhanced its stature in the short run.[38]

ACRS members recognized that short-term enhancements in their authority might spell long-term disaster. According to the AEC attorney assigned to provide the ACRS with independent legal advice, most of the members were not at all happy about this turn of events. Establishing and maintaining consensus was uppermost in the minds of many ACRS members. In October 1957, Ruel Stratton, a chemical engineer at the Hartford Insurance Company, wrote to McCullough on the topic of routinizing the committee's submissions to the commission. Stratton urged that "committee members, once the report establishes a consensus to which all agree, should under no circumstances indicate their personal deviation from the consensus publicly. To do so, even if under subpoena, could defeat the purpose of the Committee or at least would seriously detract from the value of the report in question and possibly from all other reports which have been or will be produced by the ACRS."[39]

The full Advisory Committee on Reactor Safeguards discussed this matter the following month in order to formalize what had previously been an unwritten law: no disputes in public. That Abel Wolman had submitted a minority report raising safety questions about the proposed National Advisory Committee for Aeronautics reactor undoubtedly sharpened the debate. The ACRS concluded unanimously that the final report must reflect the opinion of the ACRS as a whole, or at least a consensus of the committee. Minority opinions could be appended – although they were clearly discouraged – but once members agreed

37 Subcommittee Minutes, June 21, 1956, p. 2.
38 Ibid., pp. 210–11; Amendment to the Atomic Energy Act of 1954, 85th Cong., 1st sess., September 1957, Public Law 85–256, secs. 5, 6.
39 Interview with Lee Hydeman, January 21, 1986 [hereafter Hydeman Interview]. Stratton to McCullough, October 16, 1957, p. 1, "Dr. Abel Wolman" folder, "ACRS 1957–60" box, Wolman Papers.

to a ruling, they should refrain from "personal public expression of minority opinions." Although the same amendment that provided a statutory basis for the ACRS also mandated that future reports to the commission be made public, the hope was that they would not be widely distributed, thus containing the scope of debate. The committee also agreed to control tightly the distribution of its minutes: only committee members would receive copies regularly. Others would be allowed to review the minutes only with the explicit permission of the committee. Differences of opinion betweeen committee members should be protected from public or, for that matter, AEC or Joint Committee on Atomic Energy officials. Consequently, the minutes blanked out the names of the specific individual committee members commenting on safety issues.[40]

QUIETLY PROTECTING THE PUBLIC

While the Advisory Committee on Reactor Safeguards reestablished a consensual decision-making process, it also worried about its obligations to the public. Should the ACRS rule only on those cases referred to it, or did it have an obligation actively to pursue safety matters, whether referred by the Atomic Energy Commission or not? Concerned about its own access to information, the committee sought copies of all hazard summary reports submitted to the AEC as well as all applications for reactors submitted to or generated by the developmental arm of the Atomic Energy Commission.[41]

By far the most sweeping proposal for ACRS initiative came from Mark Mills of North American Aviation Corp. Mills urged his ACRS colleagues to use their new statutory authority to undertake a broad review of the committee's techniques, philosophy, and policies regarding reactor hazards. He proposed engaging a number of consultants, including Edward Teller. Mills also insisted that safety precautions take precedence over contractors' concerns about public relations, noting that when the ACRS had urged contractors to educate the public about reactor hazards and to conduct training analogous to fire drills, the nuclear community resisted, fearing that such steps might alarm the public.[42]

In practice the ACRS's role continued to be narrowly defined. Nonetheless it was critical to safety determinations. As McCullough anticipated in an August 1957 memo, the major responsibility of the ACRS was to review the safety of

40 Minutes of ACRS Meeting #2, November 1–3, 1957, ACRS File, Nuclear Regulatory Commission, Washington, D.C.
41 ACRS Minutes #5, March 6–8, 1958; ACRS Minutes #2, Append. B, ACRS File. Despite the ACRS requests, at times it was caught by surprise. In the instance of the NACA reactor, the Safety Staff presented new technical information that had not been considered by the ACRS in its deliberations leading to the issuance of its report. This, according to the minutes, represented "a dangerous potential of the staff report burying and superseding the Committee's recommendation." The ACRS was also concerned that the plutonium recycle reactor at Hanford had been selected, and construction started, without referring the reactor to the ACRS (ACRS Minutes #5).
42 Mills to Members of the ACRS, September 5, 1957, pp. 3–4, "ACRS 1957–60" box, Wolman Papers.

specific reactors. Although the committee also reviewed standards and regulations, it spent far less of its time on that task.[43]

Increasingly, the Division of Civilian Application interfered with the ACRS quest for independence. According to one observer, Civilian Application's director Harold Price expressed absolute hostility toward the ACRS. Price and the director of reactor development, Frank Pittman, viewed the ACRS members as "meddlers." They were "a little furry headed,... professorial." This, in spite of the industrial backgrounds of many of the ACRS members. Price objected to any requests that might give the ACRS greater independence. He was particularly eager to retain control over its budget, to limit access to consultants, and to influence its agenda. A behind-the-scenes tug-of-war raged over these issues during the winter of 1957–8. It took a commission meeting, and subsequent meetings with the general manager, to confirm the ACRS's right to set its own agenda and to deny the Safety Staff a veto over the ACRS budget.[44]

Summarizing an earlier meeting with Price in a memorandum to the files, McCullough noted that "Mr. Price openly charged in this meeting that I was attempting to take over and operate the Hazard Evaluation Branch and this was the basis of his objection to any action on the part of the Office of the ACRS." The two men differed on whether the ACRS should "dig out facts" on its own. "Mr. Price does not seem to recognize," McCullough continued, "that over and above a legal responsibility there may be an implied or generally understood responsibility and a moral obligation." Under siege for ten years, the Advisory Committee's chair – who the AEC hoped would bring the ACRS down to earth – continued to defend his committee's independence and to remind AEC officers of their moral obligation to consider safety first.[45]

The Advisory Committee on Reactor Safeguards performed a delicate balancing act. It hoped to retain its independence while preserving an insulated consensual forum for its own discussion within the Atomic Energy Commission. It was hesitant to rely on sources of support outside the AEC for fear of destroying that insulated forum, yet outside alliances were one obvious source of support. The ACRS continued to complain about access to information and consultants. In October 1958, one member of the ACRS noted that this was such a long-standing problem that applicants were beginning to comment on it. The member was "concerned that it will come to the attention of influential outsiders who will eventually complain to Congress and may eventually (if lack of cooperation continues) result in public hearings leading to establishment of an independent

43 "The ACRS and Its Relation to the AEC," Draft #1, August 13, 1957, "ACRS 1957–60" box, Wolman Papers.
44 Hydeman Interview, January 21, 1986. The ACRS Safety Staff debate can be followed in McCullough to Stratton, April 3, 1958, McCullough to Wolman, April 18, 1958, "ACRS" box, Wolman Papers; and Conner to McCullough, May 2, 1958, "ACRS 1957–60" box, Wolman Papers; for Commission–ACRS meeting and final agreement, see ACRS Minutes #6, May 8–10, 1958, and McCullough to Strauss, May 12, 1958, ACRS File.
45 McCullough to the Files, May 6, 1958, "ACRS 1957–60" box, Wolman Papers.

body by Congress not related to the AEC on which members of the Committee may have to serve."[46]

After the PRDC incident, the ACRS was tempted to abandon its consensual forum within the AEC, and in the fall of 1957 the committee certainly had an attentive and willing potential ally in the Joint Committee on Atomic Energy. Ultimately the ACRS preferred the limitations that life within the AEC imposed to the reaction that debating safety matters in a more public fashion might provoke. Despite the acrimonious debate within the AEC, when members of the ACRS testified before the Joint Committee in May 1958 everything was sweetness and light. Chairman McCullough stated that the ACRS was having no trouble receiving information from the AEC. Even when Chet Holifield invited the JCAE to be sure to ask for more staff if needed, McCullough could not be induced to expose any rifts publicly.[47]

Nearly a year later, preparing to testify again before the Joint Committee on Atomic Energy, the ACRS reviewed the preceding year's AEC testimony. One member commented that the material presented by Price and the General Counsel made it sound as if everything had been resolved – as if the committee were experiencing no administrative difficulty and would be fully informed on all reactor safety matters. The reason for the review, no doubt, was that the Advisory Committee's relations with the commission had further deteriorated. In January 1959, McCullough complained to the commission that the ACRS still did not have discretion over its own budget, that it was inhibited in its use of consultants, and that it was not receiving prompt notification of accidents. To pressure the AEC on accident notification, the ACRS had prepared a report on the dates of accident occurrences and the time elapsed before the ACRS was notified. Once again, the ACRS failed to make its problems known to the Joint Committee.[48]

The ACRS resisted temptation. In preparing for the 1959 hearings, the ACRS learned that the JCAE was considering introducing legislation that would give the Advisory Committee veto power over reactor operations. A majority of the Advisory Committee voted against this increase in its own authority, but the minutes report that there were many differing opinions about this option. Another indication that the Advisory Committee's role in the safety process had grown increasingly controversial could be found in the instructions its members received before testifying. "It may not be appropriate for the Committee to invite

46 ACRS Minutes #10, October. 15–17, 1958, ACRS File.
47 U.S. Congress, Joint Committee on Atomic Energy, "Hearings on Operation of AEC Indemnity Act," 85th Cong., 2d sess., 1958, pp. 55–6.
48 ACRS Minutes #14, March 12–14, 1959, p. 29, ACRS File; ACRS Minutes #12, December 11–13, 1958, and ACRS Minutes #13, January 8–10, 1959, p. 34, ACRS File; see "List of Incidents Reported to ACRS Since Nov. 1958," attached to ACRS Minutes #20, October 8–9, 1959, ACRS File; U.S. Congress Joint Committee on Atomic Energy, "Hearing on Indemnity and Reactor Safety," 86th Cong., 1st sess., 1959. Once again the commission sought to smooth over differences in preparation for testimony before the JCAE. McCone promised prompter reporting and more *administrative* (as opposed to technical) staff. But he warned that the ACRS must stay out of operations (ACRS Minutes #13).

clarification of its function or procedures by the Joint Committee," the memo warned.[49]

The following year the Atomic Energy Commission upped the ante. It sought to undermine the ACRS's one procedural source of power – the requirement that an ACRS letter be on file before public hearings could proceed. The AEC proposed new procedures that would schedule hearings before ACRS action. The ACRS was offended at this attempt to pressure it and resented the obvious implications that the Advisory Committee was to blame for reactor delays. As one member pointed out, the committee received the Pathfinder reactor application fourteen months after it was proposed; this was an issue the ACRS had been complaining about with regularity. Faced with stiff ACRS resistance, the commission backed away from the procedural change.[50]

Even during this fracas, the Advisory Committee on Reactor Safeguards had not sought outside support. It was apparently still fearful of broadening the scope of debate about technical matters. The PRDC dispute was probably still on the committee's mind – the appeals court had recently ruled against the PRDC and the AEC. The ACRS was not pleased with this ruling. In fact, its members speculated among themselves about how the flap might have been avoided. Some felt that the infamous ACRS letter had been unclear because it was written while opinion was still divided. In any case, there was agreement that the problem could have been avoided if the letter had remained confidential.[51]

There was another reason why the JCAE rejected a political alliance with the Joint Committee on Atomic Energy: Increasingly, the JCAE was pressing for standardizing site selection and reactor hazard criteria. This quest to standardize directly threatened the Safety Committee's ability to exercise professional discretion in safety questions that required the personal attention of committee members and consultants. Eager to press forward, states and localities urged the Joint Committee to brush aside the safety questions being raised by the ACRS.

In *Controlling the Atom*, Mazuzan and Walker provide a carefully documented account of the struggle to standardize reactor site criteria. They recount how Clifford Beck of the Safety Staff developed the criteria that ultimately formed the basis of the 1962 regulations. The criteria were built upon the AEC's prior experience. They relied on isolation from population centers, modified by the hope – emphasized not only by industry but by the ACRS as well – that engineered safeguards could replace distance as the primary protection for population centers

49 ACRS Minutes #15, April 16–17, 1959, ACRS File, p. 14; Grahm to all members of the ACRS re "April 15, 1959 Whaley Eaton Service Excerpt," April 23, 1959, Grahm to ACRS re "Draft #4 of ACRS Testimony to JCAE," April 23, 1959, "ACRS" box, Wolman Papers.

50 ACRS Minutes #23, January 28–30, 1960, ACRS File.

51 ACRS Minutes #27, July 20–2, 1960, ACRS File. Four months earlier, the ACRS had demonstrated its concern about such exposures. The committee agreed to take out all references to safety reports cited in its letters to the commission (which were made available to the JCAE). It took this action because these listings might lead to "undesirable" probing into AEC files (ACRS Minutes #24, March 10–12, 1960, ACRS File, p. 2).

should an accident occur. The criteria evolved from fixed population exclusion distances to three "benchmarks": exclusion distance, evacuation distance, and population center difference. These were calibrated according to potential radiation doses projected to escape reactor containment as adjusted for variables such as the meteorology of the site.[52]

The Advisory Committee on Reactor Safeguards thought that the state of the art was not sufficiently developed to support Beck's criteria, especially if promulgated as regulations. In October 1960, ACRS chair Leslie Silverman offered the definitive critique of Beck's proposal, writing that it was premature to establish quantitative limits on the variables involved in site evaluation. Minimum standards, he said, would stifle the kind of innovation required for the continuous improvement of safety features.[53]

At the heart of the ACRS's response lay its concern over premature standardization that would eliminate expert discretion. As Silverman put it: "Site selection is still largely a matter of judgment." Abel Wolman, for instance, stressed that as society became more complex and sophisticated, the urge to standardize restraints would be inevitable. "The hope that an authoritative rule may be substituted for thought and judgment is an ever present desire of some administrators, of the quantitative-minded precisionist, of the lawyer and the distressed citizen," Wolman testified before the Joint Committee. "All are beleaguered by the multiplicity of problems. To resolve these by formula has always been the will-of-the-wisp hope of workers in every field of human application." The development of criteria for the protection of health has invariably preceded full scientific understanding, cautioned Wolman. He reminded the JCAE that where criteria were frozen into law, revision required heroic efforts.[54]

Although nothing changed in the AEC's ability to quantify safety variables, the Advisory Committee on Reactor Safeguards eventually approved the criteria after changes largely related to presentation. Beck agreed, for instance, to call them "guides" rather than "rules." The criteria were approved despite resistance from the nuclear industry. It was rumored that even the point man for standardizing site evaluations had misgivings about them. The September 1960 ACRS minutes noted that "Dr. Beck was reported unofficially to be against firm

---

52 Mazuzan and Walker, *Controlling the Atom*, pp. 214–45; Okrent, *Nuclear Reactor Safety*, pp. 12–56. Another factor that was considered in developing the new criteria – and one the ACRS may have been sensitive about – was whether reactor sites already approved would qualify under the new criteria (see, for instance, Grahm to Files, September 20, 1960, and McCullough to Conner, Williams, and Wolman, August 15, 1960, "ACRS" box, Wolman Papers).

53 Silverman to McCone, October 22, 1960, Append. C-2 to "Reactor Site Criteria," December 10, 1960, AEC R-2/19, *CTA* File.

54 Ibid.; Abel Wolman testimony before U.S. Congress, Joint Committee on Atomic Energy, Special Subcommittee on Radiation, "Hearings on Radiation Protection Criteria and Standards," 86th Cong., 2d sess., pp. 30–1. In a memo to McCullough on the proposed criteria, Wolman complained that the ACRS was being worn down in a war of attrition. He also wondered where all the pressure was coming from, pointing out that the industry didn't want the regulation (Wolman to McCullough, January 4, 1960, "ACRS Minutes 1960" box, Wolman Papers).

regulations and to feel better able to carry out his Hazard Evaluation Branch duties without such rules."[55]

Despite a response that was at best lukewarm, the Atomic Energy Commission pressed for standardization: it feared that without such standards the ACRS would retard development. The AEC raised the issue in a letter to the ACRS that was discussed at the October 1958 committee meeting. The ACRS responded by agreeing to cooperate with the Safety Staff in attempting to articulate the criteria used to evaluate the Elk River, Vallecitos, Piqua, Santa Susana and Shippingport sites. John McCone, who had succeeded Strauss as AEC chair, considered the ACRS to be "over-conservative and contributing to the slowdown of nuclear power due to financial and safety restrictions." He said there was a need to protect the public, but without initiating hysteria. More to the point, he asked the ACRS to look carefully at its decision to limit power levels at the Hanford reactor. He pointed out that many groups disagreed with the ACRS's conservative approach. At the industrywide meetings held at the end of 1957, one utility executive expressed the prevailing view on safety: "Private utilities will be able to substantially reduce high capital costs as safety measures are learned by actual operation."[56]

The commission pressed the ACRS, using the survey of criteria in an attempt to establish some baseline that would assure site approval for reactor proposals. The obvious implication of the commission's request for explicit criteria was that ACRS decisions were not consistent. The ACRS itself was well aware of its vulnerability on this front. As ACRS chairman McCullough said, "There is a need to publish some numbers since operating without numbers, as has been done in the past, implies operating in an arbitrary manner."[57]

Concerned about the commission's responsibility to promulgate regulations in compliance with the 1954 amendments, the AEC's legal staff also pressed for standardization. When Harold Price submitted Beck's 1959 criteria, he emphasized that the commission was obligated by law to publish criteria. Counsel Robert

---

55 On the approval, see ACRS Minutes #31, January 12–14, 1961, pp. 4–16, ACRS Special Meeting Minutes, January 28, 1961, ACRS File; "Reactor Site Criteria," January 27, 1961, AEC R2/24, pp. 1–3, "Reactor Site Criteria," January 30, 1961, AEC R2/25, *CTA* File; Mazuzan and Walker, *Controlling the Atom*, pp. 234–6. On industry reaction, see ACRS Minutes #31, p. 6; for a summary of early industry reaction, see Robbins to McCullough, July 25, 1959, "ACRS" box, Wolman Papers; for a summary of later responses, see Mazuzan and Walker, *Controlling the Atom*, pp. 225–7. Quotation from ACRS Minutes #28, September 22–4, 1960, ACRS File.

56 ACRS Minutes #10, October 15–17, 1958, ACRS File. At the previous ACRS meeting, the committee tentatively had agreed that the site submitted for the City of Piqua reactor – part of the JCAE-promoted round of small, cooperatively owned reactor invitations – was not a suitable one. Also, the committee approved the Elk River reactor – again a second-round proposal – with a strong warning: The ACRS did not consider it desirable to locate a nuclear reactor of this power level so close to a growing community, but given containment, it would allow it be operated without undue hazard (ACRS Minutes #9, August 4–5, 1958, ACRS File).

Private utility representative quote from Fairman, Con Ed, in "Highlights of Industry Meeting, December 3, 1957 (Utility Organizations – Public and Private)" attached to Zehring to Strauss, [misdated 12/3/57], "Power General Dec. 1957" file, Box 84, LLS, HHPL, p. 12.

57 ACRS Environmental Subcommittee Minutes, February 18, 1959, ACRS File, p. 2.

Lowenstein elaborated on this argument in September 1960. Without regulations the AEC could not fulfill its objective of ensuring fairness to applicants and affording opportunities for public participation.[58]

It is difficult to determine the degree to which an ideological commitment to a public basis for evaluating proposals motivated the commission, in light of its past – or, for that matter, prospective – record on this issue. Price was particularly concerned with the problems created by public disclosure and often took any opportunity to prevent it. Lowenstein and General Counsel Loren K. Olson, on the other hand, seemed genuinely committed to establishing some kind of solid basis for determining siting safety. Lee Hydeman, who served in the counsel's office and helped draft regulations during the late fifties, later provided a convincing explanation of the "regulatory mentality": "They don't like anything that throws it into the judgmental area....Lawyers like to have rules to follow." This, of course, was precisely the pattern that Wolman cautioned the Joint Committee about and that his colleagues on the ACRS resisted, insisting that expert judgment was required.[59]

Although the commission had its own reasons for standardizing site review, there is strong evidence that it, too, was under pressure. The source was the Joint Committee on Atomic Energy. Congressman Carl T. Durham was particularly outspoken on this issue. In May 1958 he delivered a speech calling on the AEC to establish firm criteria for siting. It is important to note that Durham's call coincided with one of the ACRS's most difficult periods in its struggle for resources and access to information. Speeches like Durham's could not have encouraged ACRS members to turn to the JCAE for support.[60]

Like the AEC, the Joint Committee on Atomic Energy seemed to spring into action whenever ACRS siting concerns threatened to delay development. The Advisory Committee on Reactor Safeguards was well aware that the JCAE was particularly sensitive about reactors submitted under Round II of the Power Development Reactor Program. This round of public incentives to construct reactors was directed toward public power cooperatives and the smaller reactors that they were likely to consider. The ACRS felt the political fallout from its Piqua rejection when McCullough reported that "the JCAE suspects Piqua was delayed deliberately by the AEC because it was a public [power] plant."[61]

---

58 For Price comment, see ACRS Minutes #14, March 12–14, 1959, ACRS File, p. 27; Lowenstein rationale summarized in "Ad Hoc Committee on Safety Report," September 29, 1960, Append. B, AEC R-2/19, p. 2.

59 For Price's attitude toward public hearings, see Toll to Ramey, October 9, 1956, "AEC Procedure and Organization" folder, Box 65, Records of the Joint Committee on Atomic Energy, 1947–77, Record Group 128, National Archives, Washington, D.C., and ACRS Minutes #18, August 14, 1959, ACRS File; Hydeman Interview, January 21, 1986.

60 Durham's May 21, 1958, speech to the New York City National Association of Manufacturers was discussed at some length at the following month's ACRS meeting – ACRS Minutes #8, July 10–12, 1958, ACRS File.

61 ACRS Minutes #12, December 11–13, 1958, ACRS File.

Besides the personal interest of its own members in many smaller projects, the Joint Committee undoubtedly recognized that this was one of the few areas where their congressional colleagues might intervene directly in nuclear affairs. In 1955, for instance, Senator Hubert Humphrey wrote to the Atomic Energy Commission to complain that the Elk River site had been rejected because the project was to be developed by a public power cooperative in Minnesota.[62]

Debate over the reactor proposed by the town of Jamestown, New York, further enflamed the Joint Committee. The proposal called upon the Atomic Energy Commission to build the reactor while the town supplied the site and the electricity-generating facility. The AEC argued that experience with such relatively small reactors could be used to extrapolate cost data for the industry. In its initial review, the Advisory Committee on Reactor Safeguards expressed doubts about the economic feasibility of the proposal, but conceded that this was a political question. The poor quality of the site, however, could not be ignored. Both the ACRS and the AEC Safety Staff agreed: This was an unacceptable site.[63]

The timing of this decision and the preceding meeting's decision to reject the Point Loma reactor site could not have been worse. The annual "Indemnity" hearings were scheduled for April 1960. At those hearings, Congressman James E. Van Zandt made it clear that pressure from local government was an important factor driving the JCAE. The concern was excessive safety, not a lack of it. According to Van Zandt, the congressman representing Jamestown had called and accused the AEC of changing the rules of the game. In fact, Congressman Charles Goddell was in attendance at the hearing.[64]

Chet Holifield, on the other hand, strongly defended the ACRS. He urged the committee not to "let the economic standpoint cause you to adulterate your own convictions in regard to safety." This statement was all the more impressive in light of the fact that the ACRS had questioned reactor siting in Holifield's home state.[65]

---

62 ACRS Minutes #12, December 11–13, 1958, ACRS File. The second round of the Power Demonstration Reactor Program was a pet project of the JCAE. Established in response to complaints that the first round had effectively precluded smaller, publicly owned utilities from participation, Round II was directed toward the construction of small, experimental reactors, likely to be used by public utilities. A RAND Corp. evaluation in 1977 judged the project a failure, attributing this to "treating immature reactor designs as if they were ready for near-commercial demonstration coupled with involving small, publicly owned utilities as if they were the appropriate participants in what were essentially R and D projects" (Wendy Allen, *Nuclear Reactors for Generating Electricity: U.S. Development from 1946 to 1963* [Santa Monica, Calif.: RAND, 1977], p. 66). On Humphrey, see, Hubert Humphrey to K. E. Fields, October 11, 1955, "Power General, 50–56" file, Box 83, LLS, HHPL.

63 The site would have permitted housing to be built up to 500 feet of the reactor. ACRS Minutes #24, March 10–12, 1960, and Silverman to McCone, March 14, 1960, attach. to ibid., ACRS File.

64 U.S. Congress, Joint Committee on Atomic Energy, Subcommittee on Research and Development and Special Subcommittee on Radiation, "Hearings on Indemnity and Reactor Safety," 86th Cong., 2d sess., part 2, 1960, pp. 197, 232.

65 Ibid., p. 257.

Holifield also wanted standards, but for different reasons. In his opinion, only standards could prevent the kind of pressure brought to bear in the PRDC case and in more recent siting decisions. As he put it: "What I am trying to say is that judgment should be passed upon an objective technical evaluation rather than on such pressures as might be brought to bear upon an administrative body."[66]

Feeling the heat, the AEC fell back on Beck's criteria, citing them as evidence that standards were close at hand. McCullough backed up the sentiment for standards, although he refrained from stating that they were imminent. He appreciated their political significance, noting, "As an added bonus, it should then be possible to assure the safety of a nuclear plant in a manner which can be understood by the layman."[67]

Like most developmental issues, this problem would not go away. At the subsequent ACRS meeting, the commission pressed harder for ACRS approval of site criteria. The one bright spot in the discussion was that JCAE chair Holifield seemed to accept the Point Loma rejection without much of a fight. But the Jamestown issue returned once more. At a special meeting of the ACRS in June 1960, the AEC's Oak Ridge Operations Office, which had originally promoted the proposal, defended its potential. The developers also submitted a new site. But even the Division of Reactor Development was reported to have questions about the site. The ACRS realized that those proposing the reactor were willing to do almost anything to get the project approved. Since the ACRS was not being asked to comment on the wisdom of the policy, only its safety, members feared they might have to give in.[68]

Ultimately, the ACRS decided to recommend against it on broader policy grounds. ACRS chair Silverman deplored the tendency toward siting reactors in population centers simply to demonstrate that it could be done. He noted that far safer sites existed. Although the committee could find no serious technical fault based on the partial information supplied, it recommended against building the reactor as a matter of policy. Jamestown proposed two more sites and the ACRS – still criticizing the wisdom of the whole proposal – finally conceded that the sites were acceptable.[69]

Once again, denial of a specific site led to increased pressure for criteria. At a September meeting of the ACRS, Price said that the Commission desired firm site criteria with figures that could be provided to the public. He warned, "The criticism of the AEC for its handling of the Jamestown case has focused attention on site criteria, and the Chairman believes there should be action in response to

66 Ibid., pp. 212–13.   67 Ibid., pp. 200–2, 241.
68 ACRS Minutes #25, May 5–7, 1960, ACRS File. Special Meeting of the ACRS Minutes, June 7, 1960, and ACRS Minutes #26, June 22–4, 1960, ACRS File.
69 Silverman to McCone, June 30, 1960, attached to ACRS Minutes #26; ACRS Minutes #29, November 3–5, 1960, ACRS File, and Silverman to McCone, November 7, 1960, attached to ibid.

the AEC and ACRS statements that site criteria are being developed." Price undoubtedly was referring to JCAE criticism. In November, New York State threatened to make the dispute far more public than it had been. The director of the New York State Office of Atomic Development, Oliver Townsend, challenged the arbitrariness of the ACRS decision. In a letter to Holifield, Townsend proposed hearings to establish a national siting policy that could serve as a guide to the ACRS. Chief counsel to the JCAE James T. Ramey forwarded this proposal to AEC general manager Alvin R. Luedecke. Price was able to respond to Ramey quickly: There was no need for public hearings, because the AEC had incorporated site criteria in the regulations it would soon circulate for public comment. Those criteria, after the public comment period, were enacted on June 30, 1962.[70]

The combination of independence, expertise, and specialized mission distinguished the RSC/ACRS from the other actors with which it dealt. As we have seen, the committee walked a fine political line in its efforts to maintain its independence. Reinforced by the Joint Committee on Atomic Energy after the PRDC incident, yet committed to working inside the Atomic Energy Commission, the RSC/ACRS found its insistence on expert discretion threatened from without by both the commission and the Joint Committee over the question of standardized site criteria.

70 ACRS Minutes #28, September 22–5, 1960, ACRS File. Townsend to Holifield, November 9, 1960, in AEC R-2/17, November 23, 1960, *CTA* File; Mazuzan and Walker, *Controlling the Atom*, p. 228. Ramey to Luedecke, March 8, 1961, and Price to Ramey, March 20, 1961, in AEC R-2/28, March 29, 1961, *CTA* File; Mazuzan and Walker, *Controlling the Atom*, pp. 236–7.

6

The magnetic pull of professional disciplines, issue networks, and local government

The interdisciplinary nature of the Advisory Committee on Reactor Safeguards exerted several crosscutting pressures. The most obvious was the committee members' tendency to perceive problems through the lens of their own discipline. That made achieving internal consensus more difficult. The interdisciplinary nature of the committee also eventually attracted professional networks that radiated out from each of the committee's members. In the United States, the dense organizational fabric and porous administrative structure promoted a high degree of professional cross-fertilization. Professionals were intensely committed to issues defined by their discipline. Their interaction across agency boundaries eventually subjected nuclear power to organizational missions far different from those of the AEC. Although the Atomic Energy Commission actively sought to build an iron triangle to support its developmental mission, experts crucial to its authority gently, but persistently, nudged that agency into the turbulent political currents of crosscutting issue networks. These issue networks knew no agency boundaries. A long tradition of state and local autonomy in America – particularly regarding public safety, health, and natural resources – also threatened the AEC's jurisdiction.[1]

Faced with the choice between insulated debate and the potential for highly

---

1 Hugh Heclo has distinguished issue networks from iron triangles, characterizing the former as comprising actors more likely to move in and out of government, more concerned about intellectual debates than material interests, and more likely to be politically sensitive technical experts than their counterparts in iron triangles. Issue networks on the one hand have a more widespread organizational base than iron triangles, but on the other hand concentrate on narrower and often technically more complex issues than do iron triangles (see Heclo, "Issue Networks," pp. 102–3). Heclo is one of several political scientists who have written on the evolution of an older iron-triangle model into an even denser array of issue-oriented groups (see footnote 74, Chapter 3).

   That nuclear power was developed at the very time when the predominant policymaking paradigm – iron triangles – was besieged by the more fluid and divisive issue networks, undercut its ability to maintain the kind of insulated, centralized policy control it sought to establish by helping to build a nuclear industry and containing legislative oversight to the Joint Committee on Atomic Energy.

   For an excellent case study of the tension created by two disciplines' conflicting perspectives, see Andrew J. Polsky, "The Odyssey of the Juvenile Court: Policy Failure and Institutional Persistence in the Therapeutic State," *Studies in American Political Development* 3: 157–98.

politicized congressional review, the Advisory Committee on Reactor Safeguards consistently opted for the former. It showed very little stomach for engaging in broad public debate. However, an alternative somewhere in between these two extremes linked professional discipline to organizational bases – often at the local level – through issue networks. Proselytizing on issues dear to one's own profession, far from disloyalty, might almost be seen as an obligation. At that proselytizing, one ACRS member – Abel Wolman – was a virtual prophet among his own.

THE ORGANIZATIONAL IMPLICATIONS OF EXPERTISE: ABEL WOLMAN AND HIS ISSUE NETWORK

Wolman was a walking demonstration of the integral relationship between intellectual discipline and institutional base. His experience at his first institutional base, Maryland's Department of Health, steered Wolman toward environmental engineering. Having prepared for a medical career as an undergraduate at The Johns Hopkins University, Wolman switched to engineering as a graduate student at Hopkins. He was looking for jobs in New York City as a civil engineer when he received a phone call from the chief sanitary engineer of the Maryland Department of Health. Hired in 1915, Wolman moved up rapidly in the Health Department and in 1922 was named its chief sanitary engineer, a position he held until 1940.[2]

When Wolman was in graduate school, the discipline of sanitary engineering was just emerging. At Hopkins, he was exposed to the writings of William Thompson Sedgwick, chair of the departments of Biology and Public Health at MIT and author of *Principles of Sanitary Science and Public Health*. Applying some of these principles at his Maryland post, Wolman also helped build a discipline at the university. He was appointed as a lecturer in sanitary engineering at The Johns Hopkins School of Hygiene and Public Health in 1921 and subsequently received similar appointments at Harvard and Princeton. His undergraduate preparation for medicine made him perhaps the nation's only sanitary engineer with an extensive background in biology and chemistry. In 1937, Wolman was named professor and chair of the Hopkins Department of Sanitary Engineering, with a joint appointment at the School of Hygiene and Public Health.[3]

Between his academic responsibilities and those of the chief sanitary engineer, Wolman's contacts ranged far and wide. In 1921, he was named editor-in-chief of the *Journal of the American Water Works Association* and in 1923, associate editor

---

2 Walter Hollander, *Abel Wolman, His Life and Philosophy: An Oral History* (Chapel Hill, N.C.: Universal Printing and Publishing, 1981), pp. 14–15, and curriculum vitae [hereafter Wolman c.v.], p. 1109.
3 Ibid., pp. 13–14, 69; William Thompson Sedgwick, *Principles of Sanitary Science and the Public Health with Special Reference to the Causation and Prevention of Infectious Diseases* (New York: Macmillan, 1902). Hollander, *Wolman*, p. 54, and Wolman c.v.

of the *American Journal of Public Health*. He had ties to the important foundations, as well. Describing his first lecture course at Hopkins, Wolman noted the
preponderance of Rockefeller Foundation officials among the students. "Why
in the world am I talking to *them* about environmental problems?" the young
professor asked his wife. "They've been engaged in this... responsibility all over
the world." As Wolman's career developed he, too, acquired extensive international responsibilities. During World War II he advised the surgeon general and
the U.S. Navy. He also chaired the Sanitary Engineering Division of the Advisory
Board on Health Service of the American Red Cross, and worked for the Committee
on Sanitation and the Environment, U.S. Office of Foreign Relief and Re-
habilitation Operations. Wolman summed up his responsibilities best: "During
World War Two I became the consultant to everybody that was fighting."[4]

Wolman's first contact with atomic energy was through the National Research
Council's Environment Committee. After the war, the committee worried about
the safety aspects of atomic energy and its impact on the environment. Accounts
of the Manhattan Project's disregard for basic safety precautions, such as those
that Richard Feynman documented retrospectively, demonstrate that the
committee's concerns were well founded. Feynman recounted his visit to the Oak
Ridge uranium separation plant during the war. It was very dangerous: "They
had not paid any attention to safety at all," Feynman wrote. According to Wolman,
however, the Environment Committee did not have any hard examples to go on
at the time it began probing into atomic energy; its doubts were more the
product of atomic energy's "exotic" nature. Based on these rather amorphous
doubts, the committee asked Wolman to probe the matter with Lilienthal.[5]

Wolman met Lilienthal in 1947 and told him the NRC committee felt the
AEC was doing a number of things that might be unsafe. Lilienthal expressed
doubt about this but invited Wolman to travel with the Reactor Safeguard
Committee the following weekend, as they made the rounds. Later that year
Wolman received a formal invitation to join the Reactor Safeguard Committee.
He remained a member of it and its successor, the Advisory Committee on
Reactor Safeguards, through 1960. Wolman also headed the AEC's Stack Gas
Working Group.[6]

As we have already seen, achieving a consensus was crucial to the Advisory
Committee on Reactor Safeguards. The relatively inadvertent insertion of an
environmental engineer into that body dominated by physicists and chemists
made consensus more difficult to achieve. It permanently changed the dynamics

4 Ibid., p. 70; Wolman c.v.; Interview with Abel Wolman, December 18, 1985 [hereafter Wolman
Interview]. In 1925, Wolman was named to chair the Public Health Engineering Section of the
American Public Health Association.
5 On Feynman's visit to Oak Ridge, see Richard P. Feynman, *Surely You're Joking Mr. Feynman!*, pp.
103–8; quotation from p. 104; Wolman Interview.
6 Wolman Interview; Wolman c.v.

of the committee. Wolman emerged as the strongest advocate of reactor containment. He also authored the alternative site policy, encouraging developers to submit a number of sites. The ACRS could then select the best one. Wolman was also relentless in pressing developers to submit more data on the surrounding environment, particularly on the meteorology of sites.[7]

Wolman limited his views to the organizations he served and the professions he worked with. Even when overruled, he was loath to seek broader audiences than these. The PRDC project was a good example of this. Wolman was highly critical of building a breeder reactor so close to a large population center. He confined his criticism, however, primarily to the ACRS discussions. When the ACRS was overruled by the Safety Staff and the commission, rather than going public, Wolman limited his rebuttal to the organization that housed the ACRS – the Atomic Energy Commission. Wolman's most visible protest was hardly dramatic, amounting to a lunchtime chat with an aide close to AEC chair Strauss. As that aide reported to the Admiral after lunch, Wolman took him aside and blasted the PRDC project. Wolman "thinks that the Commission made a grave mistake in granting the 'conditional' construction permit to PRDC." Wolman also had written to the AEC's director of Civilian Application, Harold Price. Despite his obvious concern, Wolman refused to take his criticisms beyond the Atomic Energy Commission. When the unions (who had filed suit against the AEC) visited Wolman, he refused to discuss the matter with them. Nor was he willing to go to Congress.[8]

Wary of what he feared might produce uninformed public debate, Wolman firmly believed that professionals from a broad cross section of disciplines should debate nuclear safety. Wolman was instrumental in bringing into the ACRS other specialists concerned with the environment, such as meteorologist Leslie Silverman, who, after consulting for a number of years, became a member and succeeded McCullough as chair in 1960. Upon his retirement from the ACRS in 1960, Wolman insisted that another environmental engineer be named to the committee. Wolman lobbied to establish an environmental engineering unit in the AEC and got one, headed by his former student Arthur Gorman. Wolman and Gorman pressed the AEC, with relatively little success, to give the question of high-level waste a higher priority.[9]

Although the quintessential insider in style, Wolman was hardly bashful about expressing his opinion once inside what he considered to be the appropriate

7 Hydeman Interview, January 21, 1986; Wolman Interview. For examples of Wolman's impact on Reactor Safeguard Committee reviews, see RSC reports, AEC 172 series, Atomic Energy Commission Files, Department of Energy, Germantown, Md; and Wolman testimony on the PRDC reactor, Document #50–16–T1, Box 5, Public Document Room, Nuclear Regulatory Commission, Washington, D.C.
8 M. Dupkin to Strauss, December 26, 1956, "Reactor Experts" file, Box 93, LLS, HHPL.
9 Wolman interviews, December 18, 1985, and January 15, 1986; Hollander, *Wolman*, pp. 275, 281, 399, 599.

forums for discussion. Perhaps his experiences as a Jewish engineer at a time when Jews were almost universally excluded from that field prepared him for his situation within the AEC, where concern with the environment was not terribly high. According to Wolman, physicists in particular viewed him as an ignorant interloper. In his words:

The stage was held throughout that early period and for many years thereafter by the superb group of theoretical physicists, chemists and the like whose concern with environmental problems at that time, and to no inconsiderable extent at this time, is rarely environmentally or ecologically positive. Their concern, professionally, as with many others, was primarily with their own discipline and its activities. For someone to intrude into that stage, with some exotic views of the fact that what you're doing may have unfortunate implications, didn't sit too well.[10]

At the Atomic Energy Commission, Wolman established a beachhead for his discipline's perspective. He urged institutions for which environmental engineering was a mainstay to play a more active role in the development and regulation of nuclear power. Wolman was uniquely qualified to achieve this goal. His institutional links to the public health community and environmental engineering community spanned the jurisdictional boundaries linking state and local agencies to federal agencies. Administratively, this shared responsibility – so typical in functional areas such as highways or education – was as alien to the nuclear community as environmental engineering was heuristically.[11]

In 1950, long before there was any public debate over reactor safety or, for that matter, the effects of fallout, Wolman delivered an address to the American Public Health Association entitled "The Public Health Aspects of Atomic Energy." Citing a history of unexpected industrial accidents, Wolman warned of the dangers of "self-policing." Turning to the field of nuclear fission, Wolman chastised local health departments for ignoring the existence of nuclear trade, its products and its wastes. He was not sure whether this was owing to the mystery that surrounded the industry, to its secrecy, or to the complexity involved in measuring its effects. But he warned, "It may be predicted that this ostrich-like attitude toward atomic energy will result either in the substitution of another official agency for the health departments in the regulatory features of this

10 Hollander, *Wolman*, p. 760, quotation from pp. 398–9.
11 On the evolution of American intergovernmental relations, see: Daniel Elazar, *The American Partnership: Intergovernmental Cooperation in the Nineteenth Century* (Chicago: University of Chicago Press, 1967); George E. Hale and Marion Lief Palley, *The Politics of Federal Grants* (Washington, D.C.: Congressional Quarterly Press, 1982), pp. 7–11, 24–6; Michael D. Reagan, *The New Federalism* (New York: Oxford University Press, 1972), pp. 3–28; David B. Walker, *Toward a Functioning Federalism* (Boston: Little, Brown, 1981), pp. 19–99; Deil S. Wright, *Understanding Intergovernmental Relations* (Monterey Calif.: Brooks/Cole, 1981), pp. 43–60; Samuel P. Hays, "The Structure of Environmental Politics Since World War II," *Journal of Social History* 14, no. 4 (Summer 1981): 726–31. On education, see Gary Orfield, *The Reconstruction of Southern Education: The Schools and the 1964 Civil Rights Act* (New York: Wiley, 1969).

industry or in increasing difficulties to the public because of the failure to exercise any control.... That such a major gap in health department responsibility should be permitted to continue is one of the mysteries of the last decade. For the welfare and health of the nation, the gap should be closed."[12]

Wolman urged local departments to develop key professionals familiar with atomic energy. Even with this modest commitment, Wolman predicted, "official health agencies would recapture a responsibility which historically has always been theirs." To judge from the reaction to Wolman's speech, however, state and local health officials were not terribly eager at that time to make even that modest commitment.[13]

Wolman persisted. He used the American Water Works Association (AWWA), an organization representing state and local sanitary engineers, to lobby the Joint Committee on Atomic Energy on behalf of local, state, and federal health officials. Again, much of this activity preceded broad public concern about the side effects of radiation. In March 1955, Henry Jordan, president of the American Water Works Association, submitted a statement to the JCAE calling for increased jurisdiction for the U.S. Public Health Service (USPHS) and state health departments. The statement, drafted by Wolman, who was chair of the Committee on Water Quality Standards for the AWWA, noted that the AEC had not yet found a feasible conventional method for disposing of dangerous wastes. In all of the voluminous promotional material released by the electric utility industry, the statement continued, nothing had been said about waste. The statement congratulated the AEC on its safety record, but insisted that protection of health and safety in licensed facilities "should be definitely assigned to lifetime specialists in public health administration." The JCAE's response was terse: Radioactive wastes would be dealt with in separate hearings.[14]

Wolman did not limit his cajoling to professional associations; he was not afraid to address directly the public agencies that housed his profession. Named officially as a consultant to the USPHS in 1939, Wolman's ties went back much farther to his days as Maryland's chief sanitary engineer. Relations between the federal and state departments of public health had been very close. Wolman, for instance, has recounted how he would borrow personnel from the surgeon general to work on industrial health problems.[15]

In his effort to enhance the Public Health Service's role in radiation monitoring and standard setting, Wolman hit an obstacle named Surgeon General Leonard A. Scheele, who wanted no part of this responsibility. Nor was the AEC anxious

12 Abel Wolman, "Public Health Aspects of Atomic Energy," *American Journal of Public Health* 40 (December 1950): 1502–7, quotations from pp. 1506–7.
13 Ibid., p. 1507; Hollander, *Wolman*, p. 404.
14 Jordan statement to Joint Committee on Atomic Energy, March 1, 1955, Durham to Jordan, March 23, 1955, "Cooperation with States" folder, Box 262, JCAE Files; Wolman Interview. Hearings on radioactive wastes were not held until 1959.
15 Wolman Interview.

to give up its jurisdiction over this area. Wolman and his colleague at The Johns Hopkins School of Public Hygiene, R. H. Morgan, struggled to get the surgeon general involved. When Wolman finally was able to arrange a meeting with Scheele and a representative of the Atomic Energy Commission, the problem turned out to be precisely what Wolman had been lecturing about. Even the U.S. Public Health Service was afraid of the esoteric skills required to pursue problems in radiation monitoring and research. According to Wolman's account of this meeting, Scheele confided: "I would just as soon have nothing whatever to do with that obligation. That's a field so exotic ... so esoteric that I don't believe that we have, within our forces, the disciplines and the level of understanding that exists in the Atomic Energy Commission. Therefore I would just as soon not exercise that responsibility." Wolman characterized this as the general reaction of the public health community to the problem of radiation. His diagnosis of this unusual example of bureaucratic restraint when faced by an opportunity for turf building was that "this was an exotic trade, with terminology so unfamiliar to most people that they felt, I'd better let it alone."[16]

Wolman and Morgan's quiet diplomacy, however, ultimately pushed the Public Health Service toward a more independent role in radiation surveillance research and regulation. They were effective in part because they capitalized on broader public concerns about rising levels of fallout resulting from weapons testing. As Francis Weber, chief of the newly established Division of Radiological Health in the USPHS, testified in 1959, interest in safety measures for radiation had a long history. "[P]rior to weapons testing the matter of radiation was not too much in the public eye nor did it get too much attention," Weber pointed out. "It reminded me somewhat of attempting to stage a play before a house of empty seats ... it was very difficult to arouse interest in the matter, except on the part of professionals in the field."[17]

After prominent experts began to debate in highly public forums the health affects of fallout and then began to repudiate previously "safe" levels that had been established, the debate played to packed houses. Concern was palpable in both public opinion polls and news coverage devoted to the topic. The public's concern over the safety of testing was distilled in hearings held by the JCAE's Special Subcommittee on Radiation in the spring of 1957, and again in May 1959. The first set of hearings revealed the limited role that the USPHS had played to date. The appropriation for radiological health, Scheele's successor, Leroy Burney, reported, was only $347,000. This paltry amount barely registered

---

16 Hollander, *Wolman*, pp. 403–6, for account of meeting and quotations; see also Hollis to Jordan, February 21, 1955, "Cooperation with States" file, Box 262, JCAE Files, for an example of the PHS declining an invitation from the American Water Works Association to testify before the JCAE, suggesting it would be more effective if PHS officials didn't testify – private groups would be more influential.

17 U.S. Congress, Joint Committee on Atomic Energy, Special Subcommittee on Radiation, "Hearings on Fallout from Nuclear Weapons Tests," 86th Cong., 1st sess., 1959, p. 164.

when compared to the $25 million the Atomic Energy Commission spent on activities related to radiation monitoring research and protection.[18]

Although neither Wolman nor R. H. Morgan were inclined consciously to stimulate such widespread popular concern, they were not about to miss the opportunity it provided for their long-term efforts to professionalize radiation monitoring and research. They continued to press the surgeon general and eventually persuaded Burney to establish a small committee of inquiry into radiological problems. The committee was headed by Morgan, with Wolman serving on its advisory board. In 1959, Burney established the National Advisory Committee on Radiation (NACOR) chaired by Morgan, to assist the Public Health Service in developing its radiological health program. Later that year, Burney created the Division of Radiological Health headed by Dr. Francis Weber. Its fiscal year 1959 budget was $635,000 and for FY 1960 was scheduled to rise to $1.4 million.[19]

Even these steps, however, did not keep pace with the mounting external pressure for an expanded role. One source was, again, R. H. Morgan. But this time his forum was public – NACOR's March 1959 report to Burney entitled "The Control of Radiation Hazards in the United States." The committee's report cited steadily increasing radiation exposure and the steady downward revision in maximum permissible levels of radiation in calling for a comprehensive program to formulate and enforce sound protection standards from all sources of radiation. Its most controversial passage concluded that it was "unwise to continue the assignment of primary authority over the public health aspects of atomic energy in the same agency that has a prime interest in the promotional aspects of the field." "Ultimate authority," it argued, "should be placed in an independent agency and preferably in one with a special interest in public health; i.e. the U.S. Public Health Service." The report also recommended that the Public Health Service train the 5,200 additional professional and technical personnel required by 1970, and that it increase spending toward reaching the $50 million annual budget it called for by 1964.[20]

At about the same time as NACOR released its report, the Consumers Union

---

18 U.S. Congress, Joint Committee on Atomic Energy, Special Subcommittee on Radiation, "Hearings on the Nature of Radioactive Fallout and its Effects on Man," 85th Cong., 1st sess., 1957, p. 1445. For a good account of the debate over radiation standards once it entered its more public phase, see Mazuzan and Walker, *Controlling the Atom*, ch. 9. On the fallout debate, see also Robert A. Divine, *Blowing on the Wind: The Nuclear Test Ban Debate, 1954–1960* (New York: Oxford University Press, 1978); Philip L. Fradkin, *Fallout: An American Nuclear Tragedy* (Tucson: University of Arizona Press, 1989); Howard Ball, *Justice Downwind: America's Atomic Testing Program in the 1950s* (New York: Oxford University Press, 1986). On the AEC program, see Hewlett and Holl, *Atoms for Peace and War*, pp. 262–304.

19 Hollander, *Wolman*, p. 405; Wolman Interview; 1959 Fallout Hearings, pp. 155, 60; "Radiation Hazards Pose Problem of How Government Can Best Be Organized to Protect Public," *Science* 129 (May 1, 1959): 1210–12.

20 The National Advisory Committee on Radiation, "Report to the Surgeon General on the Control of Radiation Hazards in the United States, March 1959," in JCAE, 1959 Fallout Hearings, pp. 2558–61, 2565.

published a study of its own milk-sampling program. Based on a study of milk purchased at stores in fifty widely distributed locations in the United States and Canada, the Consumers Union concluded that the strontium-90 content of milk had been increasing since 1954, and at a far more rapid pace than AEC estimates. It also concluded that averages might be misleading owing to the wide variation in the areas sampled; local meteorological conditions resulting in high concentrations in high-density populations might produce "hot" spots, in other words, areas where concentrations exceeded recommended permissible levels. The scope of the Consumers Union's sampling and the quality of its work proved quite embarrassing to the Public Health Service. "That a private organization of limited means," the Consumers Union observed, "can carry out such a program suggests that an expanded monitoring network should be economically feasible under Federal, state or even community or dairy auspices."[21]

In its response, the Public Health Service acknowledged weaknesses in its present program. Like the NACOR report, the Consumers Union recommended USPHS responsibility. Noting that despite recent Public Health Service staff increases, the problem had remained the province of the Atomic Energy Commission, the report cautioned, "it is hard to see why judgments on matters of public health should have to depend primarily on reports of the very agency charged with the responsibility of manufacturing weapons, rather than on those of an agency whose specific job is to safeguard the public health."[22]

The NACOR and Consumers Union reports preceded by several months another round of Joint Committee on Atomic Energy hearings on fallout. At these hearings, Dr. Francis Weber, chief of the USPHS's Radiological Division, outlined plans for an expansion in surveillance, research, and training. The question of responsibility for standard setting and regulation was left open at the hearings, pending a Bureau of the Budget study of the matter. That report, issued later in the year, recommended increasing USPHS involvement in surveillance research and training. Rather than give the USPHS "ultimate authority," however, it recommended establishing a new federal body – the Federal Radiation Council. That body would establish standards, while application of these standards would continue to rest with the various existing federal agencies.[23]

Once the debate shattered the Atomic Energy Commission's monopoly on radiation protection, public concern proved to be far more sensitive to the number of tests being conducted than the bureaucratic mechanisms for conducting surveillance, research, and standard setting. Thus, when the Soviet Union and the United States declared moratoriums on testing in early 1960, the issue faded

---

21 Committee for Nuclear Information, "The Milk We Drink," March 1959, in JCAE, 1959 Fallout Hearings, pp. 1941, 1951.

22 Ibid., p. 1941; Porterfield to Ramey, March 17, 1959, in JCAE, 1959 Fallout Hearings, pp. 2168–9.

23 JCAE, 1959 Fallout Hearings, p. 1938; U.S. Congress, Joint Committee on Atomic Energy, "Summary-Analysis of Hearings May 5–8, 1959," 86th Cong., 1st sess., p. 34. Mazuzan and Walker, *Controlling the Atom*, pp. 257–9.

from view. Soviet resumption of testing in the fall of 1961 returned it to the headlines.[24]

In May 1962, several months after President John F. Kennedy announced that the United States would also resume testing, NACOR issued its third report. It concluded that although the Public Health Service had accomplished a great deal in recent years, much more remained to be done. There were still important gaps in the surveillance system, countermeasures to combat contamination were insufficiently developed, and research remained undersupported. The report repudiated the $16 million budget request for FY 1963 pending before Congress, and called for an FY 1963 budget of $25 million, rising to $100 million by 1970.[25]

Actual funding levels fell far short of NACOR's request. It was impressive nonetheless that, despite the sharp decline in broad public interest in the fallout issue with the signing of the test ban treaty in 1963, USPHS activities continued to expand. By FY 1966, the budget for radiological health was up to $20 million.[26]

The growth of the USPHS's involvement is an important example of the role professional networks played in spreading the impact and altering the course of nuclear power. Long before that impact was visible, professionals such as Wolman were seeking more hospitable institutional bases from which to consider this problem. Wolman interested Morgan, one of the nation's expert radiologists, and both of them turned to the USPHS, the natural institutional base for both men. It was tough going, however. Precisely because the USPHS lacked expertise in the field, it was hesitant to get involved, and Morgan could not have prevailed without rising levels of public concern. He and Wolman were helped as well by more organized expressions of concern such as the Consumers Union report. Each of NACOR's reports prodding the USPHS coincided with peak levels of public concern, and twice preceded congressional hearings into the matter.

Unlike public concern, which waxed and waned on the issue, the professional impetus was a steady source of pressure. Institutionalized in the Division of Radiological Health in 1958, the radiation safety network gradually expanded until it was able to sustain its quest for an independent voice in radiological surveillance and research, regardless of the level of public involvement. In fact, through reports like NACOR's "Radioactive Contamination and the Environment" – which recommended countermeasures against radioactive contamination – it occasionally stimulated broader public concern.

Along with the expertise they brought to the AEC, professionals like Wolman also brought the definition of and approach to problem solving inculcated by

<hr />

24 Divine, *Blowing on the Wind*, provides an excellent account of shifting popular concern with danger from fallout.

25 National Advisory Committee on Radiation, "Radioactive Contamination of the Environment: Public Health Action, 1962," summary page.

26 U.S. Dept. of Health and Welfare, Public Health Service, "Radiological Health Research Summary Report, July 1965–December 1966," p. 1.

their disciplines. Professionals from disciplines affected by nuclear power established beachheads within the AEC. These disciplines – whose institutional bases lay outside the AEC/JCAE nexus – were instrumental in carrying out the process of professional–institutional cross-fertilization. Wolman's work illustrates how this process multiplied interest in the broader impact of nuclear power beyond the boundaries of the tightly knit nuclear power community. That Wolman did so before the wide-ranging effects of nuclear power were particularly visible suggests the sensitivity of these issue networks. That his "proselytizing" was directed toward his fellow professionals in the fields of environmental engineering and public health (as opposed to broader or less technically skilled audiences) also illustrates one of the most basic patterns in the politics of expertise: Professional debate, from a variety of disciplinary viewpoints, preceded and in many respects paved the way for broader public involvement.

JURISDICTION WITHOUT EXPERTISE: A VIEW FROM THE STATES

Greater U.S. Public Health Service involvement was impossible without increased participation by state and local agencies as well. Inherent in the calls for greater involvement by independent public health institutions was the acceptance of the historical administrative arrangements between the federal and state agencies that had developed since the Progressive Era. That relationship was built on cooperative federalism, with the federal government providing general guidance and resources, primarily through grants in aid, with states and local agencies administering the specific programs. It was virtually impossible to engage one partner without the other.

In FY 1954, for example, approximately two-thirds of the U.S. Public Health Service's budget was allocated to grants that went to states, private institutions, and individuals outside the federal government. Under the heading "Community Health: A Federal–State Partnership," the 1955 USPHS report emphasized that health protection, like education, "is primarily a state and local responsibility." At the JCAE's 1959 Hearings on Federal–State Relationships, Davie Price, chief of the USPHS Bureau of State Services, put it bluntly: "Primary responsibility for protection of the public health in the United States is vested in the States."[27]

Federalism, at least when it came to determining reactor safety, challenged the Atomic Energy Commission's centralized control. The original Atomic Energy Act preempted any state role in the development or control of atomic energy. The commission confronted this issue when it amended the act to provide greater incentives for private investment. The question of the AEC's position should other federal agencies and state or local governments become interested in

27 U.S. Department of Health, Education and Welfare, Public Health Service, "The Public Health Service Today," January 1955, pp. 6, 22; U.S. Congress, Joint Committee on Atomic Energy, "Hearings on Federal–State Relationships in the Atomic Energy Field" [hereafter JCAE, Federal–State Hearings] 86th Cong., 1st sess., 1959, p. 102.

nuclear power was discussed in detail in a planning session early in January 1953. The commission's position remained remarkably consistent over the next twenty years. It actively sought state and local help in developing nuclear power. Particularly in the fifties and early sixties, states were eager to promote nuclear power. On the safety question, the AEC paid lip service to the principle of federalism while stiffly resisting any substantive delegation of authority to the states. As early as May 1953, AEC chair Gordon Dean told the Joint Committee on Atomic Energy that he hoped the commission eventually would follow "state and local criteria as to safety" regarding location and siting of reactors. The hope was never realized.[28]

The Atomic Energy Commission's monopoly in the field of radiation safety embarrassed authorities traditionally charged with these responsibilities. Testifying before the JCAE Subcommittee on Radiation in 1957, Surgeon General Burney was put on the defensive by the fledgling nature of his radiological programs – particularly the gaps in surveillance. One of the few areas of activity that Burney could point to was technical assistance to the states. The surgeon general cited one state that wanted to develop a radiological health program but "had no experts." Burney lent them a professional in the field for two years, during which the state developed its own expertise.[29]

The 1959 NACOR report envisioned an important role for state and local health departments. It conceded that at the present time AEC expertise in radiological health issues far outstripped that of the states and localities, and acknowledged that "competence in radiation safety has lagged until recently in many state and local health departments." Intensive efforts, however, were being made to correct this shortcoming. The report cited a USPHS survey indicating that seventy-six radiation health specialists and technical assistants were already working for sixteen states.[30]

NACOR spelled out the philosophy behind the Public Health Service state–local arrangement and endorsed it. "History gives strong support to the concept that where regulatory controls are needed for the safety of a community, these controls may be best exercised where the authority responsible for control is not far removed from the group or groups being protected," a 1959 report stated. "This concept is likely to prove equally valid in the field of radiation protection, for many radioactive materials used in medicine and industry, even though initially regulated, eventually become a part of environmental contamination and of necessity must be evaluated at the point of human exposure as a part of a normal health assessment program." Aroused over issues such as these, administrators pressed for enhanced legislative and regulatory control over radiation sources

---

28 On commission discussion of the role of state and local government, see CM 811, January 28, 1953, AEC, DOE. Quotation from JCAE, Executive Hearings, #3516, May 26, 1953, p. 49.
29 JCAE, 1957 Fallout Hearings, p. 1445.
30 "Control of Radiation," pp. 2557–8.

within their states. They were spurred on by the American Public Health Association, which issued a suggested state radiation act in 1958.[31]

Concern about fallout also powered interest in radiation protection at the state level, even though radiation produced by the Atomic Energy Commission and the nuclear industry represented a relatively small fraction of the radiation received by the general public. One catalyst for state action was the Atomic Energy Act of 1954, which loosened the previously absolute AEC monopoly over nuclear fuels, materials, and facilities. In early 1955, only California had a detailed radiation protection code in effect. By December 1958, eight states had approved strong radiation codes and nearly two-thirds of the states had some general radiation protection rules. Most of this legislation dealt with X-ray equipment and other sources of low-level radiation. Regulations regarding the siting and safety of nuclear reactors remained the exclusive province of the Atomic Energy Commission.[32]

When it came to reactors, state officials were primarily concerned with development. Throughout the 1950s, states and localities punctured the developmental calm with calls for reactor starts that they hoped would stimulate their region's economy. In March 1953, for instance, Puerto Rico's governor Luis Muñoz Marin asked the Atomic Energy Commission to locate one or more pilot nuclear power plants on his island. The Commonwealth was even willing to consider financing the effort. The prestige of nuclear power fit nicely with Puerto Rico's larger goal of economic development, or what the director of industrial promotion called "an old lady having her face lifted on the installment plan." Several months later the American Public Power Association, representing more than seven hundred locally owned utilities, called for public development of nuclear power. Governors' messages presented in 1956 placed heavy emphasis on economic development. Because this was often state government's highest priority, state-promoted atomic energy received close attention. Several multistate advisory groups and the Council of State Governments recommended model legislation for promoting atomic energy, for coordinating all activities related to atomic energy within the state, and for liaison with other states and the federal government. By 1959, twelve states had adopted some version of this administrative structure.[33]

The activities of the Southern Regional Advisory Council on Nuclear Energy were a good example of this pattern. The council chose for its executive director G. O. Robinson, an assistant to the manager of the AEC's Savannah River

31 Quotation from ibid., p. 2558. JCAE, Federal–State Hearings, p. 125.
32 W. A. McAdams, "Radiation Protection Laws and Codes – A Scramble for Action," in U.S. Congress, Joint Committee on Atomic Energy, "Selected Materials on Federal–State Cooperation in the Atomic Energy Field," 86th Cong., 1st sess., 1959, pp. 342–3.
33 For Puerto Rican reactor, see *New York Times*, March 4, 1953, p. 29:8. On public power statement, see ibid., May 4, 1953, p. 49:5. "Trends in State Government – 1956," *State Government* 29 (March 1956): 49–50; William Berman and Lee Hydeman, "A Study – Federal and State Responsibilities for Radiation Protection: The Need for Federal Legislation," in JCAE, Selected Materials, p. 400.

operations office, granted a leave to work for the council. In the words of its chair, R. M. Cooper (director of the South Carolina State Development Board), the council's intent was to "create a wholesome and receptive climate for advancement of good from the atom – to wipe out the mental barrier which even today keeps many people from realizing that 'nuclear energy' means much more than merely atomic or thermonuclear weapons." As William Berman and Lee Hydeman, authors of a Michigan Law School study of state–federal relations, concluded: "Fundamentally, what the states and the federal government must achieve is a healthy climate of opinion on atomic energy. Such an attitude is vital to the growth of an industry.... Atomic energy still remains a mysterious and largely dangerous force in the minds of many people," Berman and Hydeman continued, "The mere fact of state participation in promoting public understanding of atomic energy will tend to dispel this unwarranted area of mystery. But that is only a beginning. The states must also be in a position to maintain public confidence by developing an ability to deal effectively with atomic energy activities within their jurisdictions."[34]

The atomic industry also understood the important role state governments had to play in promoting local acceptance. Calling for states and localities to treat nuclear power constructively, the executive manager of the Atomic Industrial Forum, Charles Robbins, noted, "It is at the local level that public understanding will be obtained or denied." Like the Public Health Service, although with very different ends in mind, the industry recognized that at some point the traditional pattern of federal–state relations might serve its purposes.[35]

At times, the mutual interest that states and the Atomic Energy Commission shared in developing nuclear power clashed directly with safety concerns and economic objectives. For instance, the Atomic Energy Commission had steered a reactor development group toward building a sodium-cooled, heavy-water moderated reactor in Chugach, Alaska. By 1959, the commission felt that this technology was not sufficiently developed to proceed with the project. The AEC staff report that urged cancellation also spelled out the developmental consequences of such an action. Chugach wanted a reactor, the report noted, "not only because nuclear energy is a symbol of progress in the public mind, but also because the high costs of conventional power could be a limiting factor on industrial growth of the new state. There will probably ... be considerable pressure to substitute another reactor for the cancelled [one]," the report warned.[36]

34 R. M. Cooper, "The South Makes Nuclear History," *State Government* 30 (July 1957): 143–4; William Berman and Lee Hydeman, "State Responsibilities in the Atomic Energy Field," *State Government* 32 (Spring, 1959); see also "Remarks of E. C. Korten to the Atomic Industrial Forum," in "Summary of Meeting on Utilization of Nuclear Standards by State Governments, Feb. 5, 6, 1957," "Cooperation with States" folder, Box 262, JCAE Files.

35 Charles Robbins, "Community Relations for Atomic Energy," in *The Impact of the Peaceful Uses of Atomic Energy on State and Local Government* (New York: AIF, 1959), p. 35, quoted in Mazuzan and Walker, *Controlling the Atom*, p. 289.

36 AEC 777/81, February 16, 1959, p. 2, AEC, DOE.

The Atomic Energy Commission took a different tack toward the Hallam, Nebraska, reactor project. This was also a sodium-cooled reactor (with a graphite moderator). Here, too, the commission wondered whether it was not best to delay the project until improvements in the technology were achieved. In Hallam's case, however, the commission ultimately decided to proceed with the project. When the AEC's director of reactor development had hinted at the possibility of cancellation, Hallam's developers resisted. As the director put it, "I got the strong feeling that...many forward-looking, publicly-minded citizens of the area would be very disappointed if it were not to be built. It is a symbol of the progressivie attitude of the State and area, and is an important factor in public power's participation in the Atomic Energy program." Although the development group did not threaten direct political pressure, as they put it, "they could not be sure that the same would be true of other interested parties, both at State and National levels." As it turned out, there was no need for the reactor group to flex its political muscles. The AEC let the project limp along even though it would not be economically competitive and failed to incorporate the latest breakthroughs in sodium-graphite technology. Writing to Clinton Anderson, AEC chair John McCone explained rather sheepishly that "The [Hallam] people recognize the advantages from a long range standpoint of having a more sophisticated plant on their system. They feel, however, that the simpler reactor now designed will have inherently simpler opeuration and maintenance problems and they would rather learn the reactor art on as uncomplicated a unit as possible."[37]

No state embraced nuclear power more enthusiastically than Massachusetts. With its dependence on imported oil, no region (with the possible exception of the West Coast) had stronger economic incentives to develop alternative energy sources than New England. A brief discussion of the Massachusetts case illustrates how national policies to develop nuclear power and local interest in economic growth and scientific progress meshed in the Commonwealth. Even before the Atomic Energy Commission opened for business, the New England Council – a group of prominent businesspeople and educators – created an "atomic energy committee." Chaired by the New England Power Association's vice-president, William Webster, the committee also included Karl Compton, president of MIT. The committee embraced nuclear power optimistically. In December 1946 it predicted that reactors could be operating in New England within five years. The committee's report reflected the private sector's tolerance for, if not outright insistence on, active federal development of nuclear power, at a time when Democrats and Republicans alike were calling for federal cutbacks. As the committee's report put it, "This is one kind of government financing that we

37 AEC 777/82, February 16, 1959, AEC, DOE, p. 6; AEC 777/83, March 2, 1959; for McCone quote, see McCone to Anderson, February 26, 1959, p. 2, attached to AEC 777/83.

cannot afford to curtail whatever may be the national policy along general lines of retrenchment and economy."[38]

That same month, Massachusetts educators tried to take some of the fear out of the atom – another campaign that mirrored efforts by the national nuclear community. The *Christian Science Monitor* reported that "selected educators and atomic energy officials today tackled in Boston the pioneering job of replacing blind fear of atomic energy with understanding in the minds of school children." The *Boston Traveler* covered the same conference, reporting that teachers sacrificed four days of their Christmas vacation to attend. Chaired by Fletcher G. Watson of the Harvard Graduate School of Education, speakers included two experts from the AEC, according to the *Traveler*. The first aim of the conference, Watson told reporters, was to set off "an educational chain reaction."[39]

Throughout the early 1950s, newspaper coverage of nuclear power was highly favorable, emphasizing that New England had high energy costs and the kind of educational infrastructure that met the rare combination of demand and supply factors required to jump-start the nuclear bandwagon in America. In February 1954, the New England Governors Conference created an atomic energy committee. The governors sought to enlarge the region's nuclear industry. Webster and Compton also served on this committee, and were joined by Shields Warren, former medical director of the AEC, Sumner Pike, former AEC commissioner, and several other representatives of business, state government, and academia.[40]

In the fall of 1954, a group of New England electric companies agreed to form the Yankee Atomic Electric Company to produce electricity from nuclear power. Yankee members accounted for 90 percent of New England's electrical ouput. William Webster headed the group. Local promoters also took heart from the signals sent from Washington. While Webster announced the formation of his group, AEC chair Lewis Strauss told a national gathering of science writers that nuclear power might one day be too cheap to meter. In October 1954, Strauss traveled to Boston to tell four hundred New England industrialists that their region had a "great historic and geographic industrial advantage" when it came to nuclear power. Strauss also emphasized that the dangers related to nuclear power were no greater than those associated with other industrial endeavors.[41]

In the spring of 1955, Yankee announced plans to build a $25 million light-water power reactor in the tiny western Massachusetts town of Rowe. Local reaction was overwhelmingly positive. A number of articles emphasized that there were no concerns about safety and not a few hopes that the massive in-

38 "A-Power Here Seen in 5 Years," newspaper article [probably the *Boston Traveler*], December 20, 1946, in Boston University School of Communications New England periodicals file, Boston University, Boston, Mass. [hereafter BU Communications].

39 Theodore N. Cook, *Christian Science Monitor*, December 29, 1948; *Boston Traveler*, December 27, 1948, BU Communications.

40 John Harriman, *Boston Traveler*, February 14, 1954, BU Communications; *New York Times*, March 19, 1954, 5:2.

41 *New York Times*, September 17, 1954, 45:3; p. 5 for Strauss; and ibid., October 21, 1954, p. 41.

vestment would significantly lower property taxes. The concerns expressed all had to do with retaining the town's rural ambience. In a series of articles on nuclear power for the *Boston Globe*, Joseph Garland reported that "New England is faced with an opportunity to lead all other regions of the United States in riding the crest of the most spectacular industrial revolution in the history of the world – the peaceful development of atomic energy." Senator John F. Kennedy predicted that the Rowe reactor "may well prove to be the most important development in our region's economy in the twentieth century."[42]

The nuclear community recognized in federalism a powerful force that could help stimulate demand yet at the same time might severely impede development should safety review decentralize. Local officials, working in conjunction with business interests and the academic community, translated insulated centralized policymaking and distant objectives into concrete action that addressed practical concerns. There was much less talk at the local level about international prestige and national security, far more discussion of economic development and lower property taxes. Most significantly, local leaders were known and trusted by the communities where nuclear power was under consideration. They added their personal authority to the rather abstract authority wielded by distant experts.

What promoters of nuclear power most feared, however, was as old as state regulation itself: conflicting standards set by each of the states and jurisdictional disputes within individual states. Although most states adopted the standards of the National Committee on Radiation Protection, states applied these standards in different ways. Within states, the potential for fragmentation among regulatory bodies was imposing. William Krebs, former AEC counsel and executive secretary of the New England Committee on Atomic Energy, warned,

At the state level, one can expect the public utilities commission, the department of public health and perhaps the state planning board, the department of conservation, and the department of agriculture to play a part in the ultimate decision. The local interests will find expression in special legislative acts, zoning regulations, building codes, private restrictions, and political mechanisms of many kinds.[43]

The traditional nexus of local and global concern in the federal government was Congress. No small wonder, then, that the Joint Committee on Atomic Energy followed the question of state regulatory responsibility closely. Congressman Durham in 1956 and Senator Anderson in 1957 introduced legislation that would have provided the mechanism for turning over areas of the AEC's regulatory responsibility to the states when they became competent to exercise that

42 *Boston Traveler*, May 1, 1955; unmarked newspaper, February 10, 1956; *Christian Science Monitor*, February 16, 1956; *Boston Globe*, May 13, 1956; February 10, 1956, for Kennedy quotation, BU Communications.
43 Gabriel Kolko has written on the role played by fears of conflicting state railroad regulations in promoting national regulation in *Railroads and Regulation*; for more contemporary concerns about conflicting environmental regulation, see Hays, "Environmental Politics," p. 729; William Krebs, "What Are You Doing About Atomic Development?" *State Government* 30 (May 1956): 79.

power. Upon introducing his legislation, however, Anderson acknowledged that he was doing so mainly to stimulate comments from state officials.[44]

Anderson's position, in particular, illustrates the crosscutting pressures placed on an elected official responsible for a program that had an extremely centralized and specialized history yet was subject to disparate localized constituencies. At the 1959 JCAE Hearings on Federal–State Relations, Roy Cleere, executive director of the Colorado Department of Health, testified on behalf of the Association of State and Territorial Health Officers that the Public Health Service would be in a stronger position to assist states than the Atomic Energy Commission. Anderson was clearly uncomfortable. Although the senator did not support Cleere's position, he recounted his personal role in establishing a state health department in New Mexico in 1919.[45]

Anderson's highly visible role in the PRDC controversy and his resentment over the AEC's disregard of state (not to mention JCAE prerogatives) in the matter provided a more timely reminder of the dangers of centralized regulation. Speaking in September 1956, Anderson noted that in the case of the PRDC reactor, state agencies were not permitted access to safety information. "How can a state agency decide whether consumers are adequately protected ... if it cannot have access to background information?" The following year, speaking to the Council of Governors, Anderson pointed out that "the fight we make in behalf of the shrinking rights of the states is of some importance.... Consider the Dresden reactor. It is to be built 45 miles southwest of Chicago. But is it of no concern to the Governor of Illinois or the mayor of Chicago? It might explode; it might run away.... The Governor of Illinois and the mayor of Chicago," Anderson continued, "ought to insist on the right to examine reactor plans." Anderson noted that states and cities did not yet have the technical expertise to do this job completely. However, "if they continue to abandon this job to the federal government on the grounds that the staffs are not yet available," Anderson warned, "then no staff will ever be available ... all checking will be in the hands of the federal government only."[46]

Although drawn to the case for greater state regulation of nuclear power, Anderson never forgot the pitfalls this might create for development. Nor did his staff. As early as January 1956, a staffer sent a clipping from the Atomic Energy Guideletter (a weekly report for businessmen) circling the phrase "confusion is well on its way" in regard to radiation standards. Anderson also heard directly from nuclear industry representatives, such as General Electric's vice-president for development, F. K. McCune. McCune reminded Anderson that "the point has

44 H.R. 8676, 84th Cong., 2d sess., and S 53, 85th Cong., 1st sess. Senator Clinton Anderson, "The States' Responsibility in the Regulation of Atomic Enterprise," in JCAE, Selected Materials, p. 498.
45 JCAE, Federal–State Hearings, pp. 430–31.
46 Clinton Anderson, "The Atom – Everybody's Business or Nobody's Business?" *State Government* 29 (December 1956): 244–5; "The States' Responsibility," in JCAE, Selected Materials, pp. 499–500.

been made many times that overlapping regulation would be quite burdensome to the industry." Anderson took these warnings to heart in pressing toward an amendment to help standardize state activity.[47]

At the AEC, there was far less concern for the states' right to regulate. Speaking to the National Governor's Conference in 1956, AEC director of licensing and regulation Harold Price suggested that the states should find ways to "encourage and promote" atomic industrial development. Through a series of devices, ranging from its first meeting with governors in the summer of 1955 to the establishment of an advisory committee of state officials, the AEC sought to provide information to and receive comments from state officials. The states had some resources that the AEC hoped to make use of – information on local conditions, for instance. The AEC also hoped to employ state personnel to inspect the growing number of facilities using radioisotopes. But the essence of the AEC's philosophy toward state regulation was spelled out in the legislation it proposed in June 1957. That bill would leave states free to adopt and enforce standards, so long as they were "not in conflict with those adopted by the Commission"; state licensing could not be imposed in areas where the AEC already had licensing responsibility. Earlier that year AEC general manager K. E. Field had spelled out just how much leeway the states had in setting standards. They should not relieve or impose additional burdens. Clearly, this did not leave much room for the states to devise their own regulations. It did, however, raise the possibility that staffs would be duplicated in a field where trained personnel were already scarce.[48]

By 1959, state and regulatory activity was nevertheless increasing, and despite the JCAE bills and the AEC bill, Congress had not approved any new legislation. In January 1959, Berman and Hydeman released their study of federal–state responsibilities, which warned that "Congressional action should be taken before interests become too entrenched."[49]

Warnings such as these, criticism of its 1957 bill, and the impending Joint Committee hearings prompted the AEC to submit new legislation. The 1959 bill offered states full responsibility for regulation, but severely restricted the areas in which they could regulate – excluding, for instance, production and utilization facilities. It also stipulated that regulatory authority would be relinquished only after the AEC determined that the state substitute was adequate.

47 Norris to Anderson, January 25, 1956, "Cooperation with States, vol. 1" folder, Box 262, JCAE Files; McCune to Anderson, June 4, 1959, "Cooperation with States, vol. 4" folder, Box 262, JCAE Files.
48 "Harold Price Remarks to the Governor's Conference," June 27, 1956, "Speeches – Harold L. Price" folder, Box 76, JCAE Files; JCAE, Federal–State Hearings, p. 26; "Remarks Prepared by Robert Lowenstein," April 16, 1956, "Remarks Prepared by Curtis Nelson," "Cooperation with States" folder, Box 262, JCAE Files; Proposed AEC Amendment to Atomic Energy Act of 1954, in JCAE, Selected Materials, p. 18. Field to Bane, February 20, 1957, "Cooperation with States, vol. 1" folder, Box 262, JCAE Files; Mazuzan and Walker, *Controlling the Atom*, p. 291.
49 Berman and Hydeman, "A Study," p. 376.

Nuclear Regulatory Commission historians Mazuzan and Walker have concluded that the bill was a compromise, satisfying most parties. "Above all," the NRC historians point out, "the 1959 bill reflected the AEC's own position.... It still wanted to contain state activities within narrow bounds because of both regulatory and promotional considerations." Despite a cool response from the advisory committee of state officials, the legislation passed, with the support of Anderson and the JCAE.[50]

The legislation reinforced the AEC's claim that it was the only agency technically competent to make certain regulatory decisions – particularly in areas such as reactor licensing. Virtually nobody disputed its superior expertise. The Berman–Hydeman study concluded that "evaluating the safety of reactors requires the combined judgment of individuals representing a variety of scientific and technical disciplines. It might be impossible, and certainly it would be inefficient use of our limited scientific and technical manpower," the study noted, "for all states in which reactor construction and operation is proposed to attempt to obtain reactor-hazards-evaluation staffs composed of individuals with requisite talents." Senator Anderson stated the problem more succinctly, reminding Dr. Maurice Vischer, chair of the Minnesota State Board of Health, "You do not have, and the individual states will not have, the technical skill necessary, particularly the poorer states."[51]

Minnesota was not just arguing for the theoretical right to regulate reactors; it had adopted regulations in December 1958, requiring explicit approval of the Board of Health before reactor construction. What's more, a reactor was planned for Elk River and the Department of Health had already informed the AEC that this project would not be exempt from the new regulations. Vischer conceded that at the present moment, Minnesota "does not have technically qualified personnel in its own department of health, to do the whole job." On the other hand, he continued, the AEC could provide technical know-how to assist state governments in arriving at decisions.[52]

Vischer added three broad justifications for his agency's actions. First, despite the AEC's superior knowledge, the State Board of Health was legally responsible to the people of the state "and cannot properly evade a responsibility in this area." Second, Vischer claimed that state agencies had access to relevant information that was not necessarily available or obvious to national bodies, such as data on underground water flows. Finally, state government was closer to grass-roots sentiment. As Vischer put it, "a regulatory body with more direct responsibility, more direct sensitivity to the needs and the points of view of the citizens of a state ... should have some definite legal authority in these matters." Clinton

---

50 Mazuzan and Walker, *Controlling the Atom*, pp. 292–3, quotation from p. 293.
51 Berman and Hydeman, "A Study," p. 431; JCAE, Federal–State Hearings, p. 275.
52 Minnesota State Board of Health Regulations, in JCAE, Federal–State Hearings, pp. 264, 269; Barr to Dunbar, March 18, 1959, "Cooperation with States, vol. 4" folder, Box 262, JCAE Files.

Anderson was sympathetic to all of these arguments but also worried that decentralizing regulation would further retard development that was already lagging in the kilowatt race. Anderson summed up his own ambivalent feelings, commenting that "I am not so sure that you have a right to say that a reactor shall not be licensed....But I do believe you have a right to take an awfully good look at it and express yourself very very forcibly on whether you think it is good in the long run for the people of your state."[53]

Perceptive observers noticed that expert consensus on a number of issues was not the monolith that many of the experts themselves struggled to maintain in public. The only instance in which the vigorous debate over the safety of power reactors was revealed publicly was the PRDC siting controversy. Here, the Atomic Energy Commission and the Advisory Committee on Reactor Safeguards, not to mention the nuclear industry, moved swiftly to seal this breach in containment. The highly publicized debate over the effects of fallout fueled a far more serious assault on the AEC's integrity and the authority of its experts. For the most part, there was remarkably little explicit linkage between the fallout debate and reactor development. Indirectly, however, there was one crucial consequence. Highly visible debate over fallout led some to question generally the authority of experts.

One journalist in Massachusetts, commenting on the newly announced reactor in Rowe, warned against complacence. Although the citizens who would be living right next to the reactor had not expressed any fears, Paul Benzequin did. "The scientists reassure us quickly," Benzaquin wrote in the *Globe*, "no, no – don't worry, it can't blow up. There's nothing to worry about. Yet the scientists also promised us that the H-bomb tests in the Pacific were perfectly controlled – and we later learned that a Japanese fishing vessel was so contaminated by fall-out that many of her crew suffered radical radiation burns. And the scientists said too," Benzaquin continued, "that radiation from an H-bomb would be so dissipated by the explosion and the wind that it wouldn't bother anyone – and then later they told us about Strontium 90, the terrible isotope which can find its way into our bones and slowly bring us to invalidism and death by shooting our blood cells as a trap shooter knocks down clay pigeons."[54]

A pluralist storm gathered strength in the states. There, a variety of interests were expressed, and unlike the AEC – where development was the primary mission – some of the oldest institutions with the most developed constituencies were concerned with protecting the public health. At the state level it was the atomic promoters and coordinators who were the Johnny-come-latelies. Slowly, the states inched into the field, complicating the AEC's already complex job. Having won limited jurisdiction over regulation through the 1959 amendment, the states were extremely slow to apply this right. Undoubtedly this was because

53 JCAE, Federal–State Hearings, pp. 273–5.
54 *Boston Sunday Globe*, December 9, 1956, BU Communications.

of the lack of expertise and resources of which the AEC had warned. But as the states moved forward, they engineered access to their own sources of expertise.

Governor Orville L. Freeman of Minnesota foreshadowed these developments and, like Benzequin, raised questions about the the authority of experts by forcefully backing up Vischer. In addition to the usual constitutional arguments, Freeman addressed the question of expertise from what at the time was an unusual angle. "It is clear even to a layman that within this field," Freeman testified,

there are vast voids of knowledge. Even as to areas in which we think we know, there are disagreements between equally eminent and responsible scientists. Thus it seems particularly appropriate that there should be some variety of viewpoint and approach in dealing with problems as to which we know so little, and as to which experimentation may be so important.[55]

By the end of the decade, many of the governors who had favored development over regulation began to shift their priorities. Within ten years, nuclear power would be debated in a new forum – in fact, fifty new forums – less predisposed toward development than the AEC. If knowledge was not as impregnable as a number of the experts publicly suggested – if, in fact, there was a variety of viewpoints – shouldn't the states have a hand in weighing some of the trade-offs inherent in these technical decisions? At a minimum, shouldn't the states be informed about the range of views that existed on most issues? Shouldn't the states then subject the question to the same kind of pluralist pressures faced by other policies that promised economic development, but at a price?

55 JCAE, Federal–State Hearings, p. 477.

7

Nuclear experts on top, not on tap: mainstreaming expertise, 1957–1970

It took more than the occasional public glimpse of expert debate, controversy, and overconfidence, if not dissembling, to shake America's faith in its experts. The cumulative impact of the most publicized controversies in the late 1950s and early 1960s – such as the health effects of fallout, the damage caused by thalidomide, the benefits and risks associated with fluoridation, not to mention far more obscure bouts between experts such as the PRDC controversy – were overshadowed by one event in early October 1957: Sputnik. Coming on the heels of a decade of military accomplishments that suggested the Soviet Union was catching up to American achievements, Soviet capacity to launch a satellite into orbit was interpreted by millions – including a broad spectrum of policymakers – as proof that the Soviets had finally surpassed America in the race for scientific and technological superiority.[1]

Sputnik was the perfect symbol of a cold war that could be won only if the nation mobilized fully all of its resources. Militarily, America's top defense personnel explained, Sputnik did not prove anything. It was passive and primitive. Or as Strategic Air Commander Curtis LeMay explained, "It's just a hunk of iron."[2] Diplomatically, the Eisenhower administration first tried to shrug off the implications of the Soviet space shot. It was not the policy of the United States, Eisenhower aide Sherman Adams told the press, to accumulate the "high score in a celestial basketball game."[3] Eisenhower damage control and military–scientific reality aside, as Walter McDougall has pointed out, "No event since Pearl Harbor set off such repercussions in public life."[4] Sputnik fused a number of powerful forces. It raised serious questions in the public's mind about America's military and fiscal policy. It was dramatic: space had long engaged America's fantasies.

---

1 McDougall, *The Heavens and the Earth*, and McDougall, "Technocracy and Statecraft" (pp. 1010–40).
2 Ibid., p. 145.   3 Ibid., p. 148.
4 McDougall's account is an outstanding portrayal of the emergence of a command economy in postwar America. However, it does overemphasize the impact of Sputnik and understate the importance of the bomb, the legacy of World War II institutions, and longer-term trends toward centralization, professionalization, and bureaucratization in American life. Quotation from p. 142.

Finally, it carved out common ground between America's staunchest supporters of the military–industrial complex on the one hand and "social–educational" activists on the other.[5]

The cold war stimulated demand for centralized, expert responses to America's threatened national security. It was now the experts – funded and empowered by the federal government – who often identified specific problems. Then came the proposed solutions to newly identified problems. Those solutions, of course, also required federal action. As the cold war spread from military to economic and social fronts, experts in the social sciences and even the humanities linked national security to solutions offered by a broader range of disciplines. In the wake of Sputnik, policies as far afield from military security as civil rights and highway construction benefited from America's mass passion for national security. As Elaine Tyler May has documented, many felt that only a radically reoriented culture could protect American home and hearth against the Soviet threat.[6] Americans should worry less about the "height of the tail fin in the new car and be more prepared to shed blood, sweat, and tears if this country and the free world are to survive," warned Republican senator Styles Bridges in October 1957.[7]

Even before Sputnik, educators claimed an increasing share of the national security budget by joining the drive for civil defense. Bert the Turtle warned schoolchildren that they not only had to learn to cross the street safely and know what to do in case of fire – they now had to know what to do in case of atomic attack. Americans turned directly to experts to soothe jittery nerves: approximately one out of six white middle-class Americans consulted a professional for emotional or marriage problems by the mid-fifties. As one attitudinal survey put it, "Experts took over the role of psychic healer.... They would provide advice and counsel about raising and responding to children, how to behave in marriage, and what to see in that relationship.... Science moved in because people needed and wanted guidance."[8]

Galvanized by a dramatic symbol like Sputnik, the federal government not

5 Ibid.   6 May, *Homeward Bound.*
7 McDougall, *Heavens*, p. 142.
8 For case studies on how the cold war stimulated demand for centralized expertise in space and weaponry, see McDougall, *Heavens*; Herken, *Counsels of War* and *The Winning Weapon*; York, *The Advisors*. Hershberg's "James B. Conant" is an excellent study of one of the nation's leading scientist administrators and his response to cold war weaponry. See also Dickson, *The New Politics of Science*, pp. 26, 119–23. Don K. Price, *The Scientific Estate* is a testimonial to the increased authority and political clout of scientists in postwar America. Price correctly attributes to the new role of science in the federal government several major changes, including the narrowing gap between the public and private sectors, greater autonomy for and more initiative by executive agencies as a result of this expertise, and the adaptation of the research contract, leading to federal support for open–ended research (pp. 15–21, 36–40, 46–51). Robert Gilpin contrasts the revolution in scientific participation in weapons development during and after World War II to the case of World War I, in *American Scientists and Nuclear Weapons Policy*, pp. 9–12.
   The best survey of the wide range of responses across disciplines is Boyer, *By the Bomb's Early Light*; see also Newman, "The Era of Expertise." On the impact of Sputnik, see McDougall,

only scrambled to procure the services of more experts, it embraced a crash program to produce more experts. The "Gaither Report," formally presented to the White House in November 1957, was perhaps the leading example of the former. In the spring of 1957 Eisenhower commissioned a group of one hundred experts – many of them veterans of Los Alamos and subsequent weapons development projects – to answer a simple question. As Jerome Wiesner, who would emerge as President Kennedy's science adviser four years later, recalled, Ike asked the panel what the president should do "if you make the assumption that there is going to be a nuclear war."[9] That simple question sent the experts scurrying in a great many directions. Spurgeon Keeny, a physicist and former head of the Air Force Special Weapons Section, noted that the panel was united in one theme: "The emphasis was on solving it technologically."[10] The Gaither committee warned that due to the unexpectedly early development of Russian intercontinental ballistic missiles (ICBM), the Soviet threat might become critical as early as 1959. It proposed urgent measures and appropriations up to $50 billion to protect the nation's security.[11] Keeny, commenting on the report with the benefit of hindsight and perhaps chastened by Eisenhower's cool response to the panel's recommendations, labeled it "the high water mark of the belief that a technological solution could be found."[12]

While veterans like Keeny, Wiesner, and Herbert York ultimately stepped back from their all-abiding faith in technical solutions, most of their colleagues, and a whole generation of would-be technical experts, were transfixed by just such a prospect. Belief in technical expertise was spreading from the scientists and technicians themselves to the institutions now vital to their support. James Conant, president of Harvard University and formerly the second-ranking civilian overseeing the Manhattan Project, commented on this startling change in a secret address to military and government officials in 1952. Before the atomic bomb, Conant noted, technological conservatism was the military's chief stumbling block. Military officials had been "perhaps unduly slow in some cases to take up new ideas developed by civilian scientists." But in the wake of the bomb and the cold war something akin to "the old religious phenomenon of conversion" had struck. "As I see it now," Conant continued, "the military, if anything, have become vastly too much impressed with the abilities of research and development." Underscoring the professional–federal union that atomic and cold wars had sealed, Conant noted wryly that "some of your colleagues have become infected with the

Heavens, chs. 6 and 7, and "NASA, Prestige, and Total Cold War." For Bert the Turtle, see Federal Civil Defense Administration, "Bert the Turtle Says 'Duck and Cover,'" quoted in Brown, "'A Is for Atom, B Is for Bomb,'" p. 84. The quotation on experts is from Veroff, Kulka, and Douvan, *The Inner American*, p. 194, quoted in May, *Homeward Bound*; p. 27; for statistics on counseling, see ibid.
9 Herken, *Counsels*, p. 113.
10 Ibid., p. 80, for Keeny background; p. 113 for quotation.
11 Ibid., p. 113.   12 Ibid., p. 117.

virus that is so well known in academic circles, the virus of enthusiasm of the scientist and the inventor."[13]

The federal government intervened directly to meet the pressing demand for technical experts and to influence the direction of basic research and its application. Sputnik, according to the Atomic Energy Commission's longest-reigning chair, Glenn Seaborg, drove home the lesson "that production of the most advanced brainpower is a national problem and federal responsibility."[14] This commitment was embodied not only in the National Defense Education Act, passed in the wake of Sputnik, but in the burgeoning fellowship programs administered by an expanding federal scientific establishment. Frank Newman, who in his Ph.D. dissertation "The Era of Expertise" documented the history of postwar shortages of highly trained experts, distinguished the post-Sputnik environment from a society already dependent on experts before the Soviet space shot.

Whereas the original discussion had envisioned support for...only the most highly trained ... proposals over the next decade came increasingly to stress the value to both society and the individual of supporting higher education at all levels. The greater number of experts would result not merely in better defense but in advances in medicine, adventures in space, economic advantages through industrial productivity and international competitiveness, improvements in education and a generally superior quality of life.[15]

VIGOROUS HORIZONTAL COMPETITION: THE RACE FOR SPACE

Before America's space program took off publicly, the Atomic Energy Commission responded to America's damaged international prestige by touting its program for peaceful use of nuclear power. This was no easy task in light of the growing public perception that America was losing not only the space race but the kilowatt race as well. The ideal showcase for America's nuclear technology was the second conference on the peaceful uses of atomic energy, held in Geneva in September 1958. In the words of the AEC's official historians, "The spectacular American show, set up in the shadow of *Sputnik*, which dominated the Soviet exhibit, was designed to demonstrate unqualified American leadership and preeminence in the nuclear field."[16] Particularly in the wake of America's embarrassing space failures such as the aborted Vanguard test, nuclear power remained America's most glamorous federally subsidized technology. Space, however, was an obvious candidate to succeed nuclear power in that role.

The nuclear community in the late fifties and early sixties was far more concerned about competition for resources from America's one-time beleaguered space program than domestic opposition to commercial nuclear power. It was not

13 James B. Conant, "The Problems of Evaluation of Scientific Research and Development for Military Planning," quoted in Hershberg, "Over My Dead Body," p. 50.
14 Quoted in Newman, "Era of Expertise," p. 65.
15 Ibid., p. 66.
16 Hewlett and Holl, *Atoms for Peace and War*, p. 446.

an attack on expertise but, rather, the fear of being outflanked by a flashier and more glamorous exposition of expertise that frightened the nuclear community. Despite the public's concern about fallout from weapons testing in the fifties, and the breach in political containment that had occurred over the PRDC reactor siting, it was not the gang from the other side of the tracks – those poorly informed citizens who panicked at exaggerated dangers – that threatened the advocates of commercial nuclear power. Ultimately, it was the new kid on the technological block – the space program that threatened America's aging technological child prodigy. As the *Wall Street Journal* put it in 1962, "the once glittering atom, darling of the 1950s, has been displaced by the marvels of space exploration as the major scientific enthusiasm in the nation's capital."[17]

The Sputnik factor and the race for space that followed in its wake were so engulfing that they even forged a momentary truce between two longtime adversaries within the nuclear community. One of the few issues that AEC chair Lewis Strauss and Joint Committee on Atomic Energy chair Clinton Anderson ever agreed on was that the Atomic Energy Commission was ideally suited to handle the space challenge. Well aware of the possible challenge to what both men viewed as home turf, each also looked to space for its potential to expand the use of nuclear power. The combination of threat and opportunity quickly produced JCAE hearings on space.[18]

Strauss, as a close adviser to Eisenhower, was constrained in his advocacy of AEC control of the space program by the president's position. Just one day before the JCAE hearings, the president had launched a study designed to reorganize the executive branch in order to meet the space challenge. As Strauss stated at the hearings, "It is important to us...that we be not cast in the role of special pleading."[19] Anderson entered the fray with fewer constraints. In several speeches on the Senate floor, he insisted that the crucial variable in the race to space was organization. America's ability to organize its scientific personnel would determine the outcome of this crucial competition. The Atomic Energy Commission, according to Anderson, was the ideal candidate to run that organization. Its national labs were the logical breeding ground for space development, Anderson continued. Not surprisingly, the JCAE chair also thought that his own congressional committee fit the bill neatly when it came to the important task of oversight. Anderson introduced legislation to this effect, and the Joint Committee on Atomic Energy established a subcommittee on space.[20]

Space, however, was hotly contested turf, both on the Hill and in the bureaucratic trenches. The powerful Senate majority leader, Lyndon B. Johnson, ultimately had far more influence on the organizational structure of the program

---

17 Jonathan Spivak, *Wall Street Journal*, March 8, 1962.
18 U.S. Congress, Joint Committee on Atomic Energy, "Hearings on Outer Space Propulsion by Nuclear Energy" [hereafter Space Hearings], 85th Cong., 2d sess., 1958.
19 JCAE, Space Hearings, p. 179.
20 Ibid., pp. 207–12; *Congressional Record*, January 23, 1958, 85th Cong., 2d sess., pp. 813–16.

than did Anderson. Johnson used his perch on the Preparedness Subcommittee of the Senate Armed Services Committee, to shape the space program's organizational future. The new organization and oversight arrangements left little room for AEC/JCAE control over space policy.[21]

Relegated to a subordinate role in the development of the nation's space program, the AEC and JCAE nevertheless struggled to keep a foot in the door. Since 1956, the Los Alamos laboratory had been working on a reactor that could be used to propel rockets. The first test of Project Rover occurred in July 1959.[22] Sputnik also rekindled support for a radioisotope-heated generator to power space vehicles once they had reached orbit. Announced by Eisenhower in January 1959, the Systems for Nuclear Auxiliary Power (SNAP) program soon led to the development of small reactors that by November 1959 were able to provide three kilowatts of electricity that operated continuously for up to a year.[23] Glenn Seaborg outpaced even Kennedy's enthusiasm for space, promoting nuclear power's contribution to the space race as a bold new challenge. As the AEC chair put it, when the uses of nuclear energy for rocket propulsion become feasible, "what is now science fiction may become scientific *reality*."[24]

From a budgetary perspective, this strategy proved salutary in the short run. Appropriations for reactors with space applications tripled from 1961 to 1963. But it seemed that these increases came at the expense of civilian reactor development, where appropriations declined to zero during the same period. A February 1962 *New York Times* editorial found it "ironic" that spending cuts for civilian reactors were providing funds for the increased development of spacebound reactors.[25]

Despite the flurry of AEC space-related budget increases in FY 1962 and FY 1963, the AEC's space program ran into an old nuclear nemesis – softening demand. Even though the Atomic Energy Commission lost the battle to administer the space race, nuclear advocates had high hopes that the National Aeronautics and Space Administration (NASA) would at least contract for their services. For NASA, however, nuclear-powered rockets and power sources were only one of several alternatives. The rapid development of conventional propulsion systems

21 Richard Hirsch and Joseph Trento, *The National Aeronautics and Space Administration* (New York: Praeger, 1973), pp. 22–32. For an excellent summary of the origins of NASA, see McDougall, *Heavens*.

22 Hewlett and Holl, *Atoms for Peace and War*, p. 518.

23 Ibid., p. 519. For Eisenhower's enthusiastic reaction to the AEC's January demonstration, see January 20, 1959, Memorandum of Conference with the President, January 16, 1959, AEC, Vol. II, White House, Office of Staff Sec. Subject Series, Alapha subseries, Box 3, Eisenhower Presidential Library, Abilene, Kans. [hereafter Eisenhower Library]. I would like to thank John Yoo for bringing this material to my attention.

24 Glenn Seaborg, "The Atom in Space," May 10, 1961, in "Speeches of the Chairman, Commissioners and GM, 1947–1974," Records of the Office of Public Information, Atomic Energy Commission, Record Group 326, National Archives, Washington, D.C. [hereafter "Speeches"], Box 8, p. 2.

25 *BAS* 18 (April 1962): 46; *New York Times*, February 2, 1962.

soon outpaced nuclear technology, ending the AEC's dream of a new "scientific reality."[26] Unlike the AEC, NASA was not predisposed toward the nuclear option. Nuclear-powered rockets competed and ultimately failed on their merits.

The AEC space program eventually foundered because of its inability to sustain demand – or, in the jargon of NASA, because of the "absence of a clear cut requirement." Complaining of the Johnson administration's cancellation of the flight test of a larger generation of SNAP reactors in 1964, an angry Chet Holifield expressed concern to the president that "after an expenditure of $100 million for research and development on this reactor, suddenly we are informed that 'no immediate military requirement' exists.... The Congress," Holifield continued, "will certainly begin to ask why funds should be spent for research and development if, after spending millions of dollars, we continually stop short of our goal."[27] Unlike Project Rover, which was scrapped during the 1960s, the SNAP program limped through the decade. Initially the smallest in scope, SNAP was the most successful of all of the Atomic Energy Commission's air and space projects.[28] Having lost the jurisdictional struggle of the late fifties, and feeling its foot increasingly pinched by the NASA-controlled door, the AEC/JCAE could not find the large-scale demand it sought from the space program. Meanwhile, at the heart of the AEC's primary domestic responsibility – civilian nuclear power – demand was becoming an increasingly worrisome problem.

TOWARD ECONOMIC INTEGRATION UNDER POLITICAL APARTHEID

Like many other domestic policies during the 1950s, it was the military and the requirements of the national security state that provided the demand for nuclear power as well as the resources to meet that demand. By the end of the Eisenhower administration, that nexus left civilian nuclear power technologically mature, but politically and economically underdeveloped. Shippingport – the first nuclear reactor to generate electricity commercially – was the hybrid progeny of the navy's nuclear propulsion program and Eisenhower's race for technological prestige. Shippingport, however, was far from economically competitive with non-nuclear sources of fuel. Nor were any of the other reactors that were being planned or constructed. All were the product of a highly insulated style of decision making.

Without large government subsidies no private utility would embrace the nuclear challenge. Prospects for such subsidies improved when John McCone replaced Lewis Strauss as Atomic Energy Commission chair in July 1958. A hard-driving engineer-businessman from California, and former president and director of Bechtel–McCone, the new AEC chair had solid conservative Republican

---

26 On the pace of conventional alternatives, see Hewlett and Holl, *Atoms for Peace and War*, p. 519.

27 *New York Times*, December 17, 1963; Hollingsworth to Pastore, January 5, 1973, The Papers of Chet Holifield, University of Southern California [hereafter Holifield Papers], Box 50. Holifield to President Johnson, March 12, 1964, Holifield Papers, Box 43.

28 Hewlett and Holl, *Atoms for Peace and War*, p. 519.

credentials. But as America's largest corporations embraced a postwar mixed economy that blurred distinctions between public and private sectors, particularly in policy areas where national security or research and development were involved, solid conservative credentials no longer precluded support for massive government subsidies.[29] McCone, like Eisenhower himself, proved far more flexible than Strauss on the issue of public versus private development of nuclear power. McCone was also committed to smoothing over some of the rough edges that had poisoned the relationship between the Joint Committee on Atomic Energy and the Strauss-led AEC. This made the commission more receptive to the possibility of larger subsidies and more active federal sponsorship.[30]

Although the Atomic Energy Commission and the nuclear manufacturers publicly chided utilities for their caution, the nuclear community understood well the reasons for that hesitance. McCone stated it bluntly to President Eisenhower, in a top-secret briefing in January 1959. "For our own economy, with but few exceptions, we do not need atomic energy power in the foreseeable future."[31] Nothing could be more straightforward, nor did any statement about nuclear power prove more prophetic. McCone, however, did not stop there. To do so would have radically altered the AEC's approach to development over the past decade. Rather, he returned to familiar themes that in place of concrete domestic demand substituted abstract foreign policy benefits related to military preparedness. As the AEC chair put it, "to keep the United States in the race industrially and internationally we do need to carry out development." McCone called upon the federal government to continue and even enhance its subsidies. When Eisenhower complained that this seemed to be "another venture in the direction of public power," McCone responded that "by having this program with the President's support it would be possible to head off this tendency."[32]

The public continued to fund the development of nuclear power in order to ensure that it did not become "public." By the early sixties, nuclear power was no closer to the AEC's commercialization objective. The technology remained economically isolated from America's market economy. Never one to let an issue die, Lewis Strauss reminded Bourke Hickenlooper in 1961 that nuclear power had gone nowhere since the Admiral had left office almost two years earlier.[33]

Several conditions related to its historical development left civilian nuclear

29 On McCone, see ibid., p. 491. On America's postwar mixed economy, the best summary is Ellis W. Hawley, "Challenges to the Mixed Economy: The State and Private Enterprise," in Robert H. Bremner et al., eds., *American Choices: Social Dilemmas and Public Policy Since 1960* (Columbus: Ohio State University Press, 1986): 159–86.

30 On patching up relations, see Green and Rosenthal, *Government of the Atom*, pp. 252–65.

31 Memorandum of Conference with the President, January 16, 1959, dated January 20, 1959, AEC, Vol. II, Box 3, Office of the Staff Sec., Eisenhower Papers, p. 3. McCone placed the cost of electricity in the biggest commercial reactors being constructed at the time at 12–18 mills per kilowatt hour. This was approximately two to three times the cost of conventional power in most areas of the United States.

32 Ibid., p. 4

33 Strauss to Hickenlooper, January 23, 1961, LLS, HHPL.

power politically and economically insulated. That the military remained nuclear power's primary customer throughout the fifties was the most significant of these. Although invaluable as a catalyst for technological development, the AEC's dependence on a military constituency had its drawbacks. Politically, it left the AEC grossly underdeveloped compared to other agencies experienced in dealing with domestic policy considerations. Security concerns also continued to hamper participation by technical experts in related fields. The Reactor Development Division, for instance, charged with encouraging wider participation in reactor work, reported that "the requirements of national security tend to limit the number of new participants." The report argued that out of the 100,000 active security clearances, adding several hundred more for the purpose of industrial participation would not jeopardize security.[34]

The most significant drawback presented by dependence on an entirely military constituency was its relatively minor interest in economic constraints. The military advantage offered by nuclear-powered vessels and the relative ease with which funds were appropriated for military expenditures dwarfed concerns about cost. Technologically, nuclear power had arrived by the early sixties. Economically, however – from the standpoint of the utilities that were the new consumers it courted – it had not.[35]

That the market sought through this broadened industrial participation was overseas also contributed to the AEC's insulation. It dampened domestic interest group participation. Although a handful of American utilities were instrumental in the Power Development Reactor Program, the AEC – burned by early expectations of dramatic accomplishments from nuclear power – emphasized the specialized applications, and, particularly after the announcement of the Atoms for Peace program, the foreign applications of nuclear power, not its domestic potential.

Another crucial source of insulation – both economic and political – was the technology itself. Technical experts were in short supply. As the AEC sought to expand participation by industry in the fifties it consistently ran up against this problem. For instance, a report by the Division of Reactor Development written in January 1955 encouraged the use of technical personnel already assigned by contractors to AEC projects for commercial projects of the future. The report argued that despite the potential for conflicts of interest, such measures were necessary "because of the scarcity of people with experience and up-to-date knowledge outside the AEC reactor development program."[36]

Although the course of commercial nuclear power was shaped by its technology's integral links with national security concerns, and the esoteric nature of its technology, the pattern of participation in nuclear decision making was hardly

34 AEC 655/24, January 6, 1955, AEC, DOE, pp. 3, 5.
35 Hewlett and Duncan, *Nuclear Navy*, p. 255.
36 AEC 655/24; quotation from p. 17.

unique. The case of nuclear power should be seen as an extreme example of a phenomenon that has shaped virtually all American policymaking in the postwar era. Access to policy formulation has been far more restricted than access to policy implementation. While experts often have dominated forums for policy formulation, other political actors have gained increasing access as policies are actually implemented.[37] As John L. Campbell has written in his excellent study of nuclear policymaking, *Collapse of an Industry*, "Policy formation was usually insulated and consensual, although rarely involving citizen participation. Policy implementation was accessible and adversarial."[38]

The cumulative impact of nuclear power's relationship to national security and its esoteric technological characteristics go a long way toward explaining how it could mature technologically while remaining so insulated politically during the fifties. Given broad public deference to expertise, the relative absence of public debate between experts, the esoteric nature of the technology itself, and the tendency for policy formulation to be more insulated than policy implementation in America, nuclear technology advanced without much public debate. For the most part, the lengthy planning and demonstration process and its clandestine application to military problems conspired to conceal it from public view.

It is also clear that, in part, this insulation was quite intentional. It gave the AEC, the Joint Committee on Atomic Energy, and the nuclear industry maximum discretion politically, at a time when the industry could not compete directly economically. Besides, the nuclear community was not bashful about translating the benefits of nuclear power into layman's terms, and projecting this image to a wider audience. Writing in 1955, George L. Weil, who had been chief of the AEC's reactor branch, detected a pattern in what was broadly disseminated, and what remained limited to more specialized sources, concluding that "the beneficial prospects of nuclear power have been widely publicized. On the other hand, discussions of some of the unpleasant aspects have been limited almost exclusively to technical meetings and publications."[39]

The primary challenge for the nuclear community in the sixties was to achieve economic integration while retaining political control. Economically, nuclear power would have to compete with other sources of power and convince utilities of that advantage. Politically, however, nuclear power would have to justify large subsidies at the very time it claimed to be competitive. As if that was not difficult enough, the Atomic Energy Commission and the Joint Committee on Atomic Energy faced an even stiffer political challenge. They sought to maintain centralized discretion even though nuclear power's implementation generated visible political questions that cut across well-established functional and political jurisdictions. The AEC and the JCAE sought to maintain an insulated consensual

---

37 For a good discussion of the theoretical literature on this issue, see Campbell, *Collapse of an Industry*, ch. 2. See also Benveniste, *The Politics of Expertise*.
38 Campbell, *Collapse*, p. 90.
39 Weil, "Hazards of Nuclear Power Plants," *Science* 121 (March 4, 1955): 315–17.

style of decision making normally associated with the policy formulation stage even as they pushed that policy toward implementation.

The political trump card that virtually the entire nuclear community banked on, and the card that most members hoped would sell nuclear reactors to hesitant utilities, was expertise. Expertise had always played a central political role in the nuclear community. In the sixties, however, some experts sought and many more were invited to take on, a far more visible and explicit role in selling their wares. Perhaps reactors could be sold like shoes after all, if their designers were doing the selling. This approach, espoused and exemplified by AEC chair Glenn Seaborg during the sixties, propelled experts into far more visible political roles. After all, the stuff of politics was increasingly technical. Shouldn't those who understood these issues best be given more explicit political authority? Seeking the advantages of economic integration without the costs of its political counterpart, aggressive expert leadership was loath to sacrifice the disproportionate degree of discretion and specialized perspective that the nuclear community considered to be a part of its birthright. Although critics were quick to point out this contradiction between the desire to integrate economically and the insistence on political autonomy, in the early sixties an overriding optimism and belief in expert solutions carried the day. Nuclear power, buttressed by the rosy cost projections of the Atomic Energy Commission, entered the economic mainstream with its political autonomy apparently intact.

GLENN SEABORG: AN EXPERT ON TOP

If ever there was a time to promote a more visible role for a sophisticated technology, November 1960 was it. President John F. Kennedy planned to get the country moving again, and developing nuclear power might be just the place to start – at least as far as the Joint Committee on Atomic Energy was concerned. Seizing one of the general themes of President Kennedy's campaign, the Joint Committee on Atomic Energy argued that aggressive leadership had been one of the crucial missing ingredients in the nation's quest to promote nuclear power. In a letter to the president-elect written November 21, 1960, JCAE chair Anderson laid out the committee's thoughts on this matter. Citing the first point of the Democratic party's platform – restoring nonpartisan vigorous administration to atomic energy – Anderson wrote: "Here the key requirements are 'leadership,' and 'vigor.'" Anderson noted modestly that the Joint Committee had managed to stop much of Strauss's political maneuvering. But that was not enough. Some things had to come from the executive branch. As Anderson put it, "the AEC program still lacks the qualities of vigor, and leadership, so essential to our world position."[40]

---

40 Clinton P. Anderson to president-elect John F. Kennedy, in The Papers of Clinton P. Anderson, Library of Congress, Washington, D.C. [hereafter Anderson Papers], Box 832.

Kennedy was convinced that science provided a powerful force that could be applied directly to social problems. He tried to use science to solve problems both in domestic affairs – where he felt it could be an important economic lever – as well as in international affairs. He also shared in, and encouraged, the nation's seemingly insatiable thirst for more expertise. Shortly after his election he is reported to have griped that he didn't know enough experts, and promptly set out to overcome this handicap. Calls for more aggressive leadership and greater reliance on experts dovetailed in the appointment of Glenn Seaborg – the AEC's first scientist-chair.[41] Seaborg's reign represented the apotheosis of the politically visible scientist administrator. His route to the chair was not via the explicitly political path that more outspoken members of the Federation of Atomic Scientists had traveled. Rather, it was in the tradition of Vannevar Bush, who built political power through his skill at another type of politics – bureaucratic politics.

With Seaborg's appointment, it looked as if the old administrative adage that scientists should be on tap, not on top, had been revised: scientists – at least scientists who had honed bureaucratic political skills – emerged on top. Although Seaborg was not one of the four scientists endorsed for the position by *Nucleonics Week*, as an editorial in that journal pointed out, "It is not the name that is important. It is the principle behind the selection." According to that principle, the time had arrived for "the scientist member" of the AEC to assume the chair. As the principal spokesperson for the commission, that scientist, "would have a respect that is vital to AEC's success, he would bring prestige and knowledgeability to U.S. participation in the many international nuclear parleys in which he must play a leading role," the editorial argued. "He would be able to deal with the Joint Committee on Atomic Energy in a persuasive way in matters having important technical considerations. He would have the respect of the nuclear industry" it concluded.[42]

Seaborg himself was a leading advocate of increased reliance on scientists and engineers to staff the highest levels of government. In his early speeches as chair he was fond of noting that in the top bodies of the Soviet Union there was a preponderance of men with technical and engineering backgrounds. Conceding that we should not emulate the Soviet form of government, Seaborg went on to stress that we might learn something from that nation's reliance on expertise in higher positions.[43]

In a December 1961 speech to the American Association for the Advancement of Science (AAAS), Seaborg was more explicit. "The entry of scientists into important national advisory capacities is an inevitable concomitant of the events

41 See, for instance, Walter McDougall, "NASA, Prestige," p. 11; Halberstam, *The Best and the Brightest*, p. 1.
42 George Spensler, *BAS* 18 (June 1962): 17. *Nucleonics Week* (January 5, 1960), cited in Anderson Papers, Box 834.
43 Seaborg, June 6, 1962, "Speeches," Box 8, pp. 7–8.

of the last twenty years. I believe it is a healthy and essential development, and I have advocated it for many years. It does not seem to me," the AEC chair concluded, "that the influence of scientists in this respect is greater than it should be; indeed, in the national interest, I believe it must increase."[44]

By June 1962, Seaborg was close to suggesting publicly that science should be on top, not on tap. In a commencement address at George Washington University entitled "The Third Revolution" (by which he meant the scientific revolution), he urged that "competent persons with scientific or technological training" run for legislative office, and that the government "recruit with vigor and purpose" from this same pool to fill high-level administrative positions. Noting that government had already drafted the lawyer into not only the judicial but the legislative and executive branches, Seaborg stated that "we must conscript science and technology into this service."[45]

The basic requirement for this new leadership was "the combination of scientific capability … with political capability." These requirements Seaborg fulfilled handsomely. He was an enthusiastic and aggressive promoter. Richard Hewlett, official historian of the AEC, characterized him as an entrepreneur of science. In a 1963 speech, one of Seaborg's colleagues on the commission, John G. Palfrey, distinguished between scientists who sought political influence as informed citizens and those working from within the government. Pointing to an influential group – including Seaborg, Jerome Wiesner, and Harold Brown – Palfrey noted that these men were hardly utopian about their politics (implying the contrast to the Federation of Atomic Scientists). "Some of them," according to Palfrey, "had developed as shrewd a sense of politics as anyone." And why not? As Alvin Weinberg – himself a skilled scientist insider – put it: "Today, a research scientist must be an operator as well as a scientist."[46]

Both Seaborg and Weinberg understood that the gap between science and its application was narrowing. Seaborg spelled out his thoughts on this and related issues in several public statements. In his National Academy of Sciences speech, Seaborg stated that "the use of the atomic bomb crystallized, as never before and on a world stage, the enormous power of science and technology." Sputnik, Seaborg continued, further dramatized that lesson. During the postwar period "the gulf between basic and applied science has narrowed and in some cases become imperceptible."[47]

Competition with the Soviets and the race for international prestige made the challenge more urgent. As Seaborg put it, "One day, I believe the world either

44 Seaborg, December 27, 1961, in "Speeches," Box 8, p. 7.
45 Seaborg, June 6, 1962, "Speeches," Box 8, p. 8.
46 Ibid., p. 9; Daniel Ford, *The Cult of the Atom* (New York: Simon & Schuster, 1982), pp. 25–6; John G. Palfrey, November 8, 1963, Box 76, JCAE Files; Alvin Weinberg, *Reflections on Big Science* (Cambridge, Mass.: MIT Press, 1967), p. 40.
47 For Weinberg's views, see ibid., and Weinberg, "Science and Trans-science," in *Minerva* 10, no. 2 (April 1972): 209–22; the most definitive statement in the early sixties is Price, *Scientific Estate*. Seaborg, December 27, 1961, "Speeches," Box 8, p. 5.

will be enslaved, or it will be free." Only education could save the forces of freedom. "My recent experiences have impressed upon me," Seaborg continued, "a new sense of urgency about the steps we must take to exploit to the maximum our intellectual resources in what may be the decisive conflict for civilization."[48]

At times the challenge, as Seaborg saw it, transcended military and economic considerations and ranged from matters psychological to something approximating the mystical. Thus, in Seaborg's view, nuclear developments had "psychological and other impacts so extensive as to be difficult to evaluate," while space was "lifting our eyes to strange horizons that will affect our orientation in important ways."[49]

Given Seaborg's enthusiasm for science, it is easy to overlook the caveats he raised. Although not nearly as prominent as the references to the promise that science offered, warnings about its negative consequences do appear in his talks. Three that recurred were the unintended social and economic consequences of technological developments – increased rates of unemployment, for instance – the increasing estrangement between the physical sciences and the humanities – C. P. Snow's two cultures – and the role of the nation's citizenry in policies that increasingly required expert technical judgment.

It is not surprising that the former chancellor of the University of California, Berkeley, found in education the solution to all three of these problems. The nation's educational institutions would produce the experts required to tackle difficult social and technological problems. Education was also Seaborg's solution to the question of citizen participation in increasingly complex issues. Representative of his thinking is the following statement from his October 1962 speech entitled "Education for a Democratic-Scientific Society." "'To participate in the conduct of his affairs, a citizen in today's democracy must know more and he must know it better," Seaborg told the audience. "To take advantage of the enrichment of life society now offers, he must have knowledge and skills that are new and complex. In short, survival in our new environment depends upon the educated and what the educated can do." In other speeches he called for "creative participation," and "intelligent participation" at all levels.[50]

Seaborg's soaring rhetoric was not just hyperbole intended solely for public consumption. He believed firmly in this solution to one of the nation's most pressing political challenges. An interesting insight into his response to a specific problem that would recur time and again during his tenure as chair is gleaned from a summary of his remarks to field information officers at the AEC. When asked how to handle the matter of disagreement between two eminent scientists who had issued opposing "authoritative" statements, Seaborg responded that scientists, too, have opinions. But the whole incident illustrated the need "for

48 Seaborg, November 14, 1961, "Speeches," Box 8, p. 3.
49 Ibid., p. 4.
50 Seaborg, October 26, 1962, "Speeches," Box 8, p. 4; June 3, 1961, "Speeches," Box 8, p. 7; see also Ohio State Commencement Speech, "Speeches," Box 8.

public education in science and mathematics, for attainment of a greater degree of scientific literacy in our population so that they, themselves, can evaluate the issues. It is not as complicated as some people feel." Seaborg went on to say that he was optimistic about the future. "Pretty soon, now, people will come out of school with sufficient degree of scientific background to evaluate the issues such as this."[51]

Not since Lilienthal had an AEC chair actively urged greater participation by the broader public. But Lilienthal's prerequisites for participation could not have been farther from Seaborg's. As Lilienthal emphasized in his famous speech at Crawfordsville in 1947: "What is needed is not knowledge and judgment about scientific or engineering matters – you don't have that and couldn't take the time to acquire it of course." Lilienthal assured his audience that the AEC's technical forces could provide that. "The kind of judgments that I am talking about," Lilienthal emphasized, "is of quite a different sort – for example, a sense of what things people will accept as right and sensible and workable. What is needed is sense about human relations, about standards of fairness, about principles of self-government and of self education." Or, as Lilienthal stated earlier in the speech, "There has never been any good substitute for the all around common sense of an informed lay public."[52]

The partial fulfillment of Seaborg's vision, particularly the increased pace of professionalization and the emergence of a more highly educated middle class, led to greater demands for participation; closer scrutiny, and ultimately criticism during Seaborg's tenure. It would take a while, and it had little to do with Seaborg's appeal, but before the end of the 1960s there would be more participation than Seaborg could possibly have anticipated. What's more, it would be led by people who met Seaborg's rather stiff criteria for entry into the debate. But as we shall see, by the time it came, Seaborg had ceased inviting it, nor did he find it particularly enlightening.

THE LIMITS OF POLITICAL AUTONOMY

In 1961, Seaborg's primary concern was hardly public participation: rather, it was assuring the Atomic Energy Commission's continued dominance as a conduit for the nation's high-technology research dollars. On the surface, conditions

---

51 Seaborg, April 10, 1961, "Summary of Remarks by Chairman Seaborg at Meeting of Headquarters and Field Information Officers," "Speeches," Box 7, pp. 1–2. Seaborg's belief in the ability of education to resolve political conflict was hardly unique. Writing in 1959, *Bulletin of the Atomic Scientists* editor Eugene Rabinowitch noted that in the case of the highly publicized dispute over the effects of fallout, "if the questions put to scientists were adequately formulated and the answers received properly understood, the public and the political leadership would have easily found out that the two groups of experts [on radioactive fallout] had nearly the same answers" (cited in Gilpin, *American Scientists and Nuclear Weapons*, p. 326).

52 David Lilienthal, "Atomic Energy Is Your Business," September 22, 1947, "Atomic Energy, Radiation, 1950s" file, Wolman Papers, pp. 8–9.

appeared to be ideal for more aggressive leadership. Seaborg, after all, had been appointed by a young president who had promised to get the country moving again. That president and the majority in Congress were of the same party. Across the nation, faith in science and technology had never been stronger. Most scholars agree with David Dickson's periodization of the rise of experts in postwar American policymaking: the priesthood, or scientific estate – as Ralph Lapp and Don K. Price respectively referred to the policymaking scientists – "reached its zenith of power during the Kennedy administration."[53] The scientists, for instance, who developed America's nuclear arsenal, Gregg Herken noted, up to 1960 "had for the most part merely witnessed the making of strategy and policy on the bomb. They had remained on the sidelines of the great national debate over defense. Henceforth, they would be at its center."[54]

Seaborg sensed the opportunity. He immediately labeled the results of past AEC efforts to commercialize nuclear power disappointing. He hinted at a bold new program of incentives for reactor development. The AEC also decided that the time was right to yield to the JCAE's perennial request that power be produced at the Hanford reactor – a government-owned facility that heretofore had been used exclusively for plutonium production. Hanford-produced power would be sold commercially by the federal government in order to demonstrate the potential of the dual-purpose reactor. This was not a new idea, but AEC support for it was new. Both Strauss and McCone had opposed it in the past.[55]

Although the Atomic Energy Commission and JCAE had aggregated an overwhelming degree of authority and political discretion on issues related to nuclear power, Congress, ever since the mid-fifties, had intervened on the question of public versus private power, and 1961 was no exception. In September, Congress turned back the JCAE/AEC proposal for selling Hanford-generated power. This defeat prompted *Nucleonics Week* to open its "News of the Month" column in January 1962 with the gloomy prediction that "Another Doldrums Year Looms for Civilian Reactor Starts." As *Nucleonics* pointed out, the defeat also opened a deep rift within the nuclear community, between embarrassed Joint Committee on Atomic Energy Democratic leadership and the private utilities who had spearheaded the campaign against the Hanford reactor. Accusing longtime advocate of nuclear power Philip Sporn of asking for a fight when he could have been an innocent bystander, Senator Anderson warned that the JCAE would scrutinize Sporn's company's contract with the AEC. As Anderson put it, "I would hope that our committee would be able to engage a most militant attorney.... I try

---

53 Dickson, *The New Politics*, p. 265; Ralph Lapp, *The New Priesthood* (New York: Harper & Row, 1965); Price, *Scientific Estate*.

54 Herken, *Counsels*, p. 134.

55 Seaborg, testimony at U.S. Congress, Joint Committee on Atomic Energy, hearings on AEC Authorizing Legislation [hereafter JCAE Authorization Hearings], FY 1962, 87th Cong., 1st sess., 1961, pp. 161–3, 238–41. On McCone's opposition, see Memorandum of Conference with the President, February 11, 1959, dated February 18, 1959, AEC, Box 3, Eisenhower Library.

not to enter many fights, but when I am invited in ... I will make every reasonable effort to see that it is no sham battle."[56]

In light of the congressional defeat, the Atomic Energy Commission delayed its program of incentives. But Seaborg's retreat only increased pressure from manufacturers and the Joint Committee on Atomic Energy. More distressing than congressional intervention into the relatively insulated enclave of nuclear policymaking was White House review. Rather than increasing its commitment to developing nuclear power, the Kennedy administration drastically reduced appropriations for the Power Development Reactor Program. From the $45 million appropriated in FY 1961, the FY 1962 appropriation declined to $12 million. For FY 1963, the administration proposed a budget of zero. As Chairman Holifield pointed out in a March 15, 1962, letter to Seaborg, the status of previously authorized projects was not promising either: The process heat reactor had been canceled: the organic reactor, delayed indefinitely; and contracts for advanced boiling-water reactors were still being negotiated. This was vigorous leadership?[57]

In fact, it was. But the Kennedy administration's infatuation with cutting-edge technology carried it far beyond the now-aging light-water reactor technology – the same technology developed by the Truman administration to power the nation's nuclear submarine fleet. Scientists and engineers connected with the race for space applauded the administration's vigor while those in the nuclear community feared they would be left in its plume. A March 1962 *Wall Street Journal* headline captured their problem perfectly, proclaiming: "Atom Apathy – Space Exploits Push Nuclear Power Plants into Background." As one disgruntled congressional advocate of nuclear power commented in the article, "Everything is [John] Glenn, Glenn, Glenn. The United States will spend millions to put a man on the moon, but I couldn't get $50,000 authorized to build one atomic reactor."[58]

If space was the new faith, light-water reactor technology had, by comparison, become increasingly secularized. Skeptical of fanaticism by nature and training, budget analysts were among the first to defrock the federal role in the development of commercial nuclear power. In a memo to the files, written in February 1962, Atomic Energy Commission staffer Howard Brown summarized Bureau of the Budget analyst Fred Shuldt's reasons for rejecting the AEC's requested appropriation. Shuldt ran through several familiar refrains: the substitution of

56 *Nucleonics* 19 (October 1961): 19; and *Nucleonics* 20 (January 1962): 17. Anderson to Sporn, August 9, 1961, Box 839, Anderson Papers.
57 *Nucleonics* 19 (November 1961): 37. The Atomic Industrial Forum immediately embarked on a study of the nuclear future; see also Holifield to Kennedy, February 13, 1962, cited in Mazuzan and Walker, *Controlling the Atom*, p. 410. Holifield to Seaborg, March 15, 1962, in U.S. Congress, Joint Committee on Atomic Energy, "Hearings on Development, Growth and State of the Atomic Energy Industry Pursuant to Section 202" [hereafter "202" Hearings], 87th Cong. 2d sess., March 1962, p. 7.
58 Jonathan Spivak, *Wall Street Journal*, March 8, 1962.

space projects for civilian reactors and the priority given to defense were part of his reasons for slicing the AEC's appropriation to subsidize development of commercial nuclear power. He added a new reason to these older refrains, however. The goal of economically competitive nuclear power, Shuldt stated bluntly, had been largely reached; consequently there was no rationale for building more prototypes until more experience had been gained from the current generation of reactors.[59]

The AEC was caught in a bind that in part it had created by its own strategy. To sell nuclear power to utilities, the AEC increasingly had insisted that it was on the brink of economic competitiveness. The commission had long argued that reactors were safe and efficient. A White House more inclined to consider nuclear power in comparative terms and less willing to grant the AEC/JCAE the political autonomy it had enjoyed for so long (as long as it didn't visibly cross the public–private threshold) increasingly viewed light-water reactors precisely as the AEC had presented them – a mature technology requiring little more in the way of development. Yet consumers – utilities in this case – found few advantages in investing in that technology.

Competing scientific advice, lodged closer to the president, also cramped the AEC's insular style. The institutionalization of scientific advice that began under the Eisenhower administration, and Jerome Wiesner, the man who ran that institution for Kennedy, combined to frame questions related to energy needs and supply in a comparative perspective distinctly hostile to the AEC's far more parochial outlook. Like the debate with Kennedy's budgeteers, differences between the AEC and its White House handlers, though intellectually substantial, were relatively muted and remained within the bounds of the administration. Nevertheless, review by the scientific adviser threatened to shatter the tacit modus vivendi worked out between advocates of nuclear power and defenders of fossil fuels.[60]

Like the competition from the space program, expertise organized in a separate institution that introduced a new perspective on nuclear matters challenged the AEC's historical role as the premier science and technology agency in the federal government. As experts proliferated outside the AEC at the federal policymaking level, they demonstrated that there were other ways to organize expertise, other perspectives besides the nuclear framework, and other disciplines and subdisciplines from which the federal government might draw. The president's science adviser, operating from outside and, at least according to organization charts (though hardly as measured by political power), above the Atomic Energy Commission, marshaled the political clout of expertise to eat away at the AEC's political insulation. Although it dealt with an entirely different set of issues than the

59 Howard Brown to the files, February 19, 1962, AEC, DOE.
60 For a good summary of science advice to the president, see Saltzman, "Countdown to Sputnik."

science adviser's office did, the Advisory Committee on Reactor Safeguards, which operated within the AEC and whose members served at the pleasure of the commission, ultimately found itself in a similar position. Expertise organized in discrete units and drawn from a variety of perspectives – whether subordinate to AEC control, competing directly with it, as was the case with space, or ostensibly overseeing it, as with Wiesner – proved to be powerful denominations. They all embraced, and for that matter in part owed their existence to, the "scientific estate." Their disputes with the AEC over liturgy and doctrine, however, subtly undermined the commission's self-appointed role as sole interpreter of the gospel. This, in turn, quietly eroded the AEC's self-imposed political exile.

Alvin Weinberg, reactor pioneer, director of Oak Ridge National Laboratory (ORNL), and one of the nation's most sophisticated observers of "big science" politics, underscored the trend toward broader expert review in his annual "State of the Laboratory" address. Speaking to his lab in December 1961, Weinberg predicted that a new attitude on federal spending would reduce nuclear development budgets. The source of this change would be several studies under way in Washington on the country's energy future, including one by the National Academy of Sciences and another by the Federal Council of Science and Technology. From his vantage point as a member of Kennedy's Science Advisory Committee Weinberg could glimpse the emergence of an "over-all" energy policy that he felt would result in "a unified approach to our country's energy policy." When viewed as one of many competitive systems, nuclear energy would probably get a smaller proportion of government energy funds, Weinberg warned his scientific troops.[61]

The situation that had looked so promising to nuclear power's federal promoters in November 1960 looked rather bleak by the following year. An aggressive young president, captivated by the promise of technology, embraced technologies newer and more exciting than light-water reactors. In the meantime, the proliferation of scientific and technology-related agencies had drawn the permanent attention of the president's science adviser (his office itself a relatively new creation). Wiesner was not awed by the promise of nuclear power. Rather, he insisted on evaluating it in comparative terms. It was in this context – facing eclipse by a new, glamorous technology; restrained by a Democratic Congress that balked at any hint of "public power"; pincered by an oversight committee that itself felt the hot breath of an overbuilt manufacturing industry pressing for construction starts on the one hand and, on the other, a White House whose Bureau of the Budget viewed the AEC's mission of developing light-water technology as essentially complete and whose science adviser insisted on a comparative context – that the AEC mapped its strategy.

61 Weinberg, "State of the Laboratory – 1961," ORNL 61–12–70, December 15, 1961, Oak Ridge National Laboratory Central Files, Oak Ridge, Tenn. [hereafter ORNL Files].

A FACELIFT FOR AGING EXPERTISE: THE 1962 REPORT

By the winter of 1961–2, the Atomic Energy Commission was in a quandary. The strongest defense of the AEC's development program that Director of Reactor Development Frank Pittman could muster was that "it cannot be categorically stated that the [Power Demonstration Reactor Program is] not satisfactory."[62] The Joint Committee on Atomic Energy was more direct. At the annual hearings on the state of nuclear development in 1959, Senator Clinton Anderson called upon the AEC to construct its own prototype reactors at sites selected by the commission.[63] Anderson was also willing to consider funding for 90 percent of the capital costs involved in constructing second-generation, large-scale reactors – reactors that presumably would provide crucial economic data on the commercial competitiveness of nuclear power. McCone would not go that far, but at the commission's authorization hearings that year, he did agree to provide up to 50 percent of prototype construction costs.[64] In a speech delivered on April 25, 1961, Chet Holifield blamed nuclear power's slow growth on the the AEC. By relying heavily on its "partnership" program with private industry, the commission had retarded the development of nuclear power. The high cost of nuclear power continued to impede its development, the JCAE chair argued. Holifield demanded that the AEC develop, construct, and operate a series of nuclear projects that would lead to improved technology and lowered costs.[65] Seaborg was not willing to go that far, but the pressure was mounting.

As spring arrived with little visible Kennedy administration response to their pressure, a bipartisan mix of Joint Committee on Atomic Energy members charged that the public had already provided a massive subsidy to commercial nuclear power. A tiny bit more would realize the decade-long goal of commercialization. Shutting off the federal spigot, they argued, might kill the entire decade-long effort. In a powerful letter to Seaborg, written on March 15, 1962, JCAE chair Holifield lambasted his Democratic colleagues for "downgrading" the atomic power program. Along with the usual litany of reasons for continuing the government's support, Holifield pointed to the money already spent in pursuit of the AEC's objective. "In view of the fact that we have invested almost a billion dollars in atomic power development," Holifield argued, "and have the goal of economic nuclear power almost within reach, I believe it is incumbent upon this Administration to continue efforts to foster atomic power development through the prototype and non-economic stage."[66]

It was not just liberal Democrats who argued that past government subsidies, if nothing else, should guide future federal policy. Conservative Republican JCAE member Craig Hosmer wrote to the president on April 8, 1962. The

---

62 Pittman to Seaborg, November 28, 1961, in AEC 152/150, December 5, 1961, AEC, DOE, p.2.
63 Cited in Append. A, AEC 152/150, p. 7.
64 Ibid., p. 8.   65 Ibid., p. 7.
66 Holifield to Seaborg, March 15, 1962, in AEC 152/154, March 17, 1962, AEC, DOE, p. 2.

second sentence of the angry letter pointed to the $882 million dollars the federal government had already spent developing civilian nuclear power. "Only a small extension of the power reactor demonstration program is needed to encourage U.S. public and private utilities to build... demonstration plants," Hosmer insisted.[67]

By the end of 1959 there were thirteen power reactors in various stages of development. All received government subsidies in one form or another. None were, or even promised to be, economically competitive with power produced by fossil fuels. Two benefited only from the Price–Anderson indemnification act, and government assistance with parts of the fuel cycle. Six more were being constructed privately, but received government assistance on research and development, and fuel charges. The AEC was constructing four smaller reactors for publicly owned utilities under the second round of the Power Demonstration Reactor Program. Shippingport was constructed by the federal government at a site chosen by the AEC.[68]

Pittman himself came pretty close to stating categorically that the AEC's approach was in fact bankrupt. As he pointed out, in December 1961, the FY 1962 budget had no funds for new demonstration plants. The latest commission invitation for projects brought "no response whatever" from the private sector. The invitation before that snagged only one serious proposal. Even in that instance, the proposers had to package more than fifty utilities in order to come up with sufficient financing.[69] That development had come to a grinding halt was hardly news to any member of the nuclear community.

Pittman's memo went beyond the usual box score of proposals (or lack of them) to analyze the reasons behind the dismal figures. All of the utilities who could afford to risk capital in reactors had already done so, Pittman pointed out. Of course, if nuclear power were not such an economic gamble, there would be plenty of surplus capital available. The utilities, Pittman argued, wanted to analyze the results from reactors currently being constructed before making further investments in new projects.[70] Intangible considerations such as "the rivalry between public and private power, maintenance of U.S. leadership, desire to get practical indoctrination, and the publicity value of having a nuclear plant," Pittman concluded, would not attract the huge sums required for full-scale development. "It is now apparent," Pittman warned, "that these factors do not appeal significantly in contrast to the cost aspects." Pittman summed up the nuclear community's dilemma as the first year of the Kennedy administration came to a close: "The glamor of being 'first' is no longer there but the cost of being 'among the first' is."[71]

As if the economic news were not bad enough, Pittman also cited problems that, although soon swept under the carpet, would return to haunt the development

---

67 Hosmer to Kennedy, April 8, 1962, in AEC 152/159, April 12, 1962, AEC, DOE p. 2.
68 Append. B, AEC 152/150, p. 16.
69 Ibid., p. 19.　　70 Ibid.　　71 Ibid., p. 20.

of nuclear power. Even the Atomic Energy Commission had always confessed that for the record, nuclear power was not yet economically competitive, stating in 1959 that it hoped to achieve economically competitive power in areas of the country where fuel costs were high by 1968. It had, however, steadfastly defended the safety and reliability of nuclear technology. Whereas governors competed to attract the nuclear industry to their states in the 1950s, utilities were more cautious. Ensuring nuclear power's safety and reliability might end up costing more money, making that source of power even less economically competitive. As Pittman put it, "The siting and other hazard problems on small and medium sized reactors have had what is probably a major impact on the enthusiasm of the utility industry to bear full financial burden and the full operational and economic risk of nuclear plants." Pittman understood better than anybody that the cost of ensuring safety would only rise as reactors grew in size and were sited closer to population centers. Apparently the utilities recognized this as well. At a minimum, safety and siting would greatly affect, and in many instances determine, whether nuclear power would be able to compete economically.[72]

Reliability was also linked directly to nuclear power's cost. Like the cost of safety features, which manufacturers and utilities initially hoped would decline over time as the new technology evolved, efficiency was expected to increase as the industry gained operating experience. As preliminary data started to dribble in from the field, however, the news was not promising. Like safety, actual operating efficiency and construction experience now threatened to increase costs, rather than reduce them. As Pittman noted, "The operational problems at Dresden and other plants, increased construction costs at ECGR and EOCR, delays in start-up of EBR-II, Elk River and the NS *Savannah*, have brought a general realization that construction and operation of a nuclear plant is something that cannot be undertaken lightly."[73]

Pittman had a familiar solution to the commission's problems. The AEC, he argued, should put the utilities on notice that the commission would itself construct prototypes if the private sector failed to do so. Coming in the wake of the defeated proposal to use an already existing reactor (Hanford) to produce power, Pittman's proposal was a political long shot. As Pittman himself noted earlier in his analysis, "it is apparent that interest in new incentive programs cannot be rejuvenated at this time."[74]

New incentive programs were planned, however. In March 1962, Pittman wrote to the AEC's general manager and suggested one of the staff's most ingenious proposals for bypassing the lethargic utilities on the one hand, without triggering the "public power" alarm on the other. Why not provide all of the AEC's assistance to a designer or manufacturer contractor, rather than the utility? As Pittman noted, this "may minimize adverse comments by utilities concerning

72 Ibid.   73 Ibid.
74 Pittman to Seaborg in AEC 152/150, p. 2.

so-called Government interference with investor-owned utilities."[75] Thus, utilities would not have to risk investing in a full-scale plant, but the plant would be built, encouraging utilities to build succeeding plants. On March 8, 1962, the AEC approved this approach, although the AEC Counsel noted that the AEC's support of detailed design crossed the line into "what has been accepted as the 'No Subsidy' area." The commission noted that congressional authorization would be required. It directed Pittman to discuss the matter with both the Bureau of the Budget and the Joint Committee on Atomic Energy.[76]

While the nuclear community brainstormed, trying to come up with a long-term political strategy that would break nuclear power's economic deadlock, the AEC and JCAE continued to run interference for specific reactor projects in the hope that each might prove to be the demonstration that would jump-start a reactor bandwagon. This meant taking on some of the more controllable problems that nuclear power faced. Siting was clearly at the top of this list. In Hosmer's letter to the president of April 8, 1962, the California congressman had called for a presidential proclamation setting aside "National Power Reactor Reservations," not coincidentally starting with a portion of the Marine Corps installation at Camp Pendleton, California.

In the summer of 1961 the AEC had entered into a cooperative agreement with the Southern California Edison Company and Westinghouse to construct and operate a large, pressurized water reactor in southern California. There was a problem with siting, however. For the past ten months, the utility had tried unsuccessfully to obtain a portion of Camp Pendleton to site the reactor, and now claimed that there were no other suitable sites.[77] The history of the AEC's active intervention to obtain this site dated back to January 1961, when McCone met with Eisenhower in the twilight of his presidency and asked him to force the Marine Corps to hand over the site. Both sides started with parochial defenses of their claims, McCone warning that going ahead with a reactor at Camp Pendleton would "tend to head off a decision to make nuclear power a public program," and Marine Major General David Shoup cautioning that a reactor would interfere with troop movements and that high-tension wires might make helicopter training more hazardous.[78] Shoup soon revealed what appeared to be his real concern – a concern that opponents of the decision on the Fermi reactor (the PRDC reactor outside Detroit) might have found quite interesting. As Shoup put it, "if the utility company gets [an] installation representing $400 million, they will be so powerful they can call the tune. The result will be [*sic*]

75 Report to the General Manager by the Director, Division of Reactor Development, in AEC 152/152 March 6, 1962, AEC, DOE, p. 2.
76 Report to the General Manager by the Director, Division of Reactor Development, in AEC 152/153, March 16, 1962, AEC, DOE, p. 5.
77 Report to the General Manager by the Director of Reactor Development, in AEC 152/160, May 7, 1962, AEC, DOE, pp. 2–3.
78 Memorandum of Conference with the President, December 17, 1960, dated January 12, 1961, AEC III, Box 4, Eisenhower Library, pp. 1–2.

bring in industry and people and generate pressures to chop away more and more of the land at Camp Pendleton."[79] A modern-day NIMBY (not in my backyard) representative could not have put it better.

When Shoup reiterated that "with a $400 million investment, the power utilities could dictate," Eisenhower himself turned to McCone and out of the blue asked him about the status of the Fermi decision, which at the time was pending before the Supreme Court.[80] Eisenhower cited the situation at Fermi as a great example of Shoup's contention. The power of the utilities, once construction started, could be overwhelming.[81] Apparently neither the Atomic Energy Commission nor the Marine Corps could be expected to resist the pressure created by such massive investments, once they got started. At the same time, Eisenhower was "shocked" at the need for ten miles of low population density around the facility. The retiring president had a simple solution to that problem. He suggested "that the AEC should look at its own regulations and see if it has unnecessarily made this an impossible problem."[82]

By the spring of 1962, the Atomic Energy Commission understood that neither master strategies nor blunt pressure on behalf of specific projects would get very far unless the nuclear community could recapture the Kennedy administration's attention. To do just that, the AEC drafted a presidential statement on the development of nuclear power. Although the "hook" was the opening of Con Edison's Indian Point reactor, the statement was designed to garner tacit presidential support for the AEC's development program. As Edward Brown put it in a note to Commissioner Wilson, "the need to construct prototype and demonstration reactors employing promising new concepts, is really the 'raison d'etre' for the message." Brown reiterated the AEC's dilemma: "The progress described for the first ten [reactors] is so encouraging that it might invite complacency unless this clarification is introduced earlier in the statement."[83]

Brown's suggested transition from description of glowing technical success to request for the development funds emerged as one of the crucial planks in the AEC's strategy. The commission reasoned that although light-water reactors were technologically competitive, and close to economically competitive, funding for demonstration of new concepts could produce massive savings in the long run. It stressed that economic nuclear power was "simply a point on a cost curve... not a stopping point." Additional development funds, for prototypes and demonstrations, were crucial to achieving even cheaper power.[84]

Although the AEC failed to get the presidential statement it sought, it did garner a request from the president (originally drafted by the AEC) that it "take a new and hard look" at the role of atomic energy in the economy.[85] The AEC

79 Ibid., p. 2.   80 Ibid., p. 3.   81 Ibid., pp. 3–4.   82 Ibid., p. 3.
83 Brown to Wilson, January 11, 1962, AEC, DOE.
84 Ibid.
85 Mazuzan and Walker, *Controlling the Atom*, pp. 407–10; president's letter to Seaborg, March 17, 1962, in "Civilian Nuclear Power: A Report to the President – 1962" (U.S. Atomic Energy Commission, 1962) [hereafter Report to the President].

finally had its forum. It was not one, however, that pleased the Joint Committee on Atomic Energy. The JCAE greeted the president's call for another study with skepticism. Representative Hosmer asked, "Do you have any idea how many times we have already looked into this subject?" Senator Anderson reminded Seaborg of the previous JCAE studies, the last of which was completed just a year before. Holifield pointed out that the McKinney study took a year and a half and also reminded Seaborg that the Atomic Industrial Forum had just completed its study of civilian power.[86]

The "Report to the President," issued in November 1962, addressed the massive pressure for short-term concrete results simply by asserting that nuclear power had already reached an economically competitive plateau. Anticipating criticism like Shuldt's in the Budget Bureau, however, the report argued that although nuclear power was nearly competitive with other sources of fuels, this remained to be demonstrated. It was government's role to demonstrate it. Turning to the long-run picture, the report predicted a drain on fossil fuels and proposed a bold challenge to meet the unprecedented demand for energy. That challenge was the development of breeder reactors and the long-term conversion to what Seaborg soon began to call the "plutonium economy." A substantial program of research and development on convertor reactors would help bridge the gap from uranium-powered to plutonium-powered reactors. The "plutonium economy" was the challenge with which the AEC hoped to recapture White House funds directed toward the technological cutting edge.[87]

The most striking aspect of the report was its profound optimism. The Atomic Energy Commission's prognosis for the immediate future was distinctly upbeat. It dismissed the same technical and safety problems that its own director of reactor development had so recently cited, concentrating on economic requirements as the only remaining obstacle. After acknowledging that past efforts to meet these requirements had led to "many problems," "disappointments," and "frustrations," the report continued, "Happily, more recently much progress had been made toward solutions of these problems." In the sentence that captured the nation's attention, the report crowed, "Nuclear power is believed to be on or near the threshold of competitiveness with conventional power for large plants, in areas of the country where fossil fuel costs are high." It added that "further cost reductions are definitely in sight, provided an aggressive program is continued."[88] This optimism and enthusiasm was not based on any new operating experience or actual technological breakthroughs. The report simply asserted that the goal of reaching economically competitive nuclear power in high-cost areas had been met and would soon be surpassed.

The report also moved away from the longstanding justification of enhancing international prestige and bolstering defense, placing greater emphasis on economic

86 U.S. Congress, Joint Committee on Atomic Energy "202" Hearings, March 1962, pp. 9, 15, 19.
87 Report to the President, pp. 35–7.
88 Ibid., p. 4.

benefits – whether directly through lower costs such as the $32 billion in savings it predicted by the year 2000, or indirectly by absorbing the loss of fossil fuels.[89] If Pittman had been correct and it was no longer the "intangible" considerations such as the maintenance of U.S. leadership that swayed utilities, as opposed to "the cost aspects," the report was the perfect antidote.[90] The "aggressive" short-range program that it proposed focused squarely on overcoming what might best be described as marketing obstacles. "Unlike such revolutions as those introducing the railroad, the automobile ... and indeed electric power itself," the Report to the President argued, "the large-scale use of nuclear power will not result in qualitative new capabilities. Its public marketability will be based almost completely on economic factors." The report argued that working within America's market economy, it was the federal government's responsibility to assure widespread use of nuclear energy. The Atomic Energy Commission could achieve that "by fostering developments that make such use economically attractive."[91]

The report proposed modest additional incentives to assure the construction of plants incorporating the most competitive reactor types. Here the primary government objective was to assure that a technology which the AEC now proclaimed to be economically competitive was actually put into place. The report also advocated a program of government-financed demonstrations and prototypes for other promising reactor types in order to increase nuclear power's economic advantage.[92]

A major portion of the report was devoted to demonstrating the need for nuclear power. This was the Atomic Energy Commission's response to Jerome Wiesner's request that the nuclear program be studied in light of "the Nation's prospective energy needs and resources." Here, just as elsewhere in the report, the AEC exuded unbounded optimism. The report projected a steadily increasing rate of energy consumption well beyond the year 2000, exhausting low-cost fossil fuels in a century or less and requiring the substitution of other fuels before that. The heady projected increases in demand added a new sense of urgency to the longstanding case for development of replacement fuels.[93]

It was in the context of this overwhelming demand that the report presented a bold new program to develop breeder reactors. Many familiar with the breeder's history in the United States considered this to be the report's most optimistic and aggressive proposal. As Seaborg's cover letter pointed out, the AEC had to shift its attention from narrow short-term objectives to larger challenges. "[F]or the long-term benefit of the country, and indeed of the whole world," Seaborg

89 Earlier in the year, the AEC had released far more optimistic figures based on General Electric estimates, but none of these were based on experience (*Nucleonics* 20 [September 1962]): 17. Mazuzan and Walker, in *Controlling the Atom*, mention the shift from prestige; the Report to the President didn't entirely neglect international prestige, but it is interesting that the JCAE had to tell the AEC to put it in. See Bauser to files, July 26, 1962, Box 839, Anderson Papers, Library of Congress. Quotation from Report to the President, p. 27.

90 For Pittman, see AEC 152/150, pp. 19–20.

91 Report to the President, p. 27.   92 Ibid., pp. 49–51.   93 Ibid., pp. 16–23.

wrote Kennedy, "it was time we placed relatively more emphasis on the ... more difficult problems of breeder reactors.... Only by the use of breeders would we really solve the problem of adequate energy supply for future generations."[94]

Even critics who had tried to inject more vigor into the AEC's development program were a bit astonished by the report's unbridled optimism. Responding to a July 31 AEC briefing, Chet Holifield – no shrinking violet when it came to enthusiasm for nuclear power – told Seaborg, "While I do not question that our requirement for electric power will increase, it is hard to believe that it would continually increase at anywhere near the rate which was assumed in the presentation. If these assumptions are correct," an astonished Holifield continued, "in less than forty years we would have to be building the equivalent of two Yankee reactor plants every day in the U.S. alone."[95]

Holifield, originally skeptical of the whole idea of a study, acknowledged that the attention given to the resources problem and the role of breeders together with the interim role of converters represented a "somewhat different emphasis than heretofore expressed." But besides his doubts about the long-range electrical demand projections, he wondered why the outlook for breeding was suddenly so rosy. As he put it, "the breeder approach has now been talked about and worked on by the commission for over 15 years, yet the major problems remain to be solved. The British too have been pursuing this approach for many years with limited success to date."[96]

In the Report to the President, the Atomic Energy Commission, a model of the post–World War II high-technology agency, tried to come to grips with its ultimate nightmare. Its light-water reactor technology was aging and might possibly expire before making it into commercial production. Ten years and a billion dollars had not overcome nuclear power's economic exile. Resuscitating it required expertise far afield from the core of physicists, engineers, and chemists who had designed the original technology. It required marketing and economic skills, not to mention the steady drumbeat of encouraging statements and the political savvy to steer between the "public power" boundary Congress had declared off-limits and the aggressive short-term program the JCAE had been promoting.

The report acknowledged these short-term responsibilities; but it was far more excited about the more distant prospects of breeding. This, it was hoped, might return the AEC to the technological cutting edge and reconstruct its relative monopoly on expertise. The short-term strategy, though crucial to the commercialization of light-water reactors, most likely would have failed had the report not linked it to the larger commitment entailed in developing breeders. It was that "bold challenge," not mopping up after light-water reactors, that helped capture what had proven to be quite elusive presidential support. By portraying the program as bold, as challenging, as forward-looking, the report

94 Seaborg to the President, 11/20/62, in ibid.
95 Holifield to Seaborg, August 31, 1962, *Controlling the Atom* [CTA] File, AEC, DOE.
96 Ibid.

reflected not only the transition to the more aggressive leadership of scientists on top, it captured the spirit of the White House and the times as well.

Oak Ridge National Laboratory director Alvin Weinberg's State of the Laboratory speech, which antedated even the president's request for a study, foreshadowed just such a strategy. Although Weinberg had predicted that a more fully coordinated White House energy policy would pare nuclear budgets, he also recognized that Oak Ridge might well benefit from that development. As the director told his colleagues, "Any unified study of our energy resources will bring out again [that] our coal reserves ... are finite; that when we need nuclear energy we shall need it on a very large scale; that we are therefore justified in spending an appreciable fraction of our country's budget on continued development of long-range nuclear energy systems.... The position of ORNL with respect to such an overall trend in allotting our country's efforts to various energy systems would be very good since long-range systems – controlled thermonuclear energy and high-conversion-ratio reactors – are such important parts of our programs."[97]

The report practiced what it preached about marketing skills. Its distribution and the accompanying publicity – not technological breakthroughs or even a new operating basis for revised estimates – had a marked impact on the attitudes of those long-courted consumers, the utilities. The once wary Joint Committee on Atomic Energy emerged as its most satisfied reader. Newly appointed JCAE chair John O. Pastore called it "an outstanding job" and stated that he and his colleagues wholeheartedly endorsed the study's conclusions. Besides issuing the report, the Atomic Energy Commission had restored some of the funds previously dropped from the development program. It had also reduced, for the second time in a year, the prices it charged to utilities for AEC-processed fuel, a subsidy that all reactors benefited from. More evidence that the administration would back up the report with dollars was provided at JCAE hearings. Here, the AEC presented a twelve-year program that scheduled many of the projects it had called for in the Report to the President.[98]

The nuclear industry reacted favorably in general, although it questioned the heavy reliance on fuel conservation as the justification for the government's program. Economic marketability should have been even more heavily emphasized, the testimony of several manufacturers stressed. Manufacturers differed in their assessment of which technologies were most likely to lead the way to greater economic efficiency and the likelihood of achieving successful breeding. These differences, needless to say, reflected their companies' research commitments.

97 Weinberg, "State of the Laboratory," pp. 22–3.
98 U.S. Congress, Joint Committee on Atomic Energy, "Hearings on Development, Growth and State of the Atomic Energy Industry Pursuant to Section 202," 88th Cong., 1st sess., 1963, p. 2. The AEC proposed subsidized design funds for the demonstration program and restored a proposal for subsidizing an organic reactor; *Nucleonics* 20 (June 1962): 17. Design funds subsequently were removed by JCAE for fear that Congress would challenge current nuclear technology as being oversubsidized; *Nucleonics* 21 (August 1963): 44–5. See also *Nucleonics* 20 (July 1962): 26, and *Nucleonics* 21 (April 1963): 22.

Philip Sporn declined to testify publicly, but in a letter to Commissioner Haworth continued to stress many of the points he had raised in the preceding year's testimony: He was still not convinced that there was any basis for the AEC's newfound optimism.[99] Such concerns, however, seemed trivial in light of the tremendous enthusiasm that surrounded the report. If nothing else, the report had at least brought squabbling members of the nuclear community together.

With the emergence of a more confident Budget Bureau and a national science adviser committed to restoring political control over the nation's scientific bureaucracy, autonomy – even within the Kennedy administration – could no longer be taken for granted.[100] Consensus within the nuclear community used to ensure political autonomy on virtually any issue short of direct public construction and sale of nuclear power. Now, however, all eyes were trained on Jerome Wiesner. Wiesner did not disappoint, although he hardly embraced the report's unbridled optimism. He endorsed the report but refused to make long-term commitments to nuclear power as the only alternative energy source.[101]

MANAGEABLE POSSIBILITIES, NOT GLOBAL SOLUTIONS

One critic of the report was far more vocal in his skepticism. In a series of lectures delivered at Princeton University in February 1963, former AEC chair David Lilienthal challenged the framework in which the atom had been developed following Hiroshima. It was his concluding lecture entitled "Whatever Happened to the Peaceful Atom?" that immediately drew the attention of the nuclear community. In this lecture Lilienthal flatly claimed that "the glamour, the excitement of the boundless possibilities of power from the peaceful atom is gone. The sooner we face up to this the better.... We have failed as a nation to recognize ... that the 'profound changes' arising out of a revolution in atomic-energy supply just aren't in the cards." Despite this fact, Lilienthal continued, "we still have an organization – the AEC – that in magnitude of expenditures and personnel is geared to the objective of 1946: a revolution to bring this magic into reality, to bring on a new world."[102]

Lilienthal prescribed instead that America bring the atom "fully into the mainstream of the scientific effort of the country." In one sense, that was precisely what the AEC sought to do. Its short-term objectives directed attention toward light-water technology's problems of economic competitiveness and marketability.

99 JCAE, "202" Hearings, 1963, pp. 56–7. On industry response, see, for instance, statement of Louis Roddis, President, Atomic Industrial Forum, in ibid., pp. 794–5; and *Nucleonics* 21 (May 1963): 19–20. For Sporn's position, see Sporn to Pastore, March 13, 1963, in JCAE, "202" Hearings, 1963, p. 866; and Sporn to Haworth, January 22, 1963, AEC, DOE.
100 On assertion of centralized White House control, see Crenson and Rourke, "By Way of Conclusion," in Galambos, ed., *New American State*, p. 173.
101 JCAE "202" Hearings, 1963, Wiesner testimony.
102 Stafford Little Lectures, Lecture III, February 19, 1963, in JCAE, "202" Hearings, 1963, pp. 705–14; quotation from p. 707.

But to Lilienthal, "getting the atom fully back into the stream of American life" meant that "where there is not present or prospective economic need for a product or service," it didn't make sense for the government to continue spending public dollars to create demand. "If there is a real need, it will be met by the utility and manufacturing industries.... We should stop trying to force-feed atomic energy." Lilienthal hit the AEC where it was most sensitive, on the issue of demand.[103]

Lilienthal went beyond questions of economic demand and, through the concept of mainstreaming, foreshadowed many of the charges the AEC would face as it sought to move into the economic and political mainstream while retaining its political control over all nuclear-related matters. "Mainstreaming," Lilienthal anticipated, meant more than economic demand. Conceding that electricity from atomic power might be called "just as good" as electricity from fossil fuels if the costs were comparable, the former AEC chair warned that "the potential hazards to life and health of hundreds of thousands of people in densely populated areas adjacent to power plants (such as that project in New York City's Borough of Queens [Ravenswood]) make it inaccurate to label atomic power plants as 'just as good' as conventional power plants, even when the cost is virtually the same." The waste from nuclear plants, the ex-AEC chair reminded his audience, was "furiously" radioactive. "After all these years no entirely satisfactory technical way has been found to treat them so they will be safe."[104]

Lilienthal concluded with an example of the harm caused by the AEC's "artificial apartness." It was the case of low-level radioactive waste products that, with AEC permission, had been dumped off the coast of Massachusetts. Lilienthal warned,

By this sense of atomic apartness the State and Federal health services were deprived of the opportunity to become fully knowledgeable of the radiation hazards – or absence of hazards – in such wastes and how to protect against them. Today more and more the functions of AEC that involve existing technological agencies are being slowly transferred to those agencies. This is a move in the direction I urge: putting the atom into the mainstream of men's affairs.[105]

Beyond the specific criticism of nuclear power, the lecture series – published as *Change, Hope and the Bomb* – attacked overcompartmentalization and specialization of expertise on the one hand, while questioning the grand solutions and overselling of programs that so often accompanied these efforts. Having led the federal government into its most grandiose technological undertaking – the Tennessee Valley Authority – and its most ambitious peacetime application of scientific expertise – the Atomic Energy Commission – Lilienthal embraced another powerful political current slightly before its time. That was man's limited ability to eradicate entirely the problems of society.

Nowhere was this conviction articulated more clearly than in the new environmental consciousness expressed by several pioneering scientists and publicists.

103 Ibid., pp. 710, 712, 714.   104 Ibid., p. 706.   105 Ibid., p. 714.

As Donald Fleming has noted in exploring the "Roots of the New Conservation Movement," postwar pioneers as disparate as René Dubos and Rachel Carson shared a common belief that problems would never vanish. In Carson's case, "pests did not become extinct, but... became manageable for those who could bear to acknowledge that they never could and never should achieve absolute control over their environment."[106] For René Dubos, even when the target was the world's most dreaded microbes, absolute mastery was a foolish objective. Those who sought such goals had been seduced by "illusions and mirages."[107] Not only was this impractical, as Fleming writes, Dubos saw efforts to eradicate and exterminate even deadly microbes as "the culminating expression of a will toward 'almost complete mastery' over nature, total control over one's own destiny." Like Rachel Carson, Dubos regarded this as the "worst flaw in the modern temperament."[108] Dubos, Carson, and subsequently Barry Commoner, like Wiesner and York in the field of weaponry, arrived at their relatively more humble relationships to science through direct experience with the same forces that their colleagues with more hubris hoped to master.

Lilienthal followed a similar course. In the Princeton lectures, he renounced any remaining hubris. As he put it,

Many scientists in the course of the last fifteen years have given reinforcement to the naive point of view about the Atom and human affairs that has frustrated and confused our thinking. More than any other influential group, they have succeeded in espousing the doctrine of panacea: that in human affairs "A *solution* can be found." Time and again they have discovered such solutions, or lent the weight of their technical prestige to solutions proposed by nonscientists.[109]

Lilienthal's recommendation, that "we think in terms not of solution but of manageable possibilities," stood in stark contrast to Seaborg's "solution" to an impending energy crisis.[110]

Delivered just as America's confidence in expertise was cresting, Lilienthal's broadside provoked a quick response. The Joint Committee on Atomic Energy helped coordinate this action, requesting a detailed AEC reply to each of Lilienthal's charges and extending its annual state-of-the-industry hearings to cross-examine Lilienthal and provide a public forum for rebuttals. One of the most thoughtful responses to Lilienthal's charge that science had oversold the atom came from James Newman. Newman conceded that "the growing tribe of *savants officiels* was a nuisance"; he was particularly critical of the growing dependence of some universities on "Defense Department baksheesh." Yet Newman felt that the blame lay not with the scientists but with the politicians. "When

---

106 Cited in Donald Fleming, "Roots of the New Conservation Movement," *Perspectives in American History* 6 (1972): 34.
107 Ibid., p. 38.　　108 Ibid.
109 Lilienthal, *Change, Hope and the Bomb* (Princeton, N.J.: Princeton University Press, 1963), particularly pp. 9–26, 59–91; quotation from p. 66.
110 Ibid., p. 149.

the atom business began," Newman pointed out, "Congressmen and Government officials were overcome with admiration and awe every time a leading atomic physicist appeared before them.... The Congressmen never learned the substance, but in time they learned to use the jargon. We now have a group of so-called experts in the legislature who are no better fitted to weigh scientific evidence, no more able to distinguish professional knaves and mountebanks from disinterested and honest men than were the legislators of 1946." Newman concluded by urging Americans to listen more – not less – to capable scientists.[111]

Newman's effort to distinguish between scientists and politicians was also representative of the times. The distinction had earlier been taken for granted; now, it was becoming more difficult to tell the difference. Had not Glenn Seaborg actively called for scientists to run for elected office? Was it not essential that scientists running large operations develop political savvy to protect their budgets? Did not committees like the JCAE depend on their colleagues' deference to expertise for much of the discretion granted to that committee in its oversight and legislative capacities?

Increasingly, the various components – political, managerial, and expert – were indistinguishable. Appeals from men like Lilienthal for managerial solutions, or Newman, for scientific solutions, ignored this important development. Science had reached the top in the early 1960s, and by a route far different from that of the zealots who had sought to orchestrate an explicitly political campaign for civilian control several decades earlier. Nevertheless, although far less dramatically and over a much longer period of time than for the scientists who mobilized in more explicitly political ways, the price paid for political access was the same. It was the scientists' greatest political asset – the appearance of disinterestedness.

THE PARADOX OF POWER

Disinterested or not, the Atomic Energy Commission remained the only authoritative source of cost estimates other than nuclear manufacturers. Nor was there any operating experience on which to base these increasingly optimistic projections. The Report to the President essentially stamped the industry's cost projections with the AEC's seal of approval. A flurry of reactor starts followed on the heels of the report's release. In December 1962 three new starts were announced: The Los Angeles Department of Water and Power stated that it was entertaining bids from Babcock and Wilcox, GE, and Westinghouse, and would decide by mid-January. Connecticut Yankee announced it would contract with Westinghouse to build a pressurized water reactor. And in the biggest headline-grabber, Con Edison announced it would build the world's largest reactor in the heart of New York City at Ravenswood, just across the East River from central Manhattan.

---

111 Luedecke to Pastore, March 29, 1963, Bauser to All Committee Members, April 3, 1963, Box 75, JCAE Files; "Commentary on Lecture III," n.d., Box 42, Holifield Papers; JCAE, "202" Hearings, April 2–5, 1963. *Washington Post*, October 18, 1963, p. A4.

Con Ed said the plant would be competitive with the best conventional plant that could be built at the site.[112]

The boomlet continued into early 1963, with Minnesota Power and Light proposing a reactor for a group of rural cooperatives, and Rochester Gas and Electric announcing plans to construct a reactor. The April edition of *Nucleonics* spread the exciting news that Pacific Gas and Electric was planning to triple its capacity: two-thirds of the new capacity might be nuclear. As *Nucleonics* put it, "The plan provides another manifestation that electric utilities are beginning to regard nuclear power as conventional – as just another heat source." Niagara Mohawk Power Corp. announced in July its plans to construct a 500 MW(e) (megawatt) reactor near Oswego, New York. It, too, anticipated costs that would compete with conventional fuels in that area. In December 1963, Jersey Central Power and Light announced plans to purchase a 620 MW(e) reactor from General Electric. The Oyster Creek plant, Jersey Central Power claimed, would eventually produce power at under 4 mills per kilowatt hour (kwh).[113]

Several factors made the Jersey Central Power announcement an eye-opener. The low cost was one. Estimated costs were below projected costs for a conventional fuel plant and below the estimates for nuclear power projected by the Report to the President. As Jersey Central power proudly announced, the decision to build Oyster Creek "was based entirely on economic and engineering considerations." Once the plant was fully operational "the total cost of power from the station will be less than from any other types of plant which the company could install at this location."[114] That these low costs would be achieved without any government subsidy was a second milestone. Oyster Creek would be the first instance of unsubsidized economically competitive nuclear power. Finally, the plant would be built on a "turnkey" basis. General Electric would charge a fixed fee – an astonishingly low $60 million for delivery of a completed plant. Any cost overruns (except those resulting from monetary inflation) would be absorbed by the manufacturer.[115]

Competition between manufacturers eager to sell a technology they had already invested billions in was clearly one factor prompting the turnkey concept. To make nuclear power attractive, Westinghouse soon slashed prices to meet GE's bargain-basement price of $100 per kilowatt capacity at Oyster Creek. The "Great Bandwagon" was off and running. To the engineers charged with bringing in projects near cost, the fixed-price contracts seemed incredible. As a representative of the engineering firm working on the Oyster Creek project (Burns and Roe) told a meeting of the American Power Conference in 1966, "When you consider the fact that Oyster Creek will not be completed until mid-1967, the fixed price

112 *Nucleonics* 21 (January 1963): 17–18.
113 *Nucleonics* (March 1963): 22; (April 1963): 21, (August 1963): 43. Oyster Creek press release quoted in AEC 152/192, December 17, 1965, Attachment II, AEC, DOE, p. 12.
114 Oyster Creek press release.
115 *Nucleonics* 21 (March 1964): 17.

bidding seems even more remarkable since the true cost of constructing the large scale [reactors] is yet to be demonstrated."[116] It was not just the engineers who objected to the turnkey concept. No segment of the nuclear industry was pleased about it.[117]

Consequently, in the Great Bandwagon market that developed between 1964 and 1968, many of the reactors sold were not "turnkey" reactors. This forced greater financial risks on the utilities. Even during the height of turnkey sales, cost-plus contracts outnumbered turnkey sales.[118] Factors other than fire sale turnkey prices were at work. A RAND study pointed to the end of declining coal prices, regional pooling of power (which favored nuclear plants because of their economies of scale), increasing size of generating plants, reduced uncertainty over nuclear fuel supply, rising concern over air pollution caused by coal burners, and, most intangibly but perhaps most significantly, a growing perception among utilities that nuclear power was the wave of the future.[119]

Pausing in 1964, a year in which no new orders were placed, the move to nuclear resumed its upward momentum with a rush in mid-1965. As the AEC's director of industrial participation, Ernest Tremmel, reported to the commission's top management, "During the month of January 1966, only one fossil fuel power plant was announced ... (500 Mwe) versus four nuclear plants with a total capacity of 2,900 megawatts."[120] More good news was just around the bend, Tremmel, continued. General Electric planned to increase its estimates of projected nuclear capacity by approximately 25 percent.[121] Between August 1965 and December 1967, fifty-five orders for reactors were placed. Seven orders were placed in February 1967, the peak month in this "bandwagon" market. Even Pacific Gas and Electric's rosy estimates of 1963 were surpassed in 1966 and 1967. During those years nuclear reactors accounted for one-half of all the new power plants ordered by utilities.[122]

Long before orders tapered off in 1968, even the staunchest supporters of the AEC's developmental role conceded that the mission of commercializing light-water technology in America had been achieved. In July 1963, after the initial

116 S. Baron, "Engineering for Community Acceptance: The Architect-Engineer's Role," in AEC 152/202, May 3, 1966, AEC, DOE, p. 3.
117 Ibid.
118 Bupp and Derian, *Failed Promise*, p. 48; Gandara, *Electric Utility Decision Making and the Nuclear Option*, pp. 52–5; Mark Hertsgaard, *Nuclear Inc.: The Men and Money Behind Nuclear Energy* (New York: Pantheon, 1983), pp. 40–7.
119 Gandara, *Electric Utility Decision*, pp. 52–9.
120 Ernest Tremmel to D. A. Ink, February 14, 1966, in AEC 152/196, February 24, 1966, AEC, DOE, p. 2.
121 Ibid.
122 Ibid., p. 54. Del Sesto, *Science, Politics and Controversy*, p. 91. By 1968 the number of reactors sold declined to fourteen, dropping to seven in 1969 (Gandara, *Electric Utility Decision Making*, p. 8). The RAND analysis of utility decision-making attributes the dropoff to four factors: construction delays; little sign that these bottlenecks would be corrected in the future; rapidly escalating construction costs; and immediate demand for electricity that could not be met by slow-moving nuclear construction (ibid., pp. 60–2).

surge of orders, the Joint Committee on Atomic Energy, for instance, scaled back incentives. *Nucleonics* reported that "this approach was dictated by the mounting criticism of nuclear power programs by supporters of the new [Administration] energy study." A year later the Joint Committee on Atomic Energy justified its actions to a subsidy-hungry industry, citing Oyster Creek and another newly announced reactor, "9-Mile Point," as examples of new unsubsidized orders. For the first time, the JCAE opposed further direct assistance for light-water reactors. The JCAE took another important step toward removing economic subsidies by sponsoring the Private Ownership of Nuclear Fuels Act, which eliminated, over time, a portion of the fuel subsidy for light-water reactors. The positions of the AEC's two most formidable vertical competitors – the Bureau of the Budget and the president's science adviser – had finally prevailed.[123]

Although direct subsidies like those provided under the Power Reactor Demonstration Program were phased out for new light-water reactors by the mid-sixties, they continued for reactors already initiated under the Power Reactor Demonstration Program. So, too, did the AEC's commitment to buy back spent reactor fuel, even though there was no market for such fuel.[124] Indirect subsidies such as the Price-Anderson indemnification act remained in place. To protect these, and to protect the highly concentrated nuclear industry, from antitrust actions, the Atomic Energy Commission fought tooth and nail to preserve the industry's less visible shelters. The legal basis for much of this protection and for the AEC's direct subsidies derived from the Atomic Energy Act of 1954. Section 102 of that act provided that whenever the commission found that any type of production or utilization facility had been sufficiently developed to be of "practical value" for industrial or commercial purposes, the commission could no longer license such facilities under "demonstration" licensing procedures, procedures that waived normal antitrust scrutiny and allowed for direct AEC subsidies.[125]

Joseph Hennessey, AEC general counsel, spelled out in no uncertain terms the

---

123 Quotation from *Nucleonics* 21 (August 1963): 44. See also Del Sesto, *Science, Politics and Controversy*, p. 83; and *Nucleonics* 22 (September 1964): 21. The Atomic Energy Commission reluctantly went along with its oversight committee. Director of Reactor Development Frank Pittman, for one, had a better idea. In a November 20, 1964, memo, Pittman suggested that the AEC take support costs such as safety research and regulation out of the figures reported under civilian development (Pittman to Abbadessa, November 20, 1964, Append. D to AEC 152/176, November 23, 1964, AEC, DOE, p. 1). As another AEC memo put it, this would lead to a figure "substantially less than the figure now computed....Such a level of cost might reduce criticism that AEC is spending too much money on civilian power." There was one problem, however, as the next page of the memo made clear. "Cost of civilian power computed on such a basis would not represent the total financial resources devoted to the program" (Addendum to AEC 152/175, in AEC 152/176).

124 This plutonium buy-back, declared one government specialist, was "one of the biggest subsidies Uncle Sam gives, far more important than the advantage of leasing enriched uranium fuel" (quoted in Jonathan Spivak, *Wall Street Journal*, September 25, 1963).

125 "Determination Regarding Finding of Practical Value," in AEC 152/192, December 17, 1965, AEC, DOE, p. 2, and AEC 152/173, April 3, 1964, AEC, DOE.

dangers a practical value determination entailed for the AEC's capacity to protect the nuclear industry in 1964. Under existing legislation, should the AEC find reactors to be of "practical value" the commission would be required to give preferred consideration in licensing to applications submitted by public and cooperative utilities.[126] The AEC would be required to request that the U.S. Attorney General review the project to determine if it violated antitrust laws – a review that the concentrated nuclear industry had always been exempted from.[127] The AEC would be required to charge commercial reactors for the lease and consumption of fuel.[128] In the past, this charge had often been waived or reduced as part of the AEC's development program. A determination of practical value would also preclude the AEC from subsidizing design and construction as it had under its development program.[129] Finally, another important subsidy – guaranteed purchase prices for the plutonium created through the production of power – would be threatened if the distinction between "demonstration" and "commercial" reactors was eliminated.[130]

Offering an interpretation that adds a crucial political reason for the nuclear bandwagon's sudden departure, Hennessey concluded, "It thus appears that the types of assistance which the Act specifically limits to section 104 licensees [demonstration reactors], such as waiver of use charges, would not be disturbed by a finding of practical value ... if either a construction permit or an operating license had already issued under section 104. On the other hand," Hennessey warned, "a facility which had not obtained a section 104 construction permit or license before the finding was made would have to be licensed under section 103 [commercial reactor] and the prohibition against waiver of fuel use charges ... would appear to be applicable, even if the facility is the subject of a power demonstration contract containing a provision for waiver of such charges."[131] According to the AEC's general counsel, reactor developers could assure that their subsidies would continue if they made it under the wire before the commission determined that reactors had some "practical value."

There was no need to rush as it turned out. As long as even the AEC publicly acknowledged that nuclear power was far from competitive, the question of "practical value" remained moot. Nor was the AEC in any hurry to make such a determination. More than a year after the Atomic Energy Commission declared in its Report to the President that nuclear power was on the threshold of economic competitiveness, there were still no procedures in place governing the process by which "practical value" might be determined.[132] Fortunately, from the AEC's

126 Hennessey to Commissioners, February 12, 1964, in AEC 152/173, p. 23.
127 Ibid., p. 24.   128 Ibid.   129 Ibid., p. 25.
130 Hennessey to Commissioners, July 7, 1966, in AEC 152/203, AEC, DOE, p. 2, and "Responses to Questions Relative to Statutory Determination of Practical Value," in AEC 152/203, p. 14.
131 AEC 152/173, p. 25.
132 "Review of Commission's Position on a Finding of Practical Value," in AEC 152/173, AEC, DOE, p. 1.

perspective, the statute left the details of that process to the commission. As Hennessey interpreted it, "The absence of clear definition and limiting criteria in the statute and its legislative history leaves the Commission with considerable latitude in exercising ... discretionary authority."[133] Again the AEC would write the rules, present the arguments, weigh the evidence, and issue the judgment.

Frank Pittman, the AEC's director of reactor development, who would be responsible for providing the information on which practical value determinations would be made, was the first to offer his suggestions for the criteria to be used in determining "practical value." Pittman warned Hennessey not to jump to conclusions. Reactors should be reviewed on a case-by-case basis. "A general positive finding for a reactor class as a whole would presumably be made after firm evidence is accumulated to prove that, beyond reasonable doubt, the nuclear plants included in that class of reactors are indeed capable of competing with a reasonable share of the overall market for new electric plant capacity in a significant sector of the country." Pittman laid out the effective principle on which the AEC ultimately determined practical value. It was fine for the industry to invest billions of dollars on "extrapolation" of current knowledge," Pittman argued, which, because of lack of operating experience, involved "considerable conjecture" or "expectancy." But before the AEC eliminated subsidies and subjected the nuclear industry to the indignity of antitrust review, it needed far firmer evidence. As Pittman put it in 1964, "what we are suggesting is that prior to making a positive finding of practical value, AEC should experience a few years of operation with nuclear plants which are demonstrated to be economic or we should otherwise be in a position to conclude on the basis of facts rather than predictions that competitiveness has been achieved."[134] Quick to market commercial nuclear power as economically competitive, the Atomic Energy Commission was in no hurry to rule on its "practical value."

One unintended consequence of the AEC's successful marketing strategy, however, was to increase pressure on the AEC to rule that reactors did have practical value. Why else, asked the coal boys, were utilities spending hundreds of millions of dollars to construct them?[135] Thus when head coal boy Stephen Dunn, president of the National Coal Association, wrote to Glenn Seaborg in December 1963 and insisted that the AEC perform its statutory duty by issuing a written finding as to whether commercial power had "practical value," much of Dunn's evidence rested on statements made by the AEC and the Joint Committee on Atomic Energy.[136] Testifying at JCAE authorization hearings several months later, Joseph Moody, president of the National Coal Policy Conference, presented a powerful case, again built upon AEC and JCAE statements, of the competitive nature of nuclear power. To this rhetorical evidence, he added some

133 Hennessey to Commissioners, February 12, 1964, in AEC 152/173, p. 12.
134 Pittman to Hennessey, February 7, 1964, in AEC 152/173, pp. 27–8.
135 See AEC 152 series, and Johnson, "Coal Boys."
136 Moody to Seaborg, December 13, 1963, in AEC 152/171, March 9, 1964, AEC, DOE, p. 34.

impressive behavioral data: The utilities were investing heavily in nuclear power.[137] Several, Moody pointed out, had publicly stated that their reactors, when completed, would compete favorably with conventional fossil fuel–burning plants.[138] "The current sense of 'urgency' in expanding the commercial atomic industry," Moody told the JCAE, "represents a new, and inexplicable, phase in the various justifications that have been used for the government's program." Moody then succinctly summed up how the AEC's rationale for government subsidies had not so subtly shifted over the past two decades. "In the early stages we were assured that electric power made from atomic fission would be practically costless," Moody reminded the committee. When that failed, according to Moody, the AEC turned to "national prestige." Finally, Moody continued, "in the Report to the President in November, 1962, the Atomic Energy Commission projected a new 'justification' – clearly implying that there was an immediate and urgent need to develop a commercial nuclear power industry to meet projected national energy requirements."[139] Citing the Jersey Central estimates for unsubsidized competitive nuclear power as well as three other reactors that, with $42 million of AEC subsidies, were already under way, Moody concluded that "whether atomic power plants started now can produce electricity, with or without subsidy, cheaper than plants using coal for fuel, the fact is that the expenditure of more than $1.5 billion in government revenues has put the commercial atomic power industry close to being competitive."[140]

Moody demanded an immediate halt to the subsidies. He called on the Atomic Energy Commission to declare that reactors were of "practical value."[141] "There can be no question that they are of 'practical value' – or else they would not be built to supply commercial power," Moody fumed. He also demanded that the AEC charge the market rate for the commercial use of fuel, increase charges for other AEC-subsidized services, and eliminate government-guaranteed indemnification under the Price-Anderson Act.[142] As the National Coal Association's general counsel, Brice O'Brien, put it, "Thermal reactors have reached the point where they should be placed in the mainstream of commerce to stand on their own feet without the artificial stimulation and artificial distortion of taxpayer-financed subsidies."[143]

As the amount invested by private utilities in nuclear power climbed into the billions, the AEC's already strained efforts to argue that these were merely demonstration plants appeared increasingly comical. The joke, however, turned out to be on the utilities that invested in the great reactor bandwagon of the sixties. Like obtaining indemnification legislation, "practical value" rulings required that the AEC mount evidence that undermined the most basic assurances required for successful commercialization. Price–Anderson could not pass unless

137 "Statement of Joseph E. Moody to the JCAE," in AEC 152/171, March 4, 1964.
138 Ibid., p. 3.   139 Ibid., p. 5.   140 Ibid., p. 7.   141 Ibid., p. 11.
142 Ibid., pp. 11–15.
143 "Statement of Brice O'Brien to JCAE," March 4, 1964, in AEC 152/171, p. 17.

the Congress was convinced that reactors were so dangerous that private insurers would not take on the risk. Likewise, proving that reactors had no "practical value" required that the AEC demonstrate that its claims of economic competitiveness – claims crucial to selling nuclear power in the first place – were either false or, at a minimum, unfounded. Undoubtedly representatives from the commission, the nuclear industry, and the utilities winked as they passed around the AEC's "proof" that the reactors being constructed at that very moment were not economically viable. Expressions must have been similar when the AEC reported that catastrophic nuclear damage might reach such proportions that only the federal government could insure against potential accidents. Unfortunately for the nuclear industry, the litany of barriers to economic competitiveness cited in the AEC's "practical value" memos turned out to be far closer to the mark than the optimistic projections of the Report to the President.

Memos in the AEC 152 series, which contained no classified information but remained buried in the the Atomic Energy Commission's archives for decades protected by "official use only" stamps, constitute a veritable road map to why nuclear power failed to compete with conventional sources of power in the 1970s and 1980s. The AEC denied that nuclear power had any practical value by juxtaposing factors not previously considered in estimating costs against original AEC and industry projections. Pittman, one of nuclear power's staunchest boosters, raised some troubling questions when the topic was "practical value." He was the first to raise – albeit obliquely – a topic rarely discussed in the halls of the Atomic Energy Commission, and never mentioned publicly. Pittman urged that the utilities and manufacturers be required to provide information that would allow the commission to "formulate judgement with respect to assuring that plant costs (seller and purchaser) are based on reasonable commercial practices."[144] As the decade progressed the AEC was forced to spell out these considerations in no uncertain terms. By the middle of 1969, the AEC's general counsel, still trying to defend its failure to determine that reactors had some practical value, warned the commission that the "evidence that the San Onofre and Connecticut Yankee plants are performing in an economic fashion cannot be taken at face value because as 'turnkey' plants, the relationship between the cost to the utilities and the cost experienced by the Westinghouse Electric Corporation on Connecticut Yankee and Westinghouse and Bechtel on San Onofre is not known." The cautious general counsel was willing to venture a guess, however. "It is generally accepted," he continued, "that the 'turnkey' plant prices were not adequate to cover the cost to the supplier."[145]

By 1970, the argument was no longer oblique. In fact numbers, rather than words, stated it best. The AEC's general counsel included a remarkable chart in his April 1970 report to the commission on yet another scheduled hearing on

144 Pittman to Hennessey, February 7, 1964, in AEC 152/173, p. 29.
145 "Review of Commission's Position on a Finding of Practical Value," in AEC 152/247, August 6, 1969, AEC, DOE p. 22.

practical value. Data submitted on the capital costs of recently operational large nuclear plants was presented as follows:

Oyster Creek #1	$161/kw
9-Mile Point	305/kw
R. E. Finna.	166/kw
Dresden #2	115/kw

The report added the following note: "All of the above, except 9-mile Point, were purchased under turnkey provisions."[146]

In 1965, the year in which the Atomic Energy Commission issued its first ruling on practical value, such data did not exist. In fact the core of the commission's determination that light-water reactors had not yet demonstrated practical value rested on the absence of data. As a subsequent report to the commission put it, "In the 1965 determination, the Commission decided to exercise its discretion to await a reliable estimate of the economics based upon a demonstration of the technology and plant performance."[147] The AEC decided there had not yet been sufficient demonstration of "the cost of construction and operation" of light-water reactors to rule that they had achieved "practical value."[148] Lest the commission be misunderstood, however, the next paragraph of the 1965 determination clearly addressed utilities who might otherwise have been somewhat alarmed to discover they were investing billions in a technology that had no value. "It is entirely appropriate," the commission insisted, "for utilities and equipment companies to base their estimates of economics on something less than the demonstration of costs we require in applying section 102 of the Act."[149] The nuclear manufacturers couldn't have agreed more. As a summary of industry comment on the "practical value" question concluded, "General Electric Company, Westinghouse Electric Company, Allis-Chalmers Manufacturing Company and the Babcock & Wilcox Company testified that the uncertainties associated with the economics of nuclear power were within the realm of reasonable business risks, but that there has not been sufficient demonstration to warrant a finding of practical value."[150]

Rebuffed in 1965, the coal boys were back again in 1966. They cited another flurry of reactor announcements, and continued AEC statements citing the competitive advantages of nuclear power.[151] The AEC was not convinced, however. Its determination reiterated that the commission "should await a reliable estimate of economics based upon a demonstration of the technology and plant

---

146 "The Scheduling of a Practical Value Rule Making Proceeding," in AEC 152/265, April 7, 1970, AEC, DOE, p. 14.
147 "Review of Commission's Position on a Finding of Practical Value," in AEC 152/247, p. 4.
148 Ibid.
149 "Determination Regarding a Finding of Practical Value," in AEC 152/192, December 17, 1965, AEC, DOE, p. 5.
150 Ibid., p. 9.
151 "Review of Commission's Position," in AEC 152/247, p. 5.

performance."[152] Within days of denying that reactors had any "practical value," the AEC was circulating another document "for official use only." It was a draft report that would update the highly publicized Report to the President. Unlike the practical value determinations, which were never widely publicized, the supplement to the 1962 report would ultimately be highly publicized by the AEC. It was the optimistic statements that were more widely disseminated, statements like Alvin Weinberg's 1966 pronouncement that "nuclear reactors now appear to be the cheapest of all sources of energy."[153]

The draft report confirmed that nuclear suppliers and consumers understood that the AEC's doubts about the performance of nuclear power were for "practical value" determination purposes only. The first page of the draft report spelled this out, trumpeting that "during the four years since the 1962 report the promise shown then of a near-term place for nuclear power has developed beyond expectations. The trend in size of installed plants and industry confidence in its ability to construct economically competitive plants have led to a remarkable growth in orders for nuclear plants in the latter part of 1965 and throughout 1966."[154] Large nuclear units were being scheduled in the United States at "an astounding rate," according to the draft report.[155] The economic projections for these plants cited by the report might easily lead one to wonder why utilities would ever invest in a fossil fuel plant again. The largest reactors currently being scheduled projected costs far below Oyster Creek's already competitive costs. Investor-financed reactors projected costs ranging from "3.5 to 4.2 mills/kwh for baseload operation."[156] Publicly financed reactors were projected to produce power at the remarkably low cost of 2.6 to 3.2 mills/kwh.[157] This range of costs corresponded to the cost of power from large coal-burning plants paying from 17 to 25 cents per BTU (British thermal unit) for delivered coal at a time when the average U.S. price of coal was about 24 cents per million BTU.[158] The draft report cited another extremely promising economic indicator as well. "Since 1962 there have been a considerable number of repeat orders for nuclear electric plants."[159] By virtually any indicator imaginable, utilities could not have been more confident in nuclear power.

The U.S. Court of Appeals for the District of Columbia Circuit agreed. It was hearing a group of cases in which municipal power cooperatives charged that private utilities had excluded them from participation in the benefits of nuclear power. According to a memo written by the AEC's general counsel in the middle of 1969, that court "seemed impressed by the large number of nuclear plants for which utilities have obligated themselves and the impact which nuclear

152 Ibid.
153 Alvin Weinberg and Gale Young, *Proceedings of National Academy of Sciences* 57, no. 1 (January 15, 1967).
154 "1967 Supplement to the 1962 Report on Civilian Nuclear Power Program," AEC 152/212, December 23, 1966, AEC, DOE, p. 1.
155 Ibid., p. 18.　　156 Ibid., p. 19.　　157 Ibid.
159 Ibid.　　158 Ibid.

212 The Chain Reaction

power now has and is expected to have on the structure of the utility industry."[160] Listening to oral argument in litigation involving reactors being constructed by Duke Power, Vermont Yankee, and Philadelphia Electric, the court "appeared to have difficulty in accepting the proposition that the tremendous nuclear plant investment by business concerns (based on reasonable business risks) and its effects on the shaping of the electric utility industry did not warrant a finding of 'practical value.'"[161] As the memo noted, the AEC had promised the court to conduct another review of practical value by June 30, 1970.[162] Nor did the AEC have the degree of discretion as to the outcome of that determination that it had enjoyed in the mid-sixties. "The tenor of the questions from the bench," warned the AEC's general counsel,

> seemed to indicate that even if the Court at this sitting were to leave undisturbed the Commission's earlier "practical value" determinations, the Court would be unlikely to uphold a Commission nondetermination of "practical value" in the future rule making proceeding which the Commission is committed to hold. The Court may be troubled by a feeling that since nuclear power has for some time been a fact of life in the practical world of the marketplace, the Commission should now be willing to accept a much lower standard in defining "practical value" than it did five years ago.[163]

By April 1970, as the deadline for the AEC's determination was closing in, the general counsel reported that in the Statesville decisions the court of appeals "strongly suggested, if not virtually directed" a finding of practical value.[164]

Pressure from outside the nuclear community had finally forced the AEC's hand. There was far less at stake, however, for the AEC. In contrast to the general counsel's 1965 memo, which outlined a host of subsidies that might be lost should 'practical value' be determined, in 1969, the court of appeals found that the only remaining consequence of "practical value" was the question of antitrust review, or, as it put it, "prelicensing consideration of anticompetitive factors" – the very subject of the litigation before it. Was it pure coincidence that the AEC picked the middle of 1970 as the deadline for reconsidering the 'practical value' question? After all, 1971 was the year that the Private Ownership of Fuel amendments to the Atomic Energy Act ended distribution of fuel except by sale. By that date, contracts for the subsidized lease of fuel to reactors built under the commission's development program would have expired. And 1970 was the last year for which the AEC was committed to purchase plutonium produced in the power-generating process.[165]

For some in the nuclear community, forestalling antitrust review alone was well worth fighting for. The executive director of the Joint Committee on Atomic

160 "Review of Commission's Position," AEC 152/247, p. 21.
161 Ibid.   162 Ibid., p. 6.   163 Ibid.
164 "Scheduling a Practical Value Rule Making Proceeding," in AEC 152/265, p. 5.
165 AEC 152/215, March 21, 1967, AEC, DOE, p. 6, for fuel distribution; "Responses to Questions" in AEC 152/203, p. 14, for assurances to JCAE that "practical value" determination would have no practical impact on plutonium purchase program.

Energy, Edward Bauser, for instance, wanted to know what, if any, expertise the AEC had in the field of antitrust review. Would antitrust review be so time-consuming and burdensome, Bauser asked, "as to lead utilities to prefer other forms of generation, arrangements for which may not be subject to review?"[166] The AEC dutifully answered Bauser's questions. Its heart was still not in the review, but it could not ignore the Justice Department's increasingly aggressive stance on this issue. In May 1968 the assistant attorney general for antitrust was advised by his staff to intervene against the AEC on the Duke Power and Vermont Yankee cases. The staff memo was rather blunt on the question of practical value, arguing that "the Atomic Energy Act's intent that competitive issues should be considered in the licensing process should not be avoided by declining to find that large-scale commercial plants have commercial value."[167] When the AEC assured the Attorney General that it would review the practical value question before issuing operating licenses for the Duke Power and Vermont Yankee reactors, he agreed to support the AEC's position. From that point on, however, the AEC was committed to Justice Department review of antitrust matters, once "practical value" had been determined.[168]

This is not to say that the AEC had turned into a trust-busting tiger. When Bauser asked whether "the fact that the license applicant proposes to obtain the reactor vessel from a particular supplier would tend to establish or augment a monopoly position on the part of the supplier," the commission responded that although authorized to do so, should reactors be determined to have commercial value, "we would not expect to exercise such authority."[169] As far as the Justice Department was concerned, the AEC speculated, it already had legal authority to prosecute reactor manufacturers for antitrust violations. It was unlikely that Justice would choose to disrupt the licensing process rather than exercising its direct statutory authority. As the AEC understood it, "to deny a license to the applicant on [that] basis would be to penalize the applicant for something which was beyond his control."[170]

Regardless of the stakes involved, it was becoming increasingly difficult for the AEC to cling to its original basis for denying that 'practical value' had been achieved. By the late sixties, actual construction cost data and operating experience began to trickle in. Initially, at least in the highly insulated practical value venue, the AEC was almost gleeful as it bolstered its tenuous position with reports of operating problems and cost overruns. Reactor development chief Pittman was again the first to weigh in with a potentially

166 Bauser to Seaborg, December 19, 1969, in AEC 152/256, January 21, 1979, AEC, DOE, p. 27; quotation from p. 26.
167 Hennessey to Commissioners, May 15, 1968, in AEC 152/230, May 17, 1968, AEC, DOE, p. 3.
168 See, for instance, AEC 152/256.
169 "Proposed Responses to Questions on 'Practical Value,'" January 20, 1970, in AEC 152/256, p. 4.
170 Ibid.

costly set of factors that in previous estimates had conveniently been overlooked or brushed aside. As early as 1964, Pittman pointed out that before it could be determined that light-water reactors had demonstrated "practical value," in addition to the actual construction costs, including the true cost to the supplier in the case of turnkey plants, the AEC required firm data on the reactor's performance. Reliability, ability to sustain projected power levels, and the cost of spent-fuel processing, were all crucial variables in this calculation.[171]

By 1968, some of this data began to accumulate for two reactors – Connecticut Yankee and San Onofre. There were problems. Rather than operating at the 75 percent of its capacity originally estimated, a series of plant outages reduced San Onofre's plant capacity to 60.5 percent.[172] At Oyster Creek, the story was even grimmer. There were no data available because Oyster Creek had been delayed by work stoppages and problems with the reactor's control rods.[173] Nor was the news promising from abroad. Appending an appendix labeled "Typical Difficulties in Light Water Reactors," a 1968 status report on the practical value developments concluded that "it is evident why some utilities have expressed the view that a nuclear power plant has not matured until ... many of the practical limitations of the design have been revealed."[174] The news was better from Connecticut, where the Yankee Atomic reactor achieved a plant factor of 82 percent. Nevertheless, AEC staff reported that what little operating data existed was "limited and inconclusive."[175] There was no need for the commission to reopen the practical value question.

As the pattern of operating problems and cost escalation continued, however, the practical value memos began to sound more and more like the AEC's more public utterances: they began to ignore obvious problems and instead look on the bright side. In 1969, even though additional operating and construction cost data raised serious substantive questions about the ability of nuclear power to compete economically, AEC staff for the first time made the case that actual data from San Onofre and Connecticut Yankee could be interpreted in a way that concluded pressurized water reactors did have "practical value."[176] It is important to note that the staff – employing the standard "discussion paper" format – also mounted counterarguments, based on evidence from other reactors. Never before, however, had the staff even allowed that sufficient data might exist on which to base a positive determination, let alone suggest that the data pointed to a positive determination.

Two reasons – neither having to do with an objective analysis of the situation – lay behind this sudden shift in the AEC's position. The first was spelled out clearly in the AEC's own discussion paper: "the [Appeals] Court would be

171 Pittman to Hennessey, in AEC 152/173, p. 29.
172 Review of Commission's Position on a Finding of Practical Value, in AEC 152/228, April 5, 1968, AEC, DOE, p. 7.
173 Ibid., p. 8.   174 Ibid., p. 8 and Appendix.   175 Ibid., p. 8.
176 "Review of Commission's Position," in AEC 152/247, p. 18.

unlikely to uphold a Commission non-determination of practical value in the future rule making proceeding which the Commission is committed to hold," wrote the AEC's general counsel.[177] The AEC feared that if it did not move quickly on the question of 'practical value,' a decision would be imposed from outside the nuclear community – always the commission's worst nightmare, whatever the specific isssue involved.

The second, and in the long run more consequential, reason that AEC staff suddenly found practical value in San Onofre and Connecticut Yankee was also alluded to in that August 1969 memo, but requires some reading between the lines to interpret. Laying out the arguments against proceeding with hearings to determine practical value, the general counsel pointed to data that offset the supposedly good news at San Onofre and Connecticut Yankee. Maintenance problems had increased in Westinghouse's European reactors, the staff argued.[178] Despite the supposed economic competitiveness of San Onofre and Connecticut Yankee, the memo also noted that orders for reactors in the United States had fallen off precipitously. "[T]he dearth of utility commitments during CY [calendar year] 1969 should be considered," the general counsel warned.[179] "It is evident," he continued, "that delays in nuclear power plant construction, delays in delivery of key reactor components, inadequate scheduling and planning, rising interest rates, and other cost escalations have resulted in marked increases in the capital cost of nuclear plants."

These were statements that could no longer be winked at. Nor had the industry been doing much winking ever since the great reactor bandwagon came to a grinding halt. An examination of the data on which the staff recommended that Westinghouse reactors could be considered to have "practical value" – particularly in the case of San Onofre – only raised more questions about the nature of this recommendation. San Onofre's plant factor had plunged. From the time of its initial operation through the end of April 1969, the plant had generated only 36.1 percent of the electricity it was theoretically capable of producing.[180] Even during its most trouble-free six-month period, the operating factor was slightly below the manufacturer's original target.[181]

By the AEC's own estimates, the adjusted generated costs at San Onofre came out to 7.0 mills per kwh. This exceeded by more than 1 mill per kwh the costs of all but one of the conventional stations run by Southern California Edison and San Diego Electric. The average costs for electricity produced at those stations ranged from 4.9 kwh to 6.7 kwh. As telling as the gap itself and as distorted as the conclusions drawn from it, were the factors that went into the AEC's San Onofre estimate. One reason, the AEC had argued for five years, that practical value could not be determined was the absence of operating data. San Onofre had finally begun to provide some of that data. Yet in making its estimate, the AEC

177 Ibid., p. 21.   178 Ibid., p. 24.   179 Ibid.   180 Ibid., p. 8.
181 Ibid., pp. 8 and 15 for comparison.

did not base the plant factor on the long-awaited actual data. Rather than use that data – which had averaged out to 36 percent since San Onofre's initial operation – or even an adjusted figure that fell within the range between that figure and San Onofre's top figure of 72 percent for the period September 1968 through April 30, 1969, the AEC used a figure that was even higher than San Onofre's originally projected 75 percent. It assumed that the reactor would operate at a plant capacity of 80 percent.[182] Apparently, "considerable conjecture" and a high degree of "expectancy" was back in fashion.

Had the context been the highly publicized Report to the President or some other promotional forum, this kind of sleight-of-hand would not have been surprising. In the context of the practical value debate, however, intentionally understating the true cost of nuclear power, and then interpreting those costs – even as understated – as being competitive with conventional plants when they exceeded "by more than one mill per kilowatt hour" all but one conventional source in Southern California, signaled a dramatic shift in the AEC's priorities.[183] It was one thing to fight a quiet rearguard battle against the end to subsidies when the reactor market was booming. But with reactor sales dead in the water, with utilities well aware of the operating problems and cost overruns that recent plants had experienced, and with the threat of a "practical value" determination being imposed by the courts in any case, the AEC could no longer afford to sustain the paradox it had created by the mid-sixties. Development, even in 1970, could no longer be assumed. As the AEC moved to end its mixed message about the competitive advantage of nuclear power, however, it discovered that its word no longer carried the weight it once had with utilities. The utilities themselves were once again asking whether nuclear power really did have any practical value. Nor, by 1970, did debate remain confined within the nuclear community.

Increasingly a part of the economic mainstream, nuclear power was beginning to attract more regular attention from the broader political community. Its integration hastened the diffusion and proliferation of technical expertise, extending it beyond the AEC and manufacturers to utilities which increasingly had a major financial stake in nuclear development. Implementation also broadened the type of expertise required, as fields like utility economics became crucial to decisions affecting further development. Both these trends further eroded the AEC's monopoly of expertise by creating independent bases of expertise and introducing types of expertise far afield from the AEC's original mission.

The analysis required for planning a nuclear facility, for instance, raised the general level of sophistication of the utility's planning groups. In fact, as one student of the variables that distinguished nuclear success from failure observed, "Any utility that failed to involve itself directly in every aspect of the project was

182 Ibid., p. 16, for 80 percent plant factor assumption.
183 Ibid., p. 17.

likely to end up with a mismanaged project." Of course, that did not guarantee success. "Lacking knowhow, utilities hired experts to manage nuclear projects," James Cook continued. "But the experts themselves all too often bungled the job."[184] As hard data became available, the utilities probed in greater detail. Philip Sporn, president of the American Electric Power Company, had been urging a more careful consideration of actual performance for years. As Sporn had pointed out in 1962, "Excitement and preoccupation with nuclear fission as a new energy source obscured the economics of nuclear power.... There has been a disposition to sweep certain difficult or unpleasant facts connected with nuclear technology under the rug."[185]

In testimony submitted to Joint Committee on Atomic Energy hearings in 1962, Sporn contended that although the cost of nuclear power was declining, it was competing against a "moving target." The cost of conventional fuel conversion technology was declining also, Sporn pointed out, and promised to produce electricity more economically than nuclear sources at least through the 1970s. Sporn did not hesitate to point out the political implications of his economic analyses. As he told the American Institute of Electrical Engineers, "You cannot push a particular form of energy ahead of its time or raise it to a position it cannot occupy by virtue of its own basic weight or force."[186]

Sporn was ahead of his time again in 1967, predicting an end to the great reactor bandwagon just when business had never been brisker for reactors. As he wrote the Joint Committee on Atomic Energy, "Nuclear power, despite a number of technological innovations, stands today substantially no further advanced in its progress toward lower costs than it was 2 years ago."[187] Sporn noted that some reactors had been sold because manufacturers had cut prices owing to unusually intense competitive pressures.[188] As the utility chief gently put it, "the manufacturer risked somewhat greater uncertainty in his turnkey price than was tolerable on a repeated basis ... nuclear power could not repeat the cost levels it seemed to have achieved at Oyster Creek."[189] In December 1969, Sporn again reported to the JCAE and the news was even bleaker. "During the past two years there has taken place a remarkable and ominous retrogression in the economics of our nuclear power technology."[190] Despite the fact that fossil costs had gone up,

184 James Cook, "Nuclear Follies," *Forbes* 135, no. 3 (February 11, 1985): 92.
185 Gandara, *Electric Utility Decision*, pp. 87–91; Sporn testimony in JCAE, "202" Hearings, 1962, p. 689.
186 "202" Hearings, 1962, p. 689; Speech to American Institute of Electrical Engineers, January 29, 1962, quoted in *Nucleonics* 21 (March 1962): 25.
187 Sporn to John Pastore, December 28, 1967, in AEC 152/226, February 28, 1968, AEC, DOE, p. 9.
188 Ibid., p. 11.
189 Ibid. The director of the President's Science Adviser's energy staff wrote in early 1969 that the turnkey reactors had failed to make a profit. "As losses mounted, the manufacturers abandoned the turnkey approach and sharply increased the price of nuclear reactors and fuel" (Freeman to Whitehead, March 4, 1969, in AEC 152/243, July 11, 1969, AEC, DOE, p. 29).
190 "Developments in Nuclear Power Economics January 1968–December 1969," in AEC 152/257, January 26, 1970, AEC, DOE, p. 1.

nuclear power had "lost position vis-à-vis fossil fuel," according to Sporn.[191] This made it difficult, Sporn advised the JCAE, "to accept without something more than a grain of salt the statement of the Atomic Energy Commission 'the outlook for the future for nuclear power continues to be very promising [because] of the continuing economic competitiveness of nuclear power.' "[192]

By the late 1960s, another component of nuclear power – its costs – that had remained the relatively exclusive domain of expertise within the AEC and a few manufacturers, was accessible to the most significant consumers – in this case the utilities. As costs from their initial projects came in well beyond the initial projections, the reactor 'bandwagon' came to an abrupt halt. It would take skyrocketing conventional fuel costs in the early 1970s to get it rolling again. Paradoxically, it was at this juncture that the question of "practical value" was finally resolved. The Joint Committee on Atomic Energy and the AEC had been contemplating a legislative resolution for a number of years. In December 1970 legislation was passed to eliminate the distinction between "demonstration" and "commercial" plants.

Before the issue was finally resolved, however, the leading legislative critic of the Atomic Energy Commission's position on "practical value" in the sixties had managed considerably to broaden the scope of debate. That critic was Senator George Aiken of Vermont. Compared to the delicacy with which charter members of the nuclear community danced around the issue of practical value, Aiken's approach was refreshingly blunt. In October 1969 he wrote to Glenn Seaborg to remind the AEC chair that "for more than a year I have been warning that there would be a nuclear gold rush to get as many plants as possible licensed as research projects to escape antitrust scrutiny."[193] Although Aiken introduced legislation to require such scrutiny, it did not make it out of the Joint Committee on Atomic Energy. In 1970, Aiken increased his pressure and upped the ante. He took his case beyond the nuclear community and addressed his fellow senators. Aiken argued that the total capital investment in nuclear plants built, under construction, and on order was close to $15 billion. Equally important, Aiken continued, "is the fact that more than $2.3 billion of Federal money was invested in the advance research that made atomic power reactors profitable."[194] Aiken urged the AEC to determine that light-water reactors had "practical value."

What made Aiken's criticism unusual was that he was a member of the Joint Committee on Atomic Energy. Though the JCAE had been sharply divided in the past vis-à-vis the AEC as well as within its own ranks, with few exceptions it insisted that matters related to nuclear power be settled within the nuclear community. Aiken broke with this tradition. In his speech on the Senate floor,

191 Ibid.   192 Ibid.
193 Aiken to Seaborg, October 28, 1969, in AEC 152/252, November 5, 1969, AEC, DOE, p. 3.
194 Aiken statement to press, March 4, 1970, in AEC 152/261, March 12, 1970, AEC, DOE, p. 6.

he violated the cardinal rule of nuclear policymaking. He threatened to broaden the scope of debate. "If the private utilities and their friends persist in their efforts to set up a vast nuclear monopoly," Aiken warned, "then I see no recourse but to turn the regulation of these generating plants over to the States. In the meantime, we need a thorough investigation of the antitrust aspects of all electric power generation by a Committee of the Congress that is expert in antitrust law."[195]

A relative newcomer to the Joint Committee and a defender of the cooperative power arrangements that dotted the New England countryside, Aiken was not intimidated by the AEC's aura of expertise. He was not unaware of nuclear power's history and the crucial role played by experts in its development. As Aiken told his colleagues, "We would not have competitive atomic power today had it not been for the brilliant work of the AEC scientists in close cooperation with American industry."[196] That very success required an adjustment in the way America approached this policy, however. With the emergence of competitive nuclear power, the AEC, according to Aiken, "found itself suddenly thrown into a strange new environment – the rugged American marketplace. The atomic sicentists were no longer in their cloistered laboratories at Oak Ridge and Argonne – they were in the mainstream of American competition and were not very good swimmers."[197]

Experts rose to the top in the early sixties and, in the AEC, provided important momentum for light-water reactor technology. But the political stock of expertise was rising elsewhere as well. Newer technologies such as the space program drained funding from reactor development. A newly institutionalized presidential science advisory apparatus pushed steadfastly for a more comprehensive approach to energy development. The AEC's Report to the President sought to refurbish the image of an aging technology while regaining its monopoly on expertise through the still untested field of converter and breeder reactors. The driving force behind both strategies, however, was a boundless optimism. Some critics, like David Lilienthal, raised questions about the corollary to successful commercialization – its negative externalities and effective political representation for those affected by them. But such externalities were visible to only a few, most of whom worked for the AEC, which was hardly eager to publicize them, or to issue networks increasingly concerned within professional circles, but halting in their appeal to the broad public. Caught up in the belief in new technology and soothed by industry–AEC cost projections, utilities jumped on the bandwagon. By the mid-sixties nuclear power had successfully entered the economic mainstream. Its very success, however, generated a far broader array of interests and demands for political accountability as the paradox of "practical value" demonstrated. Senator Aiken publicized that paradox because he wanted to open

195 Ibid., p. 17.   196 Ibid., p. 16.   197 Ibid.

up nuclear power to municipal utilities. The political path that he followed, however – one that relied on broadening the scope of debate to achieve policy objectives – could be used toward different ends. It ultimately was used to destroy the consensus that the development of commercial nuclear power was beneficial. Along the way, a more powerful consensus that experts should be on top was destroyed as well.

8

Nuclear experts everywhere:
the challenge to nuclear power, 1960–1975

Nuclear power's enhanced visibility in the early sixties pushed into the limelight Atomic Energy Commission experts charged with ensuring and evaluating reactor safety. It also created record demands for their services. By 1970, thanks in part to the federal government's ongoing effort to produce more technical experts and in part to the AEC's organizational specialization, there were competing centers of safety expertise within the Atomic Energy Commission. The Advisory Committee on Reactor Safeguards – the dominant source of safety expertise in the fifties – jostled with several newly developed sources of safety expertise. The regulatory staff was one. The fledgling Hazards Evaluation Branch, which was the technical core of the Division of Licensing and Regulation (Regulatory Staff), had acquired more resources and higher stature during the early sixties. Commercialization also spawned another network of safety expertise dominated by the Atomic Energy Commission comprising the national laboratories (and their contractors) that tested through engineered experiments the theoretical calculations of the AEC's safety experts.

The proliferation of expertise and expanding number of organizations charged with safety review within the Atomic Energy Commission was part of the larger evolution that the nation's nuclear promoter hoped to complete, ultimately emerging as the nation's nuclear watchdog. As Glenn Seaborg pointed out to presidential science adviser Donald Hornig in September 1968, the AEC had scaled back its efforts to assure the economic success and reliability of light-water technology. In the past few years, Seaborg continued, the AEC had emphasized "the development and establishment of criteria, codes, and standards" regarding light-water reactor safety. It had also revamped its organization to strengthen its regulatory role.[1] In a letter to the Joint Committee on Atomic Energy a year earlier, Philip Sporn, the retired director of the American Electric Power Company, anticipated Seaborg's points, stating them more emphatically. "Without question," Sporn told the JCAE, "the Commission activities have been a major factor in the by now established atomic industry and the Commission can relax about the

---

1 Seaborg to Hornig in AEC 152/238, September 30, 1968, p. 6, AEC, DOE.

future of atomic power." But the commission's job was far from done, according to Sporn. The commission "needs to build up its resistance to any appeals to weaken its concern and its still developing regulations for safety," Sporn warned. Rather, he continued, "it needs to continue to microscopically examine every possible flaw in the chain mail armor of safety and, when detected, mend it."[2] As the AEC phased out its development role in light-water technology it prepared for technical review of the commercial applications that soon flooded in by increasing the size and skill of its in-house Safety Staff.[3]

Both the Advisory Committee on Reactor Safeguards and the Regulatory Staff demanded far more data on the possible causes of accidents, their probability, their consequences, and the reliability of safeguards mandated by the AEC. The national laboratories and their contractors responded to these demands. As Frank Pittman, director of the AEC's Division of Reactor Development, put it: "The main purpose of this whole program is to take the assumptions out of our safety analysis and put into these safety analyses some factual data." It had been the vision of the original General Advisory Committee that the national labs would provide a parallel track and ultimately supplant the ACRS as the agency's safety experts. The General Advisory Committee hoped that supplanting the Reactor Safeguard Committee with an organization concerned with both development and safety would lead to more realistic safety appraisals. Or, as Pittman stated in 1964, "We want to gather information on a scale large enough to convince those who have to be convinced that we have safe systems."[4]

The General Advisory Committee, Pittman, and a host of other nuclear optimists underestimated the power of specialization – both organizational and heuristic – to fragment the kind of quiet consensus on which the authority of experts ultimately rested. The emergence of a more elaborate Regulatory Staff, and the stepped-up role of national laboratories and their consultants who were charged with experimental verification of safety assumptions, made quiet consensus more difficult to achieve. Chapter 5 demonstrated how crucial this kind of consensus was to the members of the Advisory Committee on Reactor Safeguards.

2 Philip Sporn, "Nuclear Power Economics – Analysis and Comments – 1967," in U.S. Congress, Joint Committee on Atomic Energy "Nuclear Power Economics – 1962–1967," 90th Cong., 2d sess., February 1968, in AEC 152/226, February 28, 1968, AEC, DOE, p. 18.

3 As early as 1962, AEC staffer Don Steward complained to the Joint Committee on Atomic Energy that the Civilian Power group of the Reactor Development Division was operating with twenty-two people, compared to Regulatory Staff's eighty. Civilian Power employed only fifteen technical people; the Regulatory Staff, thirty-eight (Edward J. Bauser to Files, 4/11/62, Box 99, JCAE Files). For a summary of formal JCAE complaints about development staffing and AEC estimates of total FY 1962 and FY 1963 reactor development staffing, see U.S. Congress, Joint Committee on Atomic Energy, "Hearings on AEC Authorization Legislation, FY 1963," 87th Cong., 2d sess., 1962, pp. 165–71. For an excellent administrative history of Regulatory Staff changes during this period, see Mazuzan and Walker, *Controlling the Atom*, ch. 13.

4 Pittman testimony before U.S. Congress, Joint Committee on Atomic Energy, "Hearings on Authorization Legislation, FY 64," 88th Cong., 1st sess., 1963, p. 89. Pittman was referring specifically to the Safety Test Engineering Program at the National Reactor Testing Station. Pittman testimony before U.S. Congress, Joint Committee on Atomic Energy, "Hearings on AEC Authorization Legislation, FY 65," 88th Cong., 2d sess., 1964, p. 398.

Because the ACRS so thoroughly dominated the safety debate up to the mid-1960s, the political challenge posed by the public exposure of serious safety questions revolved primarily around maintaining consensus within that body. Like the concrete shells that protected against the spread of radioactivity should an accident occur, this style of policymaking might best be described as political "containment": it was intended to keep politically damaging fallout inside as small an area as possible.[5] With a relatively small number of safety experts, and only one respected organizational forum for resolving such matters, debate was, in fact, contained.

Even in the wake of the Fermi reactor controversy and subsequent legislation passed in 1957 requiring public hearings, the ACRS retained a virtual monopoly on the safety review of specific reactor proposals. Yankee Rowe, a 135-megawatt pressurized-water Westinghouse reactor constructed in rural Massachusetts for the Yankee Atomic Electric Company, and the first reactor subjected to the public hearings process, breezed through the new review procedures with virtually no discussion of safety.[6] All serious discussion occurred within the ACRS. That advisory body issued a clean bill of health in late September 1957, and the reactor received a provisional construction license by early November.[7]

A few individuals did raise some general safety-related questions.[8] These were dismissed by the hearing board and the commission. The most telling evidence of the cursory nature of the AEC's safety review beyond the ACRS was the proceeding's timing. The first hearing was held barely two weeks after the ACRS issued its highly technical report. John Ward, a Boston consultant, had still not obtained a copy of the report three weeks after it was issued, and one week after the first hearing.[9] Nevertheless, in approving the provisional construction permit, the AEC determined that despite the short time-frame for review, the statutory "criterion of reasonableness has been met."[10] At Yankee Rowe's operating license hearing in the spring of 1960, the hearing officer, Samuel Jensch, did hold up approval for forty-five days to probe two amendments to the original plan submitted by Yankee Atomic. The delay, however, was to ensure that the ACRS had considered these changes, not to subject the reactor's safety to broader scrutiny.[11]

5 Daniel Ford has also used the term "containment" in a political sense in his *Cult of the Atom*.
6 For a discussion of the cursory nature of the AEC's early public hearings, see Steven Ebbin and Raphael Kasper, *Citizen Groups and the Nuclear Power Controversy: Uses of Scientific and Technological Information* (Cambridge, Mass.: MIT Press 1974), ch. 1.
7 Stratton to Strauss, September 16, 1957, and chronology of documents, Docket no. 50–29 file [hereafter Yankee Rowe], Public Document Room, U.S. Nuclear Regulatory Commission [hereafter NRC], Washington, D.C.
8 For a summary of questions by individuals making limited appearances (under Section 2.731 of the rules of practice), see "Memorandum of Opinion of the Commissioners," November 5, 1957, Yankee Rowe, NRC.
9 John Ward to USAEC, October 13, 1957, Yankee Rowe, NRC.
10 "Memorandum of Opinion," November 4, 1957, Yankee Rowe, NRC.
11 Mazuzan and Walker, *Controlling the Atom*, pp. 384–5.

This was characteristic. As the official historians of the Nuclear Regulatory Commission point out, "Aware of the gravity of his responsibilities, conscientious in his exercise of authority, and lacking easy access to technically qualified advisers, Jensch elected in such situations to refer the matter back to technical experts."[12] Until the mid-sixties, those technical experts were virtually all housed in the ACRS.

By the mid-sixties specialized missions and hardening organizational boundaries reshaped the dialogue within the Atomic Energy Commission's safety community. The newer actors as well as the ACRS remained commited to quiet discussion of safety problems leading to a consensus for public consumption. Only after achieving such a consensus should the decision be exposed to broader public scrutiny, they maintained. Yet each actor, for its own political reasons, on occasion sought to broaden the debate before a consensus within the AEC had been achieved. The early breaches in "containment" were not necessarily the result of explicit political maneuvering: the increasingly elaborate organizational environment created its share of crossed signals. In October 1964, for instance, the Advisory Committee on Reactor Safeguards issued a report favorable to the construction of a reactor at Bodega Head. At the same time, the Regulatory Staff's summary analysis concluded that Bodega Head was not a suitable location owing to seismic considerations. As we shall see later in this chapter, these contradictory reports – the first instance of such a public conflict between AEC safety experts since the Fermi reactor controversy – led directly to the cancellation of the project.[13]

Two substantive issues – reactor siting and size – activated the structural faults chiseled by proliferating experts and articulating organizations within the Atomic Energy Commission. Metropolitan siting and dramatically scaled-up reactors were crucial to the industry's future economic growth. Building reactors closer to the source of electrical demand, the utilities hoped, would lower transmission costs. Larger reactors would capitalize on economies of scale. On the horizon lay the mirage of diminished safety costs. To aggressive proponents of larger reactors and urban siting, the AEC's siting and engineered safety criteria were too rigid in the fifties. Relaxing the AEC's safety requirements would make nuclear power far more economically competitive, they argued.[14] James Fairman, vice-president of Con Edison, for instance, concluded that the hazy image on the horizon might well be an oasis. As he told the industry summit on nuclear power in December 1957, "private utilities will be able to substantially reduce high capital costs as safety measures are learned by actual operation. Present containment costs are high. Experience may show all of this is not needed."[15]

12 Ibid., p. 385.
13 Okrent, *Nuclear Reactor Safety*, pp. 268–72. As we shall also see, the Regulatory Staff faced challenges from outside experts serving various federal agencies and citizen groups.
14 See Chapter 5.
15 "Highlights of Industry Meeting," December 3, 1957, attached to Zehring to Strauss, December 9, 1957, "Power, General" folder, Box 84, LLS, HHPL, p. 12.

More experienced observers, however, concluded early on that urban siting and larger reactors could only increase the cost of safety measures. George Weil, for instance, delivered the following warning to the Atomic Industrial Forum in September 1954: "[N]uclear power plants can be exceedingly hazardous, much more so than any other industrial operation with which we are familiar. This does not mean that they have to or will be hazardous. It does mean that if the industry is going to develop and grow to play a major role in our national economy, the safeguards, perhaps costly, must be provided in the design of plants."[16] Philip Sporn was also realistic about the costs of larger, more urban reactors. He spelled out the economic implications of the AEC's evolution to regulatory watchdog in 1967. As he told the Joint Committee on Atomic Energy,

In effect, for the next 10 years at any rate [the AEC] needs to assign a significant portion of the gains in economic position of nuclear power – which all of those who have a solidly founded faith in the future of atomic power must believe in – as they are brought about, to safety. It needs never to lose sight of the fact that the greatest enemy to the future of atomic power is one major accident improperly handled by the safety system.[17]

As Weil's predictions proved true in the sixties, and as the nuclear industry was in fact forced to devote greater resources to safety, just as Sporn had called for, utilities committed to the vision of reduced safety costs as articulated by Fairman blamed excessive federal regulation, not the inherent precaution required by larger, more urban reactors, for nuclear power's economic crisis. Crucial to understanding the declining authority of experts in the late sixties and the seventies, however, was the process that aired publicly for the first time these conflicting perspectives.

"CONTAINMENT" BREACHED FROM WITHIN: THE ARTICULATION OF AEC SAFETY EXPERTISE

The metropolitan siting debate soon brought far more explicit demands from within the safety community for a position that could be defended in public. Metropolitan siting was the thorniest problem addressed by the safety experts, and it did not go away, even after the most controversial proposal – a huge reactor to be built across the river from Manhattan (Ravenswood) – was withdrawn. For one thing, Con Ed talked for years of resubmitting the controversial Ravenswood project, which it had dropped in the face of united Advisory Committee on Reactor Safeguards and Regulatory Staff opposition. Also, Boston Edison informally proposed in 1965 a metropolitan reactor site on the outskirts of Boston. The general issue of urban siting thus stimulated considerable debate

16 Weil, "The Hazards of Nuclear Power Plants," September 28, 1954, in "Weil, George" folder, Box 119, LLS, HHPL, p. 9.
17 Sporn, "Nuclear Power Economics – Analysis and Comments – 1967," p. 18.

among the members of the ACRS and between the ACRS, the Regulatory Staff, and the commission.[18]

Harold Price, director of the Regulatory Staff, was hardly one to preach the merits of public debate. He had a long record of opposing public disclosure, claiming that it would be misconstrued and misused by individuals outside the nuclear community. His Regulatory Staff, however, was caught between pressure from manufacturers and utilities on the one hand, and their own growing concerns about the safety of metropolitan siting on the other. They sought to retard – in some legally defensible fashion – the siting of large reactors in urban settings. Price stated that he wanted to avoid a drawn-out, detailed design review that might well end in a decision to turn the reactor down. He sought to avoid the case-by-case review that in his opinion would leave applicants in "a state of confusion and uncertainty." A publicly pronounced moratorium on metropolitan siting was Price's proposed solution to this dilemma.[19]

A subcommittee of the Advisory Committee on Reactor Safeguards opposed an outright moratorium, concluding that "location of reactors close to cities should not be categorically discouraged or encouraged by the AEC." The subcommittee, with one member dissenting, voted to reject Price's proposed public statement. Both the ACRS subcommittee and the Regulatory Staff seemed to agree that some criteria for metropolitan siting should be developed. They differed sharply, however, on how to stem the immediate pressure from manufacturers.[20]

As head of the Regulatory Staff, Price had another matter to consider: He was responsible for promulgating the formal regulatory criteria by which reactor proposals would be judged. But neither he nor the ACRS were in a position to spell these out. The ACRS, which had always relied heavily on its expert judgment as opposed to formal criteria, sought to retain as much flexibility as possible. Price, on the other hand, wanted the force of agency policy – in this case the blunt instrument of a moratorium – behind him.[21]

18 For a comprehensive account of the Ravenswood controversy, see George T. Mazuzan, "Very Risky Business: A Power Reactor for New York City," *Technology and Culture* 27, no. 2 (April 1986): 262–84. On Boston Edison site, see *Boston Herald*, July 21, 1965, "Edison Expected to Give Go Ahead on Nuclear Plan." The three possible sites mentioned publicly were Everett, Weymouth, or Squantum. See also *Christian Science Monitor*, May 10, 1966, "Hub A–plant Study Stresses Safety."

19 For a summary of Price's remarks, see the Minutes of the Subcommittee Meeting on Reactor Siting, March 20, 1965, in "The Minutes of the Advisory Committee on Reactor Safeguards," Public Document Room, Nuclear Regulatory Commission, Washington, D.C. [hereafter ACRS Minutes].

20 Ibid.

21 When the full ACRS met on March 26 and 27, 1965, to discuss metropolitan siting, one member pointed out the parallel between the current discussion of metropolitan siting and the development of site criteria several years ago (see ACRS Minutes, March 26–7, 1965; also see Chapter 5 for an account of the evolution of siting criteria). He noted that "the criteria were formulated primarily by the Regulatory Staff and agreed to, somewhat reluctantly, by the committee." The ACRS would have to devote some detailed attention to this from the formulative stage. Several members saw the present impasse as an opportunity to state that "large reactors can be built and operated safely in populated areas, but much better efforts towards safety...will be required" (ibid.).

At the next ACRS meeting Price suggested that Seaborg might make a statement on metropolitan siting at the upcoming congressional hearings. At the same meeting, the ACRS went on record in opposition to the public moratorium statement. Seaborg, too, balked at the idea of such a public action. As the meeting minutes recount, "the chairman was uncertain as to how formal any guides on such reactor locations might be – these might never be a matter of public record." In executive session, one ACRS member contended that even Clifford Beck – Price's deputy – was against any public announcement, preferring to convey information to the reactor industry informally. ACRS members suspected that Price was trying to force the commission's hand into postponing action on large reactors.[22]

Price made his reasons quite clear: The Regulatory Staff (which had privately expressed doubts about the safety of such reactors) needed a defensible public posture. The Regulatory Staff was being pressed by Westinghouse, the Boston Edison Group, and the Consolidated Edison Group (Ravenswood reactor) for advice on large reactors in metropolitan areas. Price added another political consideration. "Delaying the proposals for the large reactors in populated areas until after the Price Anderson hearings," Price told his colleagues, "might avoid the coal interest objections to the indemnity extension."[23] In other words, a moratorium on metropolitan siting might buy the nuclear industry another decade of federal protection against liability for nuclear accidents.

In defending his plea for a public statement, Price also implied that taking a policy position in public now might reduce the likelihood of public controversy later. He reminded the Advisory Committee on Reactor Safeguards that in the case of ocean waste disposal studies, near-shore sites were mentioned publicly, and although the AEC did not intend to approve any of these sites, "much reinterpretation and time consuming Regulatory Staff correspondence resulted." Price hoped to avoid replicating this sequence with the metropolitan siting issue. He was, however, overruled by the commission; no public position was taken. The Boston Edison site was withdrawn and Ravenswood was not resubmitted.[24]

The dispute between Price's Regulatory Staff and the ACRS was not a technical one. Both organizations had their doubts about the safety of metropolitan siting in the near future. But the ACRS did not feel the need to express doubts publicly or to rule out metropolitan siting categorically. ACRS member Richard L. Doan summed up his colleagues' position: "No one wanted to say that metropolitan sites were unsuitable for reactors, but metropolitan sites would come sooner if the industry recognized that present designs were not acceptable for metropolitan sites and put some effort into improving their design." Harold Price, on the other hand, preferred to protect his staff by just such a public statement, and in

22 ACRS Minutes, March 26–7, 1965.
23 Ibid. "Price–Anderson" was the short title for the indemnity legislation, passed in 1957, that provided federal indemnification for reactor accidents that caused damages above $560 million; private insurance and federal insurance covered damages up to that amount.
24 Ibid.

doing so, he hoped to avert the even wider-ranging discussion he felt a series of site rejections might provoke.[25]

There were better ways, however, to avoid wide-ranging discussion. In retrospect it is clear that the AEC/ACRS position was the politically shrewd one. That is because the AEC, in both cases, managed to block the reactor its own experts considered to be unsafe, without ever disclosing these concerns. This left the door open for future urban siting. The *Christian Science Monitor*'s coverage of Boston Edison's proposed metropolitan site, for instance, illustrates just how this politics of "containment" worked. The *Monitor* presented Boston Edison's proposed reactor as "the world's first in-city nuclear power plant."[26] Dutifully, the *Monitor* reported that a previous urban site – Ravenswood – had been proposed by Con Ed in 1962. Quoting a Con Ed spokesperson, the *Monitor* cited cheaper Canadian power – not stiff AEC opposition – as the reason the plant was not built. The article never mentioned, nor could those outside the AEC know, that there was virtually unanimous opposition to Ravenswood within the AEC's safety community.[27]

Public opposition to Ravenswood was mentioned, if only to be immediately dismissed as irrational. It was packaged in the characteristic people-versus-the-experts framework. Interviewed by *Power* magazine in 1965, Kenneth A. Roe, president of Burns and Roe (a firm that contracted at Hanford and Oyster Creek) reiterated the nuclear industry's anthem, as had Glenn Seaborg and many other nuclear boosters. "Technically knowledgeable people understand that the operating record of atomic plants shows them to be as safe, if not safer than any... conventional plant. I believe that this information can be conveyed to the public and that in time, people's confidence will grow so that nuclear plants may ultimately be erected wherever economics dictate."[28] Quoting Donald Quinn, a nuclear engineer for Boston Edison, the *Monitor* presented public opposition to Ravenswood in this context. "The basic cause of objection to the nuclear plant," Quinn told *Monitor* readers, in May 1966, "is simply public ignorance of the facts of nuclear power. It's fear based on lack of education."[29]

Six months later, in January 1967, the *Monitor* updated Boston Edison's progress. The proposed metropolitan reactor had been moved to another site – one surrounded by far lower population densities – in Plymouth, Massachusetts. Like the rest of the nation's press, the *Monitor* had not been privy to the AEC's internal debate. Thus it again turned to a Boston Edison spokesperson to explain the reason the reactor had been moved. Charles F. Avila, Boston Edison president, explained that the reason for the move was that the AEC had "not established criteria for plants in metropolitan areas."[30] The *Boston Traveler* quoted a company

25 Doan quotation from ACRS Minutes, September 20, 1965.
26 "Hub A-plant," May 10, 1966.
27 Ibid.
28 Kenneth A. Roe, "What Lies Ahead for Nuclear Power?" *Power*, July 1965, p. 58.
29 "Hub A-Plant," May 10, 1966.
30 *Christian Science Monitor*, January 5, 1967, "Plymouth Cheers Atomic-Plant Plans."

spokesperson elaborating on this point. "[W]e couldn't wait because we're going to need the power from the plant by 1971. We couldn't tell how long it might be if we waited for the AEC."

The alchemy of the politics of "containment" fused to public relations converted stiff AEC rejections and serious safety concerns into what in the public's mind appeared to be AEC bureaucratic foot-dragging. As long as the AEC was willing to absorb the blame, the substantive debate over safety might be contained. As long as utilities did not press too hard, too publicly, there was no need to state the AEC's doubts through a public device such as a moratorium on metropolitan siting. That the utilities and nuclear industry contented themselves with tactical retreats suggests that some of their more enlightened leaders appreciated the value of political "containment" as much as the experts themselves did. However, urban siting, larger reactors, and reduced safety expenditures held the key to economic competitiveness. Utilities could not be expected to acquiesce indefinitely.

One of political "containment's" strongest backers, in fact one of its original architects – Harold Price – had strongly endorsed a strategy that would have focused public scrutiny on a crucial safety debate, shattering containment. Ultimately, Price was outflanked by colleagues who shared his concerns about the safety of urban sites, but who did not feel as much political pressure to bolster their position publicly. Nevertheless, the discourse suggested that political "containment" was hardly airtight.

Concern over metropolitan siting, and in particular the lack of operating experience for the large reactors being proposed for metropolitan areas, increased ACRS discussion about the possibility of a reactor core meltdown. The committee delved into three related issues. It took a renewed interest in the integrity of the pressure vessels; it examined ways of improving emergency core-cooling systems – essential should there be a breach in the vessels carrying water to cool the reactor core; and for the first time it treated the possibility of a core melt breaching containment as a "credible" accident for which designers should develop deterrents.[31]

Advisory Committee on Reactor Safeguards member David Okrent forced a discussion within the ACRS of pressure vessels. When the ACRS refused to take action on the general problem of pressure vessels, Okrent raised the problem as part of the ACRS review of a specific reactor – Dresden II. Although the site proposed for Dresden II was not considered to be "metropolitan" (it was forty miles from Chicago and fourteen miles from Joliet), many members viewed the proposed reactors as a likely prototype for others that soon would be located in metropolitan areas. They used the Dresden review to consider several important issues affecting virtually all reactors. The Regulatory Staff, on the other hand, seemed less concerned about vessel integrity than were Okrent and his ACRS colleagues.[32]

31 Okrent, *Nuclear Reactor Safety*, pp. 85–102, 103–33, 163–78.
32 Ibid., pp. 85–7.

After meeting twice, the Advisory Committee on Reactor Safeguards approved a favorable letter for Dresden. However, it also issued a public letter to the commission urging provisions in future designs that would reduce the likelihood of pressure vessel failure and ameliorate the consequences should one occur. The nuclear industry reacted in anger to this breach of political "containment." Industry representatives were particularly upset by the lack of advance warning. An editorial in *Nucleonics* typified the industry's indignation. "The community of power reactor designers, suppliers, and operators," *Nucleonics* exclaimed, "was taken aback last month by a terse six-paragraph letter from the Advisory Committee on Reactor Safeguards ... bearing recommendations quite unexpected at this time and whose effect on the nuclear industry may take weeks to evaluate." What seemed to trouble the industry most was that "the ACRS apparently acted without prior consultation with the technical safety experts on AEC's Regulatory Staff, and indeed gave AEC only the most cursory informal advance notice."[33]

The industry also was upset that the ACRS followed up its letter by submitting written questions about vessel integrity to construction permit applicants. Roger Coe, representing the Millstone Point reactor, for instance, stated that such correspondence becomes a public document and is quickly picked up by the trade press. He predicted that making such exchanges public would reduce their frankness and lead "perhaps to intervention by opponents of nuclear power in order to delay private nuclear development."[34] Turning to the substantive question of how to implement these additional safeguards, the ACRS – well aware of the strong reaction its November 24 letter had caused in the industry, backed down. The committee agreed that its recommendations could be implemented incrementally, except in the case of reactors located in metropolitan areas.[35]

Even with this more timid implementation plan, the strategy of raising the issue in public appears to have paid safety dividends in the long run. This was the conclusion of Elizabeth Rolph, a consultant for the RAND Corp. who was quite critical of much of the AEC's research and development program. Rolph agreed, however, that the ACRS report successfully improved standards for pressure vessels. David Okrent, who did not get everything he sought, also agreed that on the issue of pressure vessels, the AEC eventually implemented most of the ACRS's recommendations.[36]

Clifford Beck of the Regulatory Staff was not as pleased: he compared the situation at the ACRS to the one that had existed following the Fermi reactor controversy. "The ACRS was heading toward a role independent of the AEC, with its own expanding staff, a proliferation of consultants, and direct lines of communication with applicants and others outside the agency." Beck need not have been so concerned. Going public was a strategy that could only be used

33 ACRS Minutes, November 24–5, 1965. *Nucleonics* 24, no. 1 (January 1966): 17.
34 Okrent, *Nuclear Reactor Safety*, p. 95; ACRS Minutes, March 10–12, 1966.
35 Okrent, *Nuclear Reactor Safety*, pp. 93–7.
36 Rolph, *Nuclear Power and the Public Safety*, pp. 85–6. Okrent, *Nuclear Reactor Safety*, pp. 221–6.

sparingly, as the industry reaction and subsequent ACRS trimming in response to the pressure vessel controversy illustrated. The minutes of the ACRS meetings do not yield a ready explanation of why the committee went public with the pressure vessel integrity issue. What is clear is that this breach of "containment" provoked a response that stifled the "public" option for some time. The ACRS had in fact begun backing away from this bold action even as it promulgated recommendations for implementing the policy. The ACRS had a steadfast commitment to maintaining a public consensus. Afraid of more public debate, that committee had shunned JCAE support at a time when it could have used some outside support.[37] Halting as it was, however, going public proved a successful strategy for the ACRS regarding pressure vessel integrity.

Even within the Atomic Energy Commission, the safety debate was no longer carried out primarily within a single unit. The Advisory Committee on Reactor Safeguards had to arrive at a settlement with the increasingly assertive Regulatory Staff. Inadvertently, conflicting positions spilled into broader forums. The debate about siting in seismically active areas such as Bodega Head was the first instance that shattered the safety community's public facade of consensus since the Fermi reactor debate. The split, however, was inadvertent. In the case of metropolitan siting, the Regulatory Staff actively sought a more public position to ease pressure from the manufacturers and to salve the staff's misgivings about the safety of such sites. That same Regulatory Staff was incensed, however, when the ACRS publicly reported its misgivings about pressure vessels without first informally airing its concerns confidentially within the nuclear community. This was no miscommunication. The ACRS took the step in an effort to outflank a Regulatory Staff and industry that was not nearly as concerned about this problem. Unlike Price's threat of going public, on pressure vessel integrity, the ACRS unilaterally did go public. Because of the esoteric nature of the problem and the limited distribution of the ACRS letter, there was little political fallout beyond the nuclear community itself. Nevertheless, the incident reminded everyone involved of the delicate nature of political "containment," and the stakes involved should it be breached.

Those responsible for engineering tests on the theoretically derived assumptions were also tempted to broaden the audience for their findings on occasion in order to fulfill their organizational mission. Nothing sums up the manner in which this halting commitment to disclosure took place – and its roots in specialized organizational needs – better than safety director Kenneth Ergen's private correspondence with Atomic International's president, Chauncey Starr. Ergen was director of safety research at Oak Ridge National Laboratory and a proponent of nuclear power. In a letter to Starr, written in March 1965, Ergen commented on the "Public Acceptance" session of a recent American Nuclear Society site selection conference and offered several suggestions for improving the public's

37 Hewlett, "Evolving Role," *CTA* File.

response. These included "missionary work" among conservationists and strategies based on the assumption that people frequently "believe in those things that fatten their pocketbooks." It might "generally be wise, to gain public acceptance first where it is easy," Ergen continued, "in the small communities." Certain changes in nomenclature, Ergen speculated, might be useful; changing "Hazards Report" to "Safety Report," for instance. In fact, Ergen appeared to embody just what the General Advisory Committee had hoped to create by assigning safety review to the national laboratories – a commitment to empirically tested safety tempered by the practical desire to ensure nuclear power's commercial success.[38]

Despite his zeal for development, Ergen insisted that the basic mission assigned to his laboratory must be pursued, even if it momentarily raised public concern. He insisted that further experimentation, including destructive tests, be pursued, regardless of the potential for failure. "One reason for this," Ergen confided to Starr, "is that we have to learn about new reactor concepts and their safety aspects, and that there remain even some open questions on 'old' concepts. There is one thing more important than convincing the public of the safety of nuclear energy," Ergen insisted, "we have to convince ourselves."[39]

Reiterating his commitment to nuclear development, Ergen insisted that if such tests were not conducted, administrators would have to assume the worst case. Further experimentation was the only way to break out of this worst-case mentality. "There have been, and will be, cases where experimentation shows unfavorable results," Ergen conceded, but "these cases are few and are a price we have to pay for increased knowledge that allows us to relax conservation on a rational basis."[40]

Some officials working for the AEC were hesitant to go even that far. When the JCAE staff director reminded reactor development director Pittman that people outside the nuclear industry frequently cited the WASH-740 report on the probability and consequences of a reactor accident as evidence that reactors were not safe, Pittman cautioned against a new analysis. In his words: "If you make a new analysis and it doesn't improve the situation it might make the situation worse."[41] Several years later, when a new study did in fact threaten to make the situation worse, the Atomic Energy Commission decided that no public report on the study was necessary.

Ergen argued that test failures were a small price to pay and insisted that test results be reported publicly, successful or not. Ergen urged that even the failures be reported to the public. His rationale had little to do with the principles of free political or scientific discourse. Rather, the Director of Safety supported

38 William Ergen to Chauncey Starr, March 1, 1965, Box 845, Anderson Papers, Library of Congress, Washington, D.C.
39 Ibid.   40 Ibid.
41 Pittman testimony before JCAE Authorization Hearings, FY 65, 1964, pp. 404–6; quote on p. 406.

disclosure for practical bureaucratic reasons. "If there were any indication that this fishbowl policy is, in fact, not quite followed, and the opponents of nuclear energy would sooner or later ferret out such indications if they existed, then the review of safety matters would very probably be entrusted to an agency representing 'broader interests,'" Ergen warned. "This would be much worse than the present situation, and to avoid such a turn of events would be well worth the limited amount of unfavorable publicity that the open-information policy occasionally might bring."[42] Like the reasoning behind a publicly stated moratorium on metropolitan reactor siting, Ergen's logic justified occasional public disclosure as necessary in order to ward off far less sympathetic probing that might retard the overall development of nuclear power.

If political "containment" were to remain intact, mechanisms designed to vent occasional expert qualms had to be built into the system. On different occasions each unit responsible for safety within the AEC attempted to use these mechanisms, gradually broadening the scope of debate. Expert debate over safety began to reach the public at an increased pace as a result of the inexorable bureaucratic tendency of each organization to reach for that slight edge in fulfilling its mission. It was not just experts at the top of the AEC who assumed a more public political role in order to promote their points of view. Experts throughout the organization were exploiting their ability to gain political leverage or at least acknowledging the need to buy off intolerable levels of scrutiny by appealing to a broader audience.

These sporadic appeals would not have been so dangerous to the political "containment" policy had broader agreement within the safety community on issues like siting and the consequences of a core meltdown existed. But a large number of experts within this community had begun to express doubts about safety at the very moment that promoters were pressing for larger reactors with fewer siting restrictions. That is why political "containment" continued to be crucial to the AEC's overall strategy. Increased reactor size coupled with proposed metropolitan siting were making the commission more vulnerable to political attack just as the initial safety questions were spawning research that raised serious questions in the minds of the AEC's safety experts. As late as 1965, while reactors in near-metropolitan sites were being approved, Price's deputy, Clifford Beck, told the ACRS that "even routine operation, testing and maintenance of engineered safeguards are not yet well developed.... Operation for several years," according to Beck, "would help to identify and correct system deficiencies."[43]

---

42 Ergen to Starr, March 1, 1965, Anderson Papers.
43 Reactor Siting Subcommittee Minutes, March 20, 1965, in ACRS Minutes.

EXPERT DOUBTS AND PUBLIC ACCEPTANCE:
THE EMERGENCE OF A SAFETY GAP

If uncertainty within the safety community about the adequacy of safeguards for proposed reactors was one reason that public disclosure and debate was such a sensitive issue, the deeply rooted concern within the nuclear community about public acceptance was clearly another. The concerns about public acceptance were inextricably linked to the way Americans had been introduced to the field of atomic energy and the mass devastation associated with the first application of atomic energy to end World War II. From the start, those seeking to develop peaceful uses for atomic energy worried that the public's fears would retard development. Public fears and the antidotes that the nuclear community applied to them have been masterfully chronicled in *By the Bomb's Early Light* and *Nuclear Fear*.[44]

What scholars have failed to explain to date is why significant public doubt about the safety of commercial nuclear power did not materialize until the early 1970s. For more than twenty years, nuclear experts fretted over public opposition to commercial nuclear power that consistently failed to materialize. For twenty years, this concern outstripped evidence of such fear in the public at large. Most significantly, those inside the nuclear community – despite sometimes vigorous debate within their own ranks about nuclear safety – systematically dismissed the few pockets of public opposition to commercial nuclear power that did appear before 1970 as the product of irrational fears. From its inception, the Atomic Energy Commission viewed public attitudes toward nuclear power as based on "half-fact, misrepresentation and rumor."[45] This fed the "fear and hysteria" that, according to the commission, were "more hazardous to the national welfare than are the actually foreseeable hazards related to atomic energy."[46] It was this cavernous divide between the nuclear experts' own technical concerns about the safety of commercial nuclear power and the experts' perception of an irrationally panicked public on the one hand, and evidence that the American public in fact had few concerns about safety, that I have labeled the "safety gap."

The safety gap plagued the Atomic Energy Commission from its inception. In the fall of 1948, perceived fear and apprehension of hazards related to atomic energy prompted Commissioner Robert Bacher to suggest that prevailing public attitudes toward radiation hazards might prove to be the "Achilles' heel" of the atomic energy program. Citing discussions with a number of qualified observers who emphasized that "there is a considerable degree of hysteria present ... regarding atomic energy hazards," the AEC contracted with social scientists from the Michigan Center for Social Research to conduct a survey in the Rochester area. Rochester was chosen because a major new facility was planned there. The survey

44 Boyer, *Bomb's Light*, and Weart, *Nuclear Fear*.
45 AEC 157, October 23, 1948, AEC, DOE, p. 12.
46 Ibid.

was eventually broadened to include a comparison of Schenectady/Albany to eleven major urban areas. One AEC commissioner felt the problem was so serious that efforts to combat public anxiety could not wait for survey research. The commission minutes from October 1948 report that W. W. Waymack "agreed that public attitude surveys were valuable and necessary but he wished to stress that the urgency in this [public education] program was so great, and the need for public education so clear, that initiation of the program should not wait for completion of attitude surveys." General Electric was even more certain that the public was on the verge of hysteria. It was so worried about public anxiety that it wanted Schenectady excluded from the survey research. GE feared that concern over radiation hazards – dormant at the moment – might be stirred up by the survey. The AEC assured GE that the very small number of persons to be interviewed in Schenectady made it highly unlikely this would happen.[47]

The report, released in 1951, found relatively low levels of apprehension about nuclear facilities and few differences between attitudes of people who lived within twenty-five miles of a facility when compared to counterparts who lived in communities with no nuclear facilities.[48] Nevertheless, about the same time, the Atomic Energy Commission's office of Public and Technical Information Service called for an increase in the number of regularly scheduled press and radio conferences. The information service cited the dearth of press conferences as one reason "why the press and radio tend to give their readers and audience a continual feeling of mystery, uneasiness and emotional reaction instead of collected thinking about atomic energy events."[49]

By the 1950s, the Atomic Energy Commission, Joint Committee on Atomic Energy, and nuclear industry were the principal sources for virtually all news accounts written about nuclear power.[50] With its monopoly on expertise, and willing to use the full force of national security provisions selectively to promote nuclear power, the nuclear community had little to complain about as far as press treatment was concerned. The Nuclear Energy Writers Association (NEWA) symbolized this cozy relationship. It was founded in 1955 by a group of professional journalists who covered nuclear power.[51] The group soon expanded to include industry public relations operatives. For a brief period NEWA operated out of the offices of the nuclear industry's trade association. Atomic Energy

47 Bacher quoted in AEC 157, October 23, 1948, AEC Files, DOE, p. 2. AEC quotation from ibid., p. 1; see also AEC 157/2, May 19, 1949, pp. 1–2, and AEC Commission Minutes, Meeting 212, October 27, 1948, in AEC Files, DOE, p. 93, for Waymack quotation.
48 AEC 157/3, September 20, 1949, in AEC Files, DOE, pp. 4–5. AEC press release, April 25, 1951, attached to AEC 157/8, April 26, 1951 in AEC Files, DOE.
49 Public and Technical Information Services to the Commission (n.d.), in Records of the Chairman, Box 4, National Archives, Washington, D.C.
50 For an excellent historical account of press coverage of nuclear power, see William Lanouette, "The Atom, Politics, and the Press," occasional paper no. 6, Media Studies Project, Woodrow Wilson International Center for Scholars, December 1989. See p. 24 for summary of press–nuclear community relations.
51 My account of NEWA is drawn from Lanouette, "Atom, Politics, Press," pp. 65–6.

Commission chair Lewis Strauss was a proud member of NEWA. Strauss, though he followed closely the daily reporting on nuclear matters, operated at a far higher echelon than the beat reporter. The AEC chair bolstered the atom's image with the fourth estate's elite, including James Reston of the *New York Times*, the head of *U.S. News & World Report*, and the president of *Newsweek*.[52]

As we have already seen, by the late 1950s, the press could be quite critical of the speed with which the Atomic Energy Commission was building the nation's nuclear power sources. This, however, only contributed to the image that commercial nuclear power was an unalloyed benefit, denied to Americans due to bureaucratic or political foot-dragging. Two statements made in 1954 and their relative coverage capture best the overwhelmingly favorable image of commercial nuclear power disseminated by the nuclear community through the press. When Lewis Strauss told a group of science writers that "our children will enjoy in their homes electrical energy too cheap to meter," it was prominently reported in newspapers across the country. As media analyst William Lanouette has noted, a speech by George Weil that only two weeks later sounded the first public alarm about nuclear hazards associated with the production of power, received virtually no coverage.[53]

Unfortunately, it is almost impossible to get an empirical fix on public attitudes toward nuclear power installations in the fifties and sixties. There are fewer than a handful of public opinion polls regarding the peaceful uses of nuclear power or the industrial production of nuclear weapons before the mid-sixties. Those few polls that were taken supported the conclusion reached in the Schenectady/ Albany (1951) study. A Gallup poll taken in 1956 found that 20 percent of the Americans queried would be afraid to have a plant run by atomic energy located in their community; 69 percent had no fear; 11 percent had no opinion. A poll conducted by the Sindlinger Company in 1960, most likely because of the positive construction of the question and because the questions did not locate the plant in the respondents' community, reported even less concern. Responses to the proposition "Atomic power should be used to produce electricity" were 6 percent negative, 64 percent positive, 12 percent undecided, and 18 percent no opinion.[54] Although scanty, public opinion surveys before 1970 indicate relatively low levels of opposition to nuclear power. Perhaps the best evidence of public acceptance, ultimately, was the virtual absence of any organized protest against

---

52 Ibid., pp. 67–8, and Press Relations file in LLS, HHPL.
53 Lanouette, "Atom, Politics, Press," p. 69.
54 American Institute of Public Opinion poll, February 25, 1956, cited in Erskine, "The Polls,"
    p. 164. Sindlinger Company poll, 1961, cited in Stanley Nealey, Barbara Melber and William
    Rankin, *Public Opinion and Nuclear Energy* (Lexington, Mass.: D.C. Heath, 1983), p. 4. On his-
    torical American public attitudes toward commercial nuclear power, see also William R.
    Freudenburg and Eugene A. Rosa, *Public Reactions to Nuclear Power: Are There Critical Masses?*
    (Boulder, Colo.: Westview Press, 1984), and William A. Gamson and Andre Modigliani, "Media
    Discourse and Public Opinion on Nuclear Power: A Constructionist Approach," *American Journal
    of Sociology* 95, no. 1 (July 1989): 1–37.

nuclear power siting in the fifties, and the limited nature of such protests during the height of the "great reactor bandwagon" in the mid-sixties.

The situation changed dramatically by the end of the 1970s, but before the accident at Three Mile Island in March 1979. A spate of public opinion polls, reflecting growing public interest in the issue, demonstrated the shift. In 1971, pollsters found that about 25 percent of those polled opposed construction of locally based nuclear installations.[55] By 1978, 45 percent of those polled opposed local nuclear facilities. For the first time in American history, more people opposed a local installation than supported it.[56]

Long before public opinion polls indicated any growing opposition to commercial nuclear power, and in fact during the period that the Sindlinger poll suggested unprecedented levels of public support, the nuclear community became particularly perturbed about public acceptance. These doubts broadened the "safety gap" in the early sixties. The catalysts were the very issues that raised concerns within the AEC's safety network. They stemmed from siting larger reactors closer to population centers. This time, however, there at least appeared to be greater justification for the nuclear community's concerns. For one thing, the fractious debate over fallout had recently sensitized the public to the potential dangers of radiation. Although public fears about fallout seemed to vanish as soon as testing was halted in 1963, the public dispute between scientific authorities (which had partitioned the AEC's responsibility for radiation standard setting and monitoring by creating the Federal Radiation Council), the nuclear community worried that a more active role by the Public Health Service and greater state regulatory action might well have eroded public confidence in the AEC. As the July *Nucleonics* editorial pointed out, "Some in the general public are already beginning to relate reactors to bombs and fall-out."[57] Although a few observers like Paul Benzequin and activists like Joel Hedgpeth did make this connection, most who linked the two (like *Nucleonics*) appear to have been members of the nuclear community deeply committed to the development of commercial nuclear power.

Those who were anxious about the public's reception to commercial nuclear power cited several other causes for worry besides the fallout controversy. In the late fifties there was a brief flurry of concern over ocean dumping of low-level radioactive wastes.[58] The Fermi reactor controversy, in which the debate over

55 Gamson and Modigliani, "Media Discourse," p. 31.
56 Ibid. Gamson and Modigliani report that following the accident at Three Mile Island, opposition to a local plant climbed to 63 percent; it reached 70 percent in the immediate aftermath of Chernobyl.
57 On rapid shifts in the public's level of concern, see Divine, *Blowing on the Wind*. For public attitudes toward fallout, see Erskine, "The Polls," and Eugene J. Rosi, "Mass and Attentive Opinion on Nuclear Weapons Tests and Fallout, 1954–1963," *Public Opinion Quarterly* 29 (Summer 1965): 280–97. For decline in confidence in the AEC, see Mazuzan and Walker, *Controlling the Atom*, pp. 275–6, and Ball, *Justice Downwind*, chs. 2–5; Fradkin, *Fallout*, chs. 1–2, 8–9. *Nucleonics* 21, no. 7 (July 1963): 17.
58 Mazuzan and Walker, *Controlling the Atom*, pp. 341–2 and 355–72.

safety had spilled into more public forums, still concerned nuclear advocates – including some members of the Advisory Committee on Reactor Safeguards. In October 1966, an accident at Fermi reactor rekindled fears about the safety of that breeder reactor. A series of mishaps resulted in a reactor shutdown and a partial fuel meltdown. Fermi, however, was quickly distinguished from other reactors because it was a breeder reactor. A *New York Times* business section story lead sentence reported that "the nation's power program has hit a snag." The article quickly distinguished breeder reactors from the light-water reactors that powered the reactor bandwagon. The snag seemed "almost unbelievable," the *Times* continued, "in light of the record $1 billion-plus orders for nuclear power in the first nine months."[59] There had been several publicized reactor accidents before Fermi. On October 8, 1957, Great Britain experienced a nuclear accident when fire broke out in plutonium production reactors at Windscale. Not discovered until several days after it started, the fire led to a large-scale release of iodine-131, contaminating milk and depositing fallout across parts of Western Europe.[60]

The most serious American accident occurred on January 4, 1961. Three workers were killed at the AEC's National Reactor Testing Station in Idaho when the SL-1 reactor – a prototype of a small mobile reactor being developed for use by the armed forces – partially melted down. The accident received front-page coverage on January 5 in the *New York Times* and the *Los Angeles Times*, and appeared on page 3 of the *Washington Post*.[61] A second *New York Times* article that appeared on January 5 graphically demonstrated the nuclear community's contribution to the safety gap. "Within the commission and atomic industry circles ... the concern today was not so much over the cause of the accident," John Finney reported, "as over the long-range repercussions it could have on the atomic power program." Finney cited official consensus, assuring readers that these experts felt reactors could be operated safely in metropolitan areas. Finney's final paragraph concluded, "Both the commission and the industry, however, have been haunted by the fear of an accident such as the one at Idaho that would alarm public opinion and thereby restrict the commission and discourage industry in the construction of atomic plants."[62]

Once again, the nuclear community's fears raced far ahead of the general public, at least judging by subsequent press coverage. The *Times*, for example, followed up the story on January 9 with a one-sentence story buried on page 23.[63] A January 20 story that appeared on page 21 (next to "Queens Policeman

59 *New York Times*, November 13, 1966, sect. 3. For a sensationalist, journalistic account of the accident, see Fuller, *We Almost Lost Detroit*. The timing of Fuller's account (1976) is significant, corresponding with the emergence of national concern about the safety of nuclear power, but almost ten years after the accident itself.

60 Gerard H. Clarfield and William Wiecek, *Nuclear America: Military and Civilian Power in the United States, 1940–1980* (New York: Harper & Row, 1984), p. 349.

61 *New York Times*, January 5, 1961; *Los Angeles Times*, January 5, 1961, *Washington Post*, January 5, 1961.

62 *New York Times*, January 5, 1961, p. 19.

63 *New York Times*, January 9, p. 23. Fortunately, my research assistant caught this tiny notice; I had missed it entirely in my first review.

Kills Hold-Up Man" – we know how much the *Times* cared about crime or Queens) reported for the first time the cause of the accident. It was "an accidental nuclear 'runaway.'"[64] The nonchalant manner in which the press followed up this nuclear runaway, suggested that it, and most likely the public, were far from terrified. AEC assurances that it could not happen in civilian plants seemed in 1961 to be sufficient.[65]

If the virtually nonexistent coverage of what today would grab headlines for days if not months was not conclusive evidence of a safety gap, two additional pieces of evidence were. The first came from the president. Accepting outgoing chair John McCone's resignation, Eisenhower lauded the "tremendous advances" made under McCone's guidance. Eisenhower pointed out that significant progress had been made in developing civilian atomic power.[66] These were not the remarks of a president concerned about the public's reaction to the dangers of civilian nuclear power. The definitive vote of confidence on reactor safety, however, came six months later in the form of a *New York Times* editorial. The *Times* congratulated the Atomic Energy Commission for its candor in reporting on the accident, suggested that the commission was preparing to profit from the experience, and concluded, "The tragedy may already have helped tighten safety procedures." The editorial demonstrated just how deep the nation's reservoir of faith in expertise and confidence in the federal government ran. The final sentence of the editorial captured perfectly the scientific optimism of the day. The *Times* urged the commission to answer fully and satisfactorily opponents who had raised questions about AEC plans "to use atomic explosives to build a harbor in Alaska.[67]

Even when the news about nuclear power was negative – and this was rare – the framework in which that news was presented remained consistently upbeat and reassuring. As one of the most prominent critical articles to appear in the fifties (entitled "A Peril of the Atomic Age") assured readers, "The general view of the experts here is that these radiation problems, if dealt with intelligently and vigilantly, need not hamper the rapid development of peacetime atomic energy."[68] Meanwhile, inside the nuclear community, less dramatic reactor "incidents" and near misses were occurring on a regular basis, yet going unreported to the public. It was these unreported "minor" accidents and the reaction of the Joint Committee on Atomic Energy – supposedly the public's watchdog – that perhaps best captures the extent of the 'safety gap' in the early sixties. On December 12, 1961, almost one year after the SL-1 accident killed three technicians, there was another "incident" at the AEC's National Reactor Testing Station. This time nobody was injured, although there was slight property damage. The event was reported to the AEC on December 18, and reported to the JCAE

64 *New York Times*, January 20, 1961, p. 21.
65 On additional AEC efforts to distinguish the accident from potential civilian accidents, see *New York Times*, June 11, 1961, p. 24. Citing AEC sources, Finney reported that "the same problem, it is believed, does not prevail with privately operated reactors under commission licenses."
66 *New York Times*, January 6, 1961, p. 8.
67 *New York Times*, June 12, 1961, p. 28.
68 *New York Times*, July 16, 1957, p. 3.

on January 31, 1962. More interesting than the matter-of-fact details contained in the memo were two notes penned in the margins. The first, from JCAE staffer Ed Bauser, stated, "Very fortunate the elements which failed were new – not loaded with fission products." The other note, presumably from chief of staff and future AEC commissioner James Ramey, suggested that Senator Dworshak of Idaho be informed of the incident. These notes juxtaposed with the final paragraph of the AEC memo summed up the safety gap. That paragraph stated: "A press release has not been issued, but our public information people are prepared to answer inquiries should they arise."[69] None arose.

CLOSING THE SAFETY GAP AT BODEGA BAY: THE PEOPLE JOIN THE EXPERTS

There were more proximate causes for the nuclear community's heightened concern in the early 1960s. Three proposed reactors – Bodega Bay, Ravenswood, and Malibu – faced grass-roots opposition. *Nucleonics* took the nuclear community to task for the way it handled the warning signs in these three cases. It cautioned that if the new era of nuclear power plant building was to be successful, the industry must do a better job of communicating with the public. "Now that. nuclear power is on the verge of economic competitiveness," *Nucleonics* cautioned, "more and more attention will be focused on siting nuclear plants in urban centers. And, as this happens, the general public will quite properly become increasingly interested in nuclear plant sites." After criticizing the way public education had been handled at Ravenswood, the editorial concluded ominously that "if the public does not accept power, there will be no nuclear power."[70]

Almost all grass-roots opposition to nuclear reactors centered on two issues in the early 1960s – both related to siting. The first, and most pervasive, concerned conservation and preservation of the natural landscape. This was the central concern in the case of opposition to the reactor planned on Bodega Head. By the

---

69 Luedecke to Ramey, January 31, 1962, "Engineering Test Reactors" folder, Box 566, JCAE Files.
70 Though of less concern to the nuclear community, there was also grass-roots opposition to the Haddam reactor, in the Connecticut Valley, and the San Onofre reactor in San Diego. Haddam opposition centered around the possible thermal pollution that might result from discharges of heated water from the reactor. This was the first publicly voiced concern about thermal pollution – an issue that would play a more prominent role in the second half of the decade. On San Onofre see *New York Times*, December 1, 1963, and *Nucleonics* 21, no. 7 (July 1963): 21–2. On Haddam, see *New York Times*, May 30, 1964, July 21, 1964, and August 2, 1964. *Nucleonics* 21, no. 7 (July 1963): 17.
   At a meeting of the American Nuclear Society in February 1965 held to discuss reactor siting, the keynote address as well as several of the papers centered on the problem of public acceptance. The consensus was that on many of the conservation-related issues, nuclear power compared favorably with its fossil fuel competitors. Those comparisons – the participants felt – should be pointed out. It was acknowledged that nuclear plants did have a unique problem with safety. This, many panel members felt, was because of the public's vague association of everything nuclear with the bomb, and general fear of the unknown. The remedy, it was agreed, was more education (W. K. Ergen, "Nuclear-Power-Reactor Siting," *Nuclear Safety* 6, no. 4 [Summer 1965]: 379–80).

second half of the 1960s, concern for preserving the landscape had evolved toward protecting the environment – particularly the aquatic environment that would have to absorb the reactor's heated waste-water. Site-related safety issues – the second issue grass-roots critics concentrated on – paralleled opposition from conservationists, and at Ravenswood and Bodega Bay eventually overshadowed it.

Opponents, regardless of the initial reason for their involvement, eventually drew upon all of the other critiques available. Seeking public attention, they capitalized on a variety of forums and drew new sources of expertise into the dispute. This expanded the scope of debate and raised doubts where previously – at least from the public's perspective – expert consensus had prevailed. Out of these early struggles emerged an issue that as much as conservation and safety became a rallying point for opponents: the issue of participation.

The first opposition to the Bodega Head site came from a landholder – Rose Gaffney – who on occasion challenged trespassers with a baseball bat. Pacific Gas and Electric's plan to construct a 325-megawatt boiling-water reactor designed by General Electric on Bodega Head, fifty miles north of San Francisco, was not formally announced until June 1961. The plant was scheduled to be completed by the end of 1965.[71] Gaffney, however, got wind of the project long before that and filed suit against Pacific Gas and Electric in 1958. The suit was joined by Pacific Marine Station director Joel Hedgpeth.

Neither Gaffney nor Hedgpeth had been concerned with safety issues to begin with. Gaffney was trying to save her land from an eminent domain takeover, and Hedgpeth opposed the plant because he felt it would destroy the natural beauty of the area. Nevertheless, Hedgpeth publicized the safety issue as a means of stopping the plant when his secretary leaked the news to the *Santa Rosa Press-Democrat* that a nuclear plant was intended for this site. Hedgpeth questioned siting a reactor so close to the San Andreas fault. To dramatize the potential danger of a reactor located so close to an active fault, he submitted to the court a map of the fallout area from Britain's Windscale reactor accident, superimposed on the Bodega Head countryside.[72] Early citizen complaints filed with the Atomic Energy Commission raised questions about possible increases in radiation and other potential safety hazards. Elizabeth and Charles Chambers, who lived only blocks from the proposed site, expressed concern about low-level emissions and the possibility of an accident. Mrs. Chambers took great pains to demonstrate her knowledge about nuclear power. Despite a firm grasp of the facts, Chambers argued, she was still opposed to it. What appeared to bother her the most was Pacific Gas and Electric's cavalier attitude toward the public. "It simply does not

---

71 For details of the PG&E proposal, see *Electrical West Industry News Letter*, June 28, 1961 (special edition).

72 This account is drawn from two sources with conflicting perspectives on nuclear power: *Nucleonics* 21, no. 10 (July 1963): 17–18; and Joel Hedgpeth, "Bodega Head – a Partisan View," *BAS* 21 (March 1965): 2–4.

impress me or the other citizens here," she protested, "when the representatives of the PG&E hold parties in town to cajole the public into acquiescence, pass off legitimate and important questions with the intimation that they reflect 'cowardice,' and announce in the face of all reason to the contrary that what is 'good for PG&E is good for the people.'"[73]

Hearings conducted by the California Public Utilities Commission (CPUC) in 1962 produced the certificate of convenience that PG&E desired, but not before spawning several by-products that were less fortunate from the utility's perspective. One was the Public Utilities Commission chair's minority opinion, in which he proclaimed to PG&E, "People fear nuclear power, and you must explain it to them. Utilities have not met their obligation to explain nuclear power and they would do themselves a service to do so." Also, it was during these hearings that David Pesonen, a young Berkeley biological scientist with considerable organizational skills, emerged as the major opposition leader.[74]

Like Hedgpeth, conservation was Pesonen's primary goal, yet he, too, raised concerns about the earthquake hazard. Pesonen, who was conservation editor of the Sierra Club newsletter, organized the Northern California Association to Preserve Bodega Head and Harbor (NCAPBHH). He claimed he was not against the development of nuclear power plants: it was a nuclear power plant at that particular location that bothered him. Others in his newly formed organization, however, voiced general concerns about the danger of radioactive releases.[75]

For both Pesonen and Hedgpeth, due process was crucial, for two reasons. First, organized opposition had begun to seek access to expertise of its own. Grass-roots groups began to seek the means by which their own experts could be admitted to the forums in which formal technical discussion took place. Thus, when the Northern California Association to Preserve Bodega Head filed a petition before the California Public Utilities Commission, it charged that the association had not been given the opportunity to cross-examine Pacific Gas and Electric experts. In denying the petition, the CPUC countered that the Association to Preserve Bodega Head was too late. PG&E had testified ten months ago, whereas only now was the association asserting it had been denied due process or alleged "that it can now produce experts on seismology and earthquake hazards." Pacific Gas and Electric thus won this skirmish. It was now clear, however, that neither side could hope to win the war without the aid of experts. Due process, opponents believed, ensured that their expert advisers would at least be heard, and perhaps even have the opportunity to dispute or at least question the previously unchallenged opinions of colleagues working for the nuclear industry or the Atomic Energy Commission.[76]

---

73 Charles Chambers and Elizabeth Chambers to AEC, April 2, 1960, "Humboldt Bay" folder, Box 574, JCAE.
74 *Nucleonics* 21, no. 10 (October 1963): 19.
75 Ibid.
76 Quotation and account of the denied petition is from the *New York Times*, July 12, 1963.

The question of due process was also crucial to mass mobilization and to expanding the range of participants interested in these highly technical issues. Access and the right to participate was an inclusive issue that tapped deeply felt American beliefs about the political process and scientific discourse. The success of both, as far as most Americans were concerned, depended on the free exchange of ideas by qualified participants. In the case of popular participation, the qualifications were those outlined by Lilienthal in his Crawfordsville speech decades before: common sense. In the case of technical access, the organized opposition had engaged expertise, seemingly qualified to join the scientific discourse. Arrayed against these principles stood a federal apparatus that – particularly when compared to other domestic policy areas – had a history of extraordinary centralization and insulation.

The volatile nature of this juxtaposition became apparent at a public meeting to discuss the Bodega Bay reactor late in 1962. The meeting dragged on without much visible enthusiasm from the audience until Alex Grendon, the governor's coordinator of atomic energy, "inadvertently introduced the issue," as Hedgpeth tells it, "which ultimately drew politicians into the fray as well as recruiting a whole new group of opponents." Grendon, Hedgpeth continued, "would be one of the first to concede that tact is not one of his strong suits ... and his blunt but essentially accurate statement concerning the ultimate power of experts alone to decide the issue ... stirred up tempers in the audience that never quite subsided." Grendon's comments were broadcast several times to an even larger audience in a radio program appropriately titled "The People vs. the Experts."[77]

In fact, it was becoming increasingly difficult to distinguish "the people" from "the experts." On the one hand, much of Pesonen's influence rested on his claim that he had "the people" behind him. Pesonen tirelessly sought to broaden his audience. Because safety was more dramatic than conservation, Pesonen quite explicitly sought to use fears stemming from the fallout controversy raging at the time and to connect these with reactor safety concerns through public relations stunts. For example, demonstrators released balloons that stated, "This balloon could represent a radioactive molecule of strontium-90 or iodine-131. It was released from Bodega Head on Memorial Day, 1963. PG&E hopes to build a nuclear reactor plant at this spot, close to the world's biggest active earthquake fault." The Northern California Association to Preserve Bodega Head and Harbor's legal actions were also motivated in large part by a desire to broaden its base of support. Despite legal setbacks, as Hedgpeth put it, "each of these actions kept the news alive, and recruited more of these quiet supporters who hesitate to be publicly counted but quietly contribute."[78]

Simultaneously, Pesonen and Hedgpeth broadened their access to expertise. The reactor opponents hired Pierre St.-Amand, a professional seismologist from

77 Hedgpeth, "Bodega Head," p. 5.
78 Balloon quotation from ibid.; also see *New York Times*, May 31, 1963. Hedgpeth, "Bodega Head," pp. 5–6.

the Naval Ordnance Testing Station in China Lake, California. Hedgpeth lobbied Berkeley faculty concerned about the effects a nuclear power plant might have on marine life near the site of a proposed University of California marine station. Scientists from the Scripps Institution of Oceanography also joined the debate. A group of them claimed to have discovered an entirely new complex of bedrock faults near the proposed reactor site.[79] None of these hired or "volunteered" experts directly altered the debate. At Berkeley, a National Science Foundation report scuttled efforts to put the entire faculty on record against the reactor. Nor did the information contributed by the Scripps Institute affect the AEC's debate. However, the "counterexpertise" that Pesonen and Hedgpeth brought to bear, and the broader network of scientists engaged or mobilized by them, did affect the outcome at Bodega Head in two important ways.

First, and perhaps most significant for future debates, Pesonen and Hedgpeth examined skeptically the supposedly authoritative statements made by experts hired by Pacific Gas and Electric, and the Atomic Energy Commission. Although not experts in reactor technology, seismology, or reactor safety, Pesonen and Hedgpeth were nonetheless trained as scientists. Confident of their knowledge in one scientific discipline, they were not hesitant to challenge the statements of supposed experts in another. They were not awed by the other side's professionals. Hedgpeth and Pesonen claimed to be speaking for "the people," but it was actually their credentials as scientists that gave them privileged access to the debate. Although their training in a scientific discipline other than physics, chemistry, or engineering obviously dictated the perspective through which each viewed reactor siting, as the debate grew public that was less important than their scientific credentials. Pesonen, Hedgpeth, and the experts they worked with were the first wave in a steady procession of consulting experts who began to turn the debate about reactor safety into an internecine battle between informed professionals. It was the increased rate at which inter- and intraprofessional disputes were aired publicly that ultimately undermined public confidence in reactor safety and in expertise in general, not a war between "scientific" experts and the "antitechnological" people.

The grass-roots citizen protest orchestrated by Pesonen coupled with the string of "expert" reports stimulated by his organization attracted the attention of highly placed officials outside the AEC. This second legacy of reactor opponents' access to expertise prolonged the debate and subjected it to the opinion of institutionalized expertise beyond the boundaries of AEC and the nuclear community. A case in point was St.-Amand's report. It influenced the subsequent actions of the U.S. Geological Survey [USGS] and its boss – Secretary of the Interior Stewart Udall. The reports produced by the USGS and Udall's highly visible intervention, in turn, substantially altered the debate within the Atomic Energy Commission.

---

79 Sheldon Novick, *The Careless Atom* (Boston: Houghton Mifflin, 1968), p. 41. *New York Times*, July 16, 1963. Hedgpeth, "Bodega Head," p. 6.

As the California Public Utilities Commission was conducting hearings, the AEC Regulatory Staff and Advisory Committee on Reactor Safeguards proceeded with their own review of the Bodega Head site. Neither body was particularly concerned about the site's proximity to an active fault. Without directly consulting professional seismologists, geologists, or seismic engineers, the ACRS issued a routine letter of approval in April 1963.[80] The letter cited seismological studies performed for PG&E. It also noted that according to these reports, the reactor would not be located on top of an active fault. It called for careful examination during construction to confirm this fact, but it gave the green light for a construction permit to be issued.[81]

One day after the ACRS issued its letter, St.-Amand, as the Association to Preserve Bodega Head's seismological expert, wrote to Udall's special assistant.[82] St.-Amand's report seemed to have a decisive impact on the Geological Survey's acting director. On May 20, 1963, he wrote to the Assistant Secretary of Interior and reported that although PG&E's consultants had concluded it was unlikely that faulting would occur in the future, St.-Amand warned that the site was unsafe for reactor construction unless exceptional precautions were taken.[83] St.-Amand's caution was hard to argue with, the acting director concluded. Although the PG&E consultants' arguments might be valid, "St.-Amand's figures are the more conservative and should probably form the basis for the design criteria."[84]

While the Department of Interior reviewed St.-Amand's report, the NCAPBHH went public with its charges that the reactor was planned for a potentially hazardous site. On May 6, it issued a press release calling for the California Public Utilities Commission to reopen its hearings. Although the press release noted St.-Amand's new findings, it featured more prominently St.-Amand's analysis of recently updated reports from PG&E's own consultants. Pesonen charged that the Utilities Commission had failed to examine the contents of this final report, which was submitted late in the process. That report, and the preliminary hazards summary report submitted to the Atomic Energy Commission, Pesonen noted, undercut PG&E's earlier testimony. As the press release put it, "The reports of PG&E's earthquake hazard consultants showed the site to be so unsuitable that the company has tried to suppress the reports, has altered some of their conclusions and has failed to submit the most recent and damaging information to the Atomic Energy Commission."[85]

80 D. B. Hall to Seaborg, April 18, 1963, "Bodega Bay," file, Box 543, JCAE. On faulting and reactors, see Richard L. Meehan, *The Atom and the Fault: Experts, Earthquakes, and Nuclear Power* (Cambridge, Mass.: MIT Press, 1984).
81 Ibid., p. 2.
82 Director, Geological Survey to Assistant Secretary of the Interior, May 20, 1963, attached to Price to Conway, May 23, 1963, "Bodega Bay" folder, Box 543, JCAE.
83 Director to Assistant Secretary, May 20, 1963, attached to Price to Conway.
84 Ibid., p. 2 .
85 NCAPBHH press release, May 6, 1963, "Bodega Bay" folder, Box 543, JCAE; quotation from p. 2.

Pesonen's press release received relatively little attention. The concerns it raised, however, worked their way to the top of the Department of Interior's bureaucracy. On May 20, 1963, Secretary of Interior Udall wrote to Seaborg. Using language that shattered the nuclear community's soothing rhetoric, Udall told Seaborg, "I feel there is reason for grave concern."[86] Shrewdly, Udall never mentioned Pesonen or the Northern California Association to Preserve Bodega Head and Harbor. Rather, he quoted from PG&E's own consultants' report, which stated that "at least one and perhaps two or more major earthquakes can be expected near the site within the next century. These may be as strong or even somewhat stronger than the California earthquake of April 18, 1906." Udall called for a thorough investigation before the AEC issued any construction permit and offered the Geological Survey to provide the AEC with expertise that seemed to be lacking in its review.[87]

It was an offer the AEC was in no political position to refuse, particularly after the *San Francisco Chronicle* ran an article on page 2, two days later, headlined "Bodega A-plant Risks Quake Peril, Udall Says."[88] The earthquake issue was reopened in the AEC safety community, which now loaded up on seismic expertise.[89] The Regulatory Staff accepted Udall's offer and engaged the U.S. Geological Survey to examine the seismic history of the site and comment on the possibility of further faulting.[90] Employing the USGS had its political advantages. Joint Committee on Atomic Energy chair Chet Holifield, for instance, informed his California constituents who were concerned about the reactor safety that the AEC now drew upon "the expert assistance available through the United States Geological Survey."[91]

Monopolizing expertise, even producing expertise through its fellowships and research programs, had been one of the Atomic Energy Commission's greatest sources of political influence. As nuclear power moved into the implementation phase, however, a number of issues arose that went far beyond the commission's technical reach. Department of Interior research geologist Robert Rose, who also recommended a comprehensive seismic study before allowing construction, summed up the AEC's dilemma:

Qualified engineering specialists and nuclear scientists must be given the responsibility for passing judgment on the types and standards of construction required to meet specified

86 Udall to Seaborg, May 20, 1963, attached to Price to Conway, p. 1.
87 Ibid., p. 2.
88 *San Francisco Chronicle*, May 22, 1963.
89 Okrent, *Nuclear Reactor Safety*, p. 264. *Nuclear Safety* 5, no. 2 (Winter 1963–4): 210. ACRS Minutes, May 1963.
90 Robert Wilson to Stewart Udall, May 23, 1963, in AEC Files, DOE. The Regulatory Staff and the ACRS engaged a host of experts over the summer. The Regulatory Staff requested the U.S. Geodetic Survey, Nathan Newmark of the University of Illinois, and Robert Williamson of Holmes and Narver Co. to advise on seismic-related matters. The ACRS engaged engineer Karl Steinbrugge to do the same. And Pacific Gas and Electric engaged three consultants to comment on seismic matters (Okrent, *Nuclear Reactor Safety*, p. 265).
91 Holifield to Green, August 20, 1963, "Bodega Bay" folder, Box 543, JCAE, p. 3.

conditions. By the same token, however, seismologists and engineering geologists must be relied upon to determine and define the origin, nature and magnitude of those conditions.[92]

In areas where the Atomic Energy Commission had little of its own expertise, political "containment" was almost impossible to establish, let alone maintain. Given the publicity that accompanied the Geological Survey's offer to help, nobody was surprised that the review was rapidly politicized by Secretary of Interior Udall (to whom the U.S. Geological Survey reported). Although the AEC could not operate without this source of institutionalized expertise, it paid a price for it. Udall's political agenda did not correspond to the AEC's. As Udall toured Africa, acting Secretary John Carver kept the political heat on. In September 1963, Carver transmitted the USGS's preliminary report. He quoted some of the report's more provocative passages, such as "Few places on the earth are exposed to more certain earthquake risk than are those along the San Andreas fault," and "Acceptance of Bodega Head as a safe reactor site will establish a precedent that will make it exceedingly difficult to reject any proposed future site on the grounds of extreme earthquake risk."[93]

Udall's personal intervention at such an early stage in his bureau's investigation may have been motivated to a certain degree by partisan politics. As the *New York Times* reported in August 1963, California Democrats had sought to make Bodega Head a statewide campaign issue in the 1964 election, with Governor Brown criticizing the site in July, and Lieutenant Governor Glenn Anderson urging the AEC to reject PG&E's application. Udall, who had to be well aware of these rumblings, assumed a prominent position on this issue by challenging the AEC.[94]

Opposing Bodega Head also appealed to an important segment of Interior's natural constituency: conservation groups. In his May 20 letter to Seaborg, along with seismic hazards Udall discussed "probable adverse effects on coastal fisheries in the event of an accident," and "possibilities of adverse effects on public use of the new Point Reyes National Seashore." More critical to our story than the specific source of Udall's intervention – partisan or bureaucratic political agendas – was the vehicle through which he readily entered an internal AEC debate – the USGS's seismological expertise requested by the AEC's Regulatory Staff. As the AEC became dependent on sources of expertise housed in other federal agencies, it was subjected to the political agendas of those institutions and their politically sensitive leaders.[95]

Udall's "grave concerns," and even the subsequent misgivings raised in the September preliminary report, were voiced before any new hard data were

92 Rose to Rettie, May 14, 1963, attached to Price to Conway, p. 5.
93 Udall's concerns are recapitulated in Secretary of Interior to Seaborg, September 25, 1963, in AEC Files, DOE. Wilson to Udall, May 23, 1963, AEC Files, DOE. Secretary of Interior to Seaborg, September 25, 1963, AEC Files, DOE.
94 *New York Times*, August 25, 1963.
95 Udall to Seaborg, May 20, 1963; see also *Nuclear Safety* 5, no. 2 (Winter 1963–4): 210.

introduced into the debate. They were based on conflicting estimates of consulting seismologists. That situation changed in early October 1963. The new data did not help those defending the site. In an ironic twist to the story, it was Seaborg who broke the news to the press: New excavations revealed that the fault extended into the bedrock underneath the proposed site of the reactor itself.[96]

The news was not good from the AEC's perspective, but at least it appeared that the Atomic Energy Commission had discovered the problem. As the lead sentence of a front-page article in the *San Francisco Chronicle* put it, "The Atomic Energy Commission reported yesterday that an earthquake fault penetrates the bedrock of the Pacific Gas & Electric Company's ... power plant."[97] Delivering the news did not mitigate its embarrassing impact. Another edition of that day's *Chronicle* quoted Pesonen. The executive secretary to the Northern California Association to Preserve Bodega Head and Harbor stated, "the discovery of the fault 'confirms what we have been saying all along.'"[98] In both editions, the text of the articles made it clear that Department of Interior expertise led to the discovery. California Public Utilities chair William Bennett, the only dissenting vote on the CPUC, was also quoted. "This demonstrates again that California Utilities should prove their cases in proper manner instead of through barrages of press releases and conducted tours," Bennett told the *Chronicle*. "I'm gratified to have a supporting opinion from so respected an authority."[99]

As the *Wall Street Journal* described it on October 7, 1963, the new discovery "is sending a tremor through the Atomic Energy Commission's deliberations."[100] Although PG&E dismissed the fault's discovery as relatively inconsequential and insisted that the company's "independent" scientists had not found any evidence that the site was unsafe, the Atomic Energy Commission requested that the U.S. Geological Survey remain on the scene, conduct further excavations, and issue a final report.[101] The issue received national attention when the *New York Times* editorialized on December 28, 1963, about the natural beauty and seismic problems at the Bodega Head site.[102]

96 Seaborg to Carver, October 3, 1963, in AEC Files, DOE. Carver was then acting Secretary of the Interior. Udall's departure and the disruptions it caused in Interior may well explain why the AEC was able to break the news, unfortunate as it was for Bodega Bay's prospects.

97 *San Francisco Chronicle*, October 5, 1963, p. 1. The *Santa Rosa Press-Democrat* did a better job unraveling the real chronology. Its lead reported, "Two U.S. Geological Survey researchers investigating the potential earthquake hazard of a proposed atomic energy power plant at Bodega Bay, Calif., have discovered a new fault extending into bedrock at the location of the plant, the office of Sen. Clair Engle, (D-Calif.), reported today" (October 4, 1963).

98 *San Francisco Chronicle*, October 5, 1963, p. 9.

99 Ibid.

100 *Wall Street Journal*, October 7, 1963.

101 Ibid., and Seaborg to Carver, October 3, 1963, attached to AEC Press Release, October 4, 1963, "Bodega Bay" folder, Box 543, JCAE.

102 *New York Times*, December 28, 1963; For PG&E response, see press release attached to Cooper to Bauser, January 6, 1964, and Stroube to Editor, December 30, 1963, "Bodega Bay" folder, Box, 543, JCAE. Among other data cited, PG&E sent the *Times* a copy of an August 19, 1963, article in the *San Francisco News–Call Bulletin* in which science editor George Duscheck called the Bodega Head area a "seaside slum."

The USGS report (issued in January 1964) raised further doubts in the minds of the Atomic Energy Commission's Regulatory Staff. Even PG&E had to concede publicly that the Interior Department's report revealed "several areas of disagreement between the findings of government geologists and the conclusions of the independent geologists and seismologic consultants retained by the... Company to evaluate the Bodega site."[103] Coupled with the Great Alaskan Earthquake in March 1964, the Geological Survey's report tipped the Regulatory Staff's position. At the Advisory Committee on Reactor Safeguards meeting in May the Safety Staff reported that the plant could be designed to meet USGS estimates of maximum ground displacement, but the risk "could be effectively eliminated by moving the plant to a location a couple of miles distant from the main fault zone. On this basis, we have concluded that the site proposed by PG&E is not suitable for a reactor of the general type and power level proposed."[104]

The Reactor Safeguards Committee, on the other hand, was prepared to write a more favorable letter allowing that the ground displacement could be accommodated in the design without changing the reactor's location. The two staffs continued to meet, and by October 6 the Regulatory Staff issued a report to the ACRS stating that the trade-off between risks and benefits was a policy issue. The ACRS, perhaps assuming that the Regulatory Staff had been persuaded, issued a favorable report to Seaborg on October 20, 1964. To their surprise, however, the Regulatory Staff's summary analysis took a stand on that policy issue, concluding that Bodega Head was not a suitable location.[105]

The AEC's public announcement of these contradictory positions represented the first instance, since the Fermi reactor incident, that two expert bodies on safety regulation within the commission had disagreed in public.[106] Their contradictory reports had an immediate impact. A Joint Committee staffer's note, penciled onto Price's memo informing the JCAE that the AEC had issued two conflicting reports, brought the curtain down on the decade's most publicized dispute between experts on reactor safety to date. "Oral notification 10/30/64," the note stated in a matter-of-fact manner, "to JCAE that PG&E had withdrawn application for Bodega Head."[107] The official PG&E release announced the withdrawal and stated that "the doubt raised by the staff, although a minority view, is sufficient to cause us to withdraw our application."[108]

103 "Statement By S. L. Sibley," January 20, 1964, "Bodega Bay" file, Box 543, JCAE.
104 Udall to Seaborg, January 16, 1964, "Bodega Bay" file, Box 543, JCAE; Okrent, *Nuclear Reactor Safety*, pp. 266–7. Regulatory Staff Report to ACRS Meeting #55 (May 6–8, 1964), cited in ibid.
105 For ACRS and Regulatory Staff reports, and final USGS report, see Price to Conway, October 26, 1964, "Bodega Bay" file, Box 543; *New York Times*, October 28, 1964; Okrent, *Nuclear Reactor Safety*, pp. 267–8.
106 The JCAE was well aware of this fact. See Conway to All Committee Members, October 27, 1964 "Bodega Bay" file, Box 543, JCAE. For public announcement, see "AEC Releases Two Reports," press release, October 27, 1964, attached to Price to Conway, October 26, 1964.
107 Marginalia on Price to Conway, October, 26, 1964.
108 PG&E press release, October 30, 1964, "Bodega Bay" file, Box 543, JCAE; *New York Times*, October 31, 1964.

There were several new elements present in the Bodega Head fray that had been missing in the licensing proceedings of the past. While states had played an increasingly active role in the fifties, no state official had openly criticized a specific reactor proposal. Leading the opposition to Bodega Bay were some of the state's most prominent public officials including the governor, lieutenant governor, the state Democratic Central Committee chairman (supported by the committee), Public Utilities chair William Bennett, and Alan Cranston.[109] Also, for the first time there was conspicuous national political opposition from outside the Joint Committee on Atomic Energy regarding the issue of reactor safety. Udall was the most prominent critic of the proposal in the executive branch. In Congress the state-wide debate captured the attention of Senator Clair Engle, a Democrat from California. Engle introduced a joint resolution into the Congress on April 13, 1964, which required firm assurances about the plant's safety before any construction was authorized.[110] Although the resolution would probably not have passed even if Engle had not died shortly after introducing it, merely the hint of legislative oversight that ranged beyond the confines of the Joint Committee on Atomic Energy implied by the resolution was a new and important phenomenon in 1964.

David Pesonen wrote to Joint Committee on Atomic Energy member Senator John Pastore hoping to persuade him that the Bodega Bay site was misguided. Pesonen placed the "Battle of Bodega Bay" in the broadest context imaginable, claiming it was "a symptom of many of weaknesses of our times."[111] The first four reasons cited by Pesonen – man's carelessness toward the "irreplaceable natural environment," the failure to plan effectively, ineffective regulation, and "the single-mindedness and 'might-makes-right' of the nation's largest private power utility" – are arguments familiar to scholars of both the environmental movement and the New Left critique of America's establishment. Both these movements incubated even as Pesonen wrote, and undoubtedly influenced this Berkeley grass-roots organizer.

Symptoms five and six cited by Pesonen have received less scholarly attention, although they have proved to be just as powerful in altering the course of American politics in the late 1960s and beyond. Pesonen described these final "symptoms" as: "The uncritical trust by the public in the omnipotence of the 'scientist' and the 'engineer' who can build anything," and "The abdication of policy-making by government to the scientific elite." Those like Pesonen who sought to challenge the omnipotence of science, or who sought to participate in policymaking involving highly technical or esoteric matters, were promptly labeled "anti" technology, "anti" science, and even "anti" rational. Undoubtedly, some of the rank and file among Pesonen's foot soldiers warranted such labels.

Ignored by scholars but illuminated by the "Battle of Bodega Bay" was the

---

109 Pesonen to Conway, July 31, 1964, "Bodega Bay" file, Box 543, JCAE.
110 Ibid.
111 Pesonen to Pastore, January 7, 1964, "Bodega Bay" file, Box 543, JCAE.

degree to which scientist–grass-roots organizers like Pesonen fused what had previously been regarded as the two separate worlds of science on the one hand, and participation on the other. Pesonen not only broadened the debate to include an audience that may well have had an unsophisticated grasp of seismology, he also drew into the debate some leading experts in that field. When the Atomic Energy Commission suggested that public concern played no role in the evolution of the AEC's safety review at Bodega Head, Pesonen was quick to challenge that assertion. The U.S. Geological Survey did not join the review until May 1963, Pesonen wrote the Joint Committee's chief of staff.[112] Interior's assessment, "significantly, relies heavily on findings by Dr. Pierre St.-Amand, a consultant retained by our citizens Association," Pesonen continued. The USGS again turned to St.-Amand's work to issue a subsequent report, Pesonen reminded the JCAE. "In light of this and the subsequent USGS report which generated major design changes by the company, we can reasonably infer that public participation has had a tangible effect on the technical evaluation of the Bodega site," he concluded.[113]

If Pesonen's recapitulation was good history, it was even better forecasting. Just as experts like Abel Wolman quietly and "professionally" linked nuclear-related issues to the organizations and disciplinary perspectives associated with their own field, organizers like Pesonen increasingly would publicly and often stridently link expertise – in such fields as seismology, ecology, and biology – to grass-roots citizen concerns powered by demands for participation and broader constellations of attitudes, ranging from environmentalism to New Left critiques. These linkages usually engaged other public officials and the agencies that represented constituencies previously excluded from nuclear decision making. Though the techniques of insiders like Wolman and outsiders like Pesonen were extremely different, they in fact shared much in common. In part because of personality but also in part because of their scientific training, neither man was afraid to challenge expert authority. Perhaps most significantly, each challenged the AEC's authority in scientific fields where the Atomic Energy Commission had little experience and even less expertise directly under its control. Wolman, employing quiet persuasion within his environmental engineering and public health community, and Pesonen, using loud publicity stunts directed toward a far larger population linked to the issue through geography, not occupation, promoted and then employed disputes between experts to demonstrate that political choice was involved in many of the day-to-day decisions made by experts.

Perhaps the greatest difference between insiders like Wolman and outsiders like Pesonen, other than the techniques used to persuade and the constituencies they directed their messages toward, was access to information. As a charter member of the Advisory Committee on Reactor Safeguards, Wolman had virtually

---

112 Pesonen to Conway, September 14, 1964, "Bodega Bay" file, Box 543, JCAE.
113 Ibid., p. 2.

unlimited access, and even a limited degree of direct influence, on the decision-making process. Outside critics like Pesonen, on the other hand, had severely circumscribed access and relied on indirect influence. Thus, broader public access and due process became one of Pesonen's greatest concerns, whereas it never particularly troubled Wolman.

The complexity of the issues involved, the number of experts engaged, the number of organizations housing these experts – even within the Atomic Energy Commission – and the porous and decentralized nature of American policy implementation allowed critics like Pesonen, who were willing to challenge the authority of experts, to penetrate with relative ease the self-consciously insulated world of nuclear policymaking. Pesonen challenged the secrecy. At the same time he used the nuclear community's own doubts and internal debates to indict nuclear power and demand greater scrutiny. Although he did not use the term, Pesonen was the first person outside the nuclear community to recognize that a significant "safety gap" existed. His was a crusade to close it.

A case in point was Pesonen's critique of PG&E's "other" reactor, built at Humboldt Bay. As Pesonen wrote to the Joint Committee's executive director in September 1964, "For a number of reasons, including a deceptive public relations campaign by the company, the Humboldt Bay installation faced no significant public scrutiny."[114] Yet inside the nuclear community, Pesonen argued, there was cause for concern. Pesonen's evidence rested on a June 1960 issue of *Nuclear Safety*, a technical journal published at Oak Ridge National Laboratory and one of the major communication links within the nuclear community. Pesonen noted that the article's author found that ambiguities existed with only one out of nineteen reactor license applicants to date: Humboldt Bay. At that reactor, Pesonen continued, the author found "substantial technical grounds for concern over the safety." Always the advocate, Pesonen hammered home his point about what I have labeled the "safety gap." "In view of the company's record with its first and only other nuclear installation," Pesonen lectured Conway, "we consider that comparable ambiguities – affecting the public safety of the San Francisco Bay Area – very likely would remain in the record of the Bodega application were it not for public participation early in the licensing procedure."[115]

Public opposition, because it prolonged the debate, introduced new experts into it, and exposed rifts between them, was a decisive factor in expanding vastly the scope of the Bodega Bay debate. The utility's decision to withdraw its reactor proposal, however, was not the direct result of public opposition. As in the past, the utility withdrew because of resistance within the AEC's safety community. The pattern at Ravenswood was similar. Publicly organized opposition, voiced in local forums, raised safety questions that were simultaneously being debated behind closed doors by experts within the AEC. The utility – in this case Consolidated Edison – pooh-poohed safety concerns raised outside the AEC. But

114 Pesonen to Conway, September 14, 1964, "Bodega Bay" file, Box 543, JCAE, p. 2.
115 Ibid.

when faced with the likelihood of Advisory Committee on Reactor Safeguards and Regulatory Staff rejection – the utility withdrew the application. By withdrawing the Bodega Bay reactor, PG&E ended the local controversy just as it was beginning to raise national questions about the safety of commercial nuclear power.[116]

Pesonen responded by pursuing a two-pronged strategy. He probed deeper into the insulated but far-flung nuclear community. At the same time he sought national forums for the bounty he brought back from these forays. Pesonen pioneered what was soon to become a growth industry in the field of nuclear opposition. He was the first nuclear opponent systematically to plumb the obscure but easily penetrated forums in which nuclear policy was discussed, and to broadcast these internal debates to a national audience. He soon expanded his sources far beyond published technical journals like *Nuclear Safety* to industry trade and professional meetings, and to dreary congressional subcommittee hearings, which in the past had been attended only by nuclear advocates and the coal industry opposition – critics easily dismissed owing to their vested economic interest in stopping nuclear power.

Pesonen's greatest harvest, although it would not ripen for years, was his discovery that the Atomic Energy Commission, with the blessing and in some instances encouragement of the rest of the nuclear community, had suppressed a study that recalculated the consequences of a theoretical nuclear accident. Originally incorporated in the WASH-740 report publicly released in 1957, the first study found that a serious accident, though highly unlikely, might kill as many as 3,400 people and damage $7 billion in property. WASH-740 epitomized the conflicting promotional and regulatory missions the Atomic Energy Commission was saddled with. Its specific political purpose was to convince the Congress that there would be no nuclear industry if the federal government did not provide insurance and indemnification against the catastrophic losses that might stem from a nuclear accident. It achieved that objective when the Price–Anderson Act sailed through Congress with virtually no opposition in 1957. WASH-740 and the Price–Anderson Act, however, made it increasingly difficult to convince Americans that the Atomic Energy Commission was doing all that it could to fulfill its regulatory responsibility. After all, the threat of crippling liability suits was one of the central ways that free market systems ensured safety.

One year after PG&E withdrew the Bodega Bay reactor, Pesonen wrote an article for *The Nation* entitled "Atomic Insurance: The Ticklish Statistics."[117]

---

116 The one exception to this pattern was at Malibu. There, reactor siting was also challenged by organized public opposition. Although the Regulatory Staff and the ACRS approved the site, the Atomic Safety Licensing Board (ASLB) – the AEC forum now responsible for public hearings – sided with intervenors, who had engaged the services of Professor H. Bernioff (the PG&E consultant at Bodega Bay). The ASLB approved construction but required that the plant be redesigned to withstand ground displacement (Okrent, *Nuclear Reactor Safety*, p. 274).

117 David E. Pesonen, "Atomic Insurance: The Ticklish Statistics," *The Nation* (October 18, 1965): 242–5.

Moving well beyond his original concerns about conservation at a specific site, Pesonen charged that the AEC had suppressed a study conducted by Brookhaven National Laboratory that updated the WASH-740 report. As was the case with the original report, the update was commissioned as part of the campaign for federal indemnification. The Price–Anderson Act expired in 1967, and the nuclear community was anxious to renew it – preferably long before the deadline.

Pesonen used hard-to-find, but public, information to piece together his charge that the AEC had suppressed the update, and that it had done so because the update threatened to intensify public concern about safety. Pesonen quoted an address by AEC commissioner John Palfrey delivered to the Atomic Industrial Forum on December 3, 1964, entitled "The Price-Anderson Law and Safety."[118] In it, Palfrey alluded to the Brookhaven update, informing the nuclear community that the report was not yet ready. Palfrey refused to talk about the report's conclusions. He did make it clear, however, that a major effort was under way.

Turning to the proceedings of a February 1965 meeting of the American Nuclear Society, Pesonen quoted the AEC's deputy director of regulation, Clifford Beck. Beck told his colleagues that what was really troubling the nuclear community was pressure to relax siting restrictions and move reactors closer to population centers at the same time that current reactors had five times the power and three times the fuel lifetime of reactors proposed just three years ago. Beck used the graphic metaphor of a thermos bottle in describing how increased size could change safety calculations. For the small reactor, Beck told the panel, "it may make good sense to surround it with a leak-proof, concrete "thermos bottle' containment, and in case of accident, just walk away and let everything inside settle down and cool off."[119] Things were different, however, with the far larger reactors currently being proposed for urban sites. "If you walk away from that 'thermos bottle,'" Beck warned, "its temperature curve ... will rise continuously and will simply heat up until it bursts, so a reliable cooling system must be added."[120] Pesonen used Beck's talk to illustrate technical concerns about larger reactors, and to imply that the consequences of an accident might be much more far-reaching than the original WASH-740 report suggested.

Ironically, it was Pacific Gas and Electric's public relations adviser, Hal Stroube, who gave Pesonen the most damaging evidence that some sort of cover-up was contemplated. Speaking at the same professional meeting as Beck, Stroube urged the AEC to cancel the "now-in-progress up-dating of the Brookhaven Report. I've eaten a steady diet of WASH-740 in the past three years as it became the bible of the anti-Bodega crowd," Stroube complained. Stroube agreed that any new report must be placed in context, but warned that doing so was "about as easy as trying to unscramble an omelet and return it to its several shells."[121] Quoting from the AEC Division of Technical Information transcript of the proceedings, Pesonen informed *The Nation*'s readers that before the meeting

118 Ibid., p. 243.   119 Ibid., p. 243.   120 Ibid.   121 Ibid.

adjourned, the "opinion was expressed that the original report had already caused serious problems in gaining public support of nuclear power and that the revision would tend to begin a new cycle of problems."[122]

Pesonen saved his best evidence for last. He quoted a June 18, 1965, letter from Seaborg to Joint Committee on Atomic Energy chair Chet Holifield, in which Seaborg danced around the issue of whether an actual report existed or not, but did report that "assuming the same kind of hypothetical accidents as those in the 1957 study, the theoretically calculated damages would not be less and under some circumstances would be substantially more than the consequences reported in the earlier study."[123] As Pesonen noted, the calculations necessary to support such a conclusion were quite complex. That Seaborg stated it suggested that he had at least a draft report to support the statement.

Frank Pittman, testifying in 1964 at the commission's authorization hearings, suggested that "if you make a new analysis and it doesn't improve the situation it might make matters worse." The events surrounding the WASH-740 update suggested this was fast becoming the nuclear community's credo.[124] Pesonen, though apparently unaware of Pittman's statement, tapped into the same specialized media used by the nuclear community. There he discovered and in turn disseminated information about an important safety review that had been stifled by the Atomic Energy Commission. Because the Price–Anderson renewal sailed through both houses of Congress in September 1965, Pesonen argued, the updated WASH-740 report was not needed to convince legislators. Congress rubber-stamped the Joint Committee's recommendation to renew federal indemnification and liability insurance. As Pesonen put it, "The AEC has proved itself most dexterous in defining the term 'subsidy.'"[125] According to the AEC, Price–Anderson was not a subsidy. Once Price–Anderson had been renewed, updating WASH-740 – though clearly useful to the questions that swirled around the siting debate – could only create political havoc, as far as the AEC was concerned. "For public relations reasons," Pesonen concluded, "as well as for legislative success in extending Price–Anderson, the AEC appears to have abandoned or suppressed the updated report – a major research project of potentially widespread public importance."[126]

Political "containment," though challenged by Pesonen, remained intact. Pesonen was reduced to piecing together scattered evidence that a study had been conducted to update WASH-740 and that the update increased the estimated damage that would occur in the unlikely event of a nuclear accident. On both scores he was correct, but he lacked the details required to convince others. The true measure of political containment can be effectively assessed today because Daniel Ford, following in the tradition of Pesonen and by obtaining thousands of pages of internal documents through Freedom of Information requests, and

122 Ibid.   123 Ibid., p. 244.   124 "Authorization Hearings."
125 Ibid.   126 Ibid.

David Okrent, a member of the Advisory Committee on Reactor Safeguards, have published accounts of the WASH-740 update.[127] Their accounts confirm the large-scale effort conducted at Brookhaven National Laboratory to update the WASH-740 report. They also explain why the nuclear community was hesitant to publish the update. When the Brookhaven team met in October 1964 with the AEC's steering committee charged with overseeing the update, it reported that as many as 45,000 people might be killed by a major reactor accident.[128] The reason for the tremendous increase in casualties was the vastly increased size of new reactors and the radioactive material they might contain at the time of an accident. Faced with such staggering projections, the steering committee headed by Clifford Beck sought at least to place them in context and to lift the "pall of gloom" by asking the Brookhaven team to estimate the probability of such an accident occurring. But the Brookhaven scientists refused to do so, citing the dearth of operating experience on which to base such projections.[129]

Having failed to budge the Brookhaven scientists, the steering committee had no choice but to present the commission with what clearly was a major political problem. As the minutes of one steering committee meeting put it, the "impact of publishing the revised WASH-740 report on the reactor industry should be weighed before publication."[130] Richard Doan of the AEC's regulatory staff was concerned that the report, if published, would provide the grounds for lawsuits that might retard future reactor construction.[131] The net result of these consultations and of the AEC's consultations with the industry was the letter Seaborg wrote to Holifield (quoted by Pesonen) in June 1965. Briefing the Advisory Committee on Reactor Safeguards on the Brookhaven update, Clifford Beck noted that because reactors had grown so in size, the hypothetical accident postulated by Brookhaven assumed that the reactor core would melt through the containment floor into the earth – a process known as "the China Syndrome." David Okrent, a member of the ACRS at the time, summed up the discrepancy between the internal discussion and the public record in characteristically muted and understated terms. "This appears to be the first unequivocal statement by the regulatory staff," Okrent wrote in his 1981 account of these discussions, "that containment failure was a likely result of core melt in large LWRs [light-water reactors]. However, this conclusion was not drawn in the letter from Chairman Seaborg to Representative Holifield."[132] Small wonder that Pesonen, the ultimate outsider, could barely chisel at the nuclear community's still formidable political "containment."

Although local resistance at Bodega Head, Ravenswood, and Malibu was intense, the concerns about an "irrational" public expressed in *Nucleonics* and the industry's

127 Ford, *Cult of the Atom*; Okrent, *Nuclear Reactor Safety*.
128 Ford, *Cult*, p. 69.　　129 Ibid., pp. 72–3.
130 Quoted in Ford, *Cult*, p. 73.　　131 Ibid., p. 76.
132 Okrent, *Nuclear Reactor Safety*, p. 102.

siting conference remained exaggerated. At least in private, many of the experts charged with assuring nuclear safety seemed far more concerned than the average citizen. With the exception of Pesonen's effort to nationalize concerns about the consequences of a reactor accident, public opposition was limited to specific site-related issues. At a number of sites, such as San Onofre, opposition did not result in project cancellation or even delays. It was the AEC Regulatory Staff, the ACRS, and, increasingly, Atomic Safety and Licensing Board (ASLB) approval that determined the fate of reactor proposals in the first half of the 1960s. Grass-roots opposition – with the exception of Malibu – did not seem to affect these processes. In fact, the other important instance of ASLB-imposed restrictions – Oyster Creek – was the product of ASLB's own probing and of state intervention.[133]

Perhaps the strongest evidence that a safety gap existed – where expert fears far outweighed public doubts – was the number of reactors approved with little or no opposition. Con Edison's Indian Point II, for example, the largest reactor licensed to date (1966) and one that caused considerable debate among safety experts at the ACRS, breezed through the construction permit licensing hearings. What's more, the utility industry as a whole did not seem disturbed by localized opposition or occasional restrictions imposed by ASLBs. As previously discussed, shortly after Ravenswood and Bodega Bay were withdrawn, orders for reactors poured in.[134]

The public, despite the industry's worst fears, and in spite of scattered, geographically concentrated opposition, continued to demonstrate relatively little concern about the safety of reactors. As ACRS member Osborn stated: "Safety questions no longer appear to be important considerations in the opposition of the general public to reactors."[135] Safety experts, on the other hand, cautioned each other that safeguards were not yet well developed. To the extent this safety gap was beginning to close, it was because debate within the nuclear community could no longer be "contained," and because a handful of outsiders who questioned the safety of nuclear power excelled at exposing internal debate to broader public scrutiny. Unlike the iron triangle forged by the nuclear community, issue networks that engaged expertise housed outside the AEC had little incentive to contain the debate. Both Abel Wolman in public health and Secretary Udall in the Department of the Interior had an active obligation to alert their constituencies. Pesonen's constituency, as he envisioned it, reached beyond those of Wolman and even Udall. Only highly public, often symbolic, communications could reach it.

Meanwhile, subdivisions within the nuclear safety community remained staunchly committed to the political strategy of "containment." But as the units matured, commitment to their organizational mission overpowered the will to "contain." Ultimately the experts' fears either seeped through to the public or were force-fed by outside challengers. As the kind of debate that had gone on

---

133 *Nucleonics* 23, no. 1 (January 1965): 26.
134 *Nucleonics* 24, no. 10 (October 1966): 27.
135 Subcommittee Meeting on Reactor Siting, in ACRS Minutes, March 20, 1965, p. 4.

within the safety community for years was brought to the public's attention in the mid-1960s, the safety gap narrowed. Harold Price disagreed with Osborn's assessment of public fears in 1965. In Price's opinion, the public's attitude "reflects the attitude of the nuclear industry which apparently has not yet convinced itself that reactors are completely safe."[136]

Whatever the exact size of the gap between public awareness of safety questions and expert debate about these issues in the mid-sixties, one thing is clear: The size of that gap narrowed significantly in the next decade. The phenomenon described thus far – the ever-broadening forums in which experts raised safety questions – remained a crucial force behind this change. But that engine was fueled by the popularization of an alternative perspective on technological development and the expertise that served it. Environmentalism was the integrating force that brought together the disparate elements – state authorities, tangentially involved federal agencies, grass-roots citizens groups, and the experts that served or in some instances motivated these parties. Eventually the set of political actors loosely faxed together by the language and perspective of environmentalism emerged as a force capable of challenging the grip on expertise and national organization monopolized by the nuclear community.

THE ENVIRONMENTAL REVOLT AGAINST SPECIALIZATION

The emergence of the environment as a permanent concern of national policymakers is a phenomenon that has just begun to be treated by historians.[137] The most important synthetic work on post–World War II environmentalism to date has been done by Samuel Hays.[138] He places the environmental movement in a

---

136 Ibid.
137 There is a far richer literature on the conservation movement rooted in the Progressive Era, preservationists such as John Muir, and the origins of ecology. See Roderick Nash, *Wilderness and the American Mind* 3rd ed. (New Haven, Conn.: Yale University Press, 1982); Hays, *Conservation and the Gospel of Efficiency*; Fleming, "Roots of the New Conservation Movement," pp. 7–94; John Higham, "From Boundlessness to Consolidation: The Transformation of American Culture 1848–1860" (William Clements Library, 1969); Lynn White, "The Historical Roots of Our Ecological Crisis," *Science* 155 (1971): 1203–7; Donald Worster, *Nature's Economy: The Roots of Ecology* (San Francisco: Cambrigde University Press, 1985, c. 1977); Alan Trachtenberg, *The Incorporation of America: Culture and Society in the Gilded Age* (New York: Hill and Wang, 1982), pp. 3–37, 140–81, 208–34; Paul Brooks, *Speaking for Nature: How Literary Naturalists from Henry Thoreau to Rachel Carson Have Shaped America* (Boston: Houghton Mifflin, 1980), pp. 1–32; Stephen Fox, *John Muir and His Legacy: The American Conservation Movement* (Boston: Little, Brown, 1981), pp. 3–102; Michael P. Cohen, *The History of the Sierra Club, 1892–1970* (San Francisco: Sierra Club Books, 1988); William G. Robbins, *Lumberjacks and Legislators* (College Station: Texas A & M University Press, 1982); Donald Nicholas Baldwin, *The Quiet Revolution: The Grass Roots of Today's Wilderness Preservation Movement* (Boulder, Colo.: 1972).
138 Hays, *Beauty, Health and Permanence* and "The Structure of Environmental Politics Since World War II," pp. 719–38. See also Rachel Carson, *Silent Spring* (Boston: Houghton Mifflin, 1987 [original ed. 1962]); Peter J. Taylor, "Technocratic Optimism, H. T. Odum and the Partial Transformation of Ecological Metaphor After World War II," *Journal of the History of Biology* 21 (1988): 213–44; Bill Devall and George Sessions, *Deep Ecology* (Salt Lake City: Gibbs Smith, 1985); Patricia Hines, ed., *The Recurring Silent Spring* (New York: Pergamon, 1989); Frank

framework that pits the developmental orientations of large-scale national institutions against local particularistic concerns with the everyday quality of life. That millions became concerned with such issues, in turn, was the product of a rapidly rising standard of living and a shift from exclusive interest in work, to greater concern about leisure time. Rising standards of living and the accompanying shift in values augmented the professional and technical resources available to local communities. In Hays's words, "Much of the struggle focused on scientific and technical information, not on just what was the 'accurate' information, but on control of the resources by which information was generated, disseminated, gathered and applied in decision-making."[139]

Environmentalism challenged the kind of specialization that was so crucial to nuclear power's technological, economic, and political development. The discipline of ecology epitomized the assault on specialized approaches that might lead to progress in one area only to create greater harm somewhere else. Ecology was "synthetic and holistic in aim," Donald Fleming has written, "rather than analytical. It broke down the isolation of scientific disciplines from one another to the common end of grasping the total stresses upon 'a total ecological web.'"[140] In *Silent Spring*, Rachel Carson did not call for an attack on science, rather, she turned to a branch of science other than chemistry – biology – to provide alternatives for controlling insects. Biologists might succeed, Carson hoped, because they appreciated "the whole fabric of life to which these organisms belong."[141]

As the environmental perspective garnered increasing political support in the late sixties and early seventies, it established powerful institutional bases. The Office of Technology Assessment's Russell Peterson, for instance, frequently attacked overly specialized approaches. As he told the American Association for the Advancement of Science in 1978, "while we pride ourselves on our perceptions of truth in our areas of specialty, we must widen them to include a holistic perspective. Otherwise what is really important is not always obvious."[142] Even President Richard Nixon by 1970 had adopted the rhetoric of environmentalism. In establishing the Environmental Protection Agency (EPA), Nixon told the nation that "despite its complexity, for pollution control purposes the environment must be perceived as a single interrelated system."[143] Environmentalists viewed nuclear power as part of a larger system, rather than as an end in and of itself.

Graham, *Since Silent Spring* (Boston: Houghton Mifflin, 1970); Walter A. Rosenbaum, *The Politics of Environmental Concern* (New York: Praeger, 1973); Alfred Marcus, *Promise and Performance: Choosing and Implementing Environmental Policy* (Westport, Conn.: Greenwood, 1980); David Vogel, "The 'New' Social Regulation in Historical and Comparative Perspective," in Thomas K. McCraw, *Regulation in Perspective* (Cambridge: Harvard, 1981): 155–86; Joseph L. Sax, *Defending the Environment: A Strategy for Citizen Action* (New York: Knopf, 1971); Joseph M. Petulla, *American Environmentalism: Values, Tactics, Priorities* (College Station: Texas A & M University Press, 1980).
139 Hays, "Environmental Politics," p. 727.
140 Fleming, "New Conservation Movement," p. 46.
141 Carson, *Silent Spring*, p. 278.
142 Peterson quoted in Dickson, *The New Politics*, p. 241.
143 Nixon quoted in Marcus, *Promise and Performance*, p. 46.

The environmental perspective, emphasizing a holistic approach, drawing on grass-roots citizen support, and operating out of increasingly powerful institutions, provided just the political force needed to make good on David Lilienthal's call for "mainstreaming the atom." Although Lilienthal had not mentioned thermal pollution or used phrases such as "environmental impact," his stinging criticism of the Atomic Energy Commission's 1962 Report to the President called for integrating the atom into the nation's other scientific endeavors and ridding the AEC of its claim to special treatment. The AEC in fact had succeeded in bringing nuclear power into the economic mainstream. It had captured the attention of the political mainstream – through its highly visible promotional campaign spearheaded by the Report to the President. But that visibility had unanticipated consequences, the most significant of which was environmentalists' demand for broader participation.

Lilienthal anticipated another crucial element that both united disparate interests that shared the environmental perspective and castigated a central belief shared by the nuclear community. Environmentalists from René Dubos to Rachel Carson insisted on scientific and technological humility. Even when it came to eradicating death-threatening microbes, Dubos argued for control rather than complete elimination. Those who sought to eradicate entirely were not only impractical, they were arrogant as well, Dubos argued.[144] Rachel Carson was moved to write *Silent Spring* in part because the massive use of insecticides convinced her that man was approaching nature with arrogance rather than humility.[145] Lilienthal had sought to inject a touch of humility into the Atomic Energy Commission's approach when he urged the commission to seek "manageable possibilities" rather than panaceas and absolute solutions.[146] Scientific humility, however, and promises of energy too cheap to meter, the plutonium economy, and transmuting science fiction into scientific reality, were mutually exclusive.

The AEC and its oversight committee, the Joint Committee on Atomic Energy, resisted and rebuffed political actors who claimed a stake in the nuclear debate as a result of their growing sensitivity to the environmental perspective. Questioning the nuclear community's arrogance and citing nuclear power's wide-ranging implications, these skeptics demanded more information and were less willing to defer automatically to nuclear insiders. The AEC and JCAE, however, were committed to the long tradition of broad discretion within their area of expertise, and were loath to concede any part of their turf. Having turned back with relative ease the disparate efforts of state agencies and of experts working with citizen groups in the 1950s and early 1960s, the AEC/JCAE alliance was confident it could hold the line in the future. That task became more difficult as those involved in the challenges that had operated in relative isolation during nuclear power's first two decades embraced a common perspective.

144 Fleming, "New Conservation Movement," p. 38.
145 Graham, *Since Silent Spring*, p. 17.
146 On Lilienthal's critique, see Chapter 7.

Inherent in all of these disputes were questions of participation and access. This sharp "insiders–outsiders" divide gave critics another issue that they shared in common. It was these issues that eventually overshadowed the more technical ones. Framing the debate as a contest between insulated decision making and free discussion among a variety of interested and informed parties, the opposition made the whole matter far more accessible to the broader public. Experts – particularly in the biological sciences – led the way.

In *Beauty, Health, and Permanence*, Hays illuminates three streams of environmental concerns that came together by the 1960s: protecting scarce natural areas, warding off threats to community degradation, and protecting health. Ironically, given the challenge that environmentalists ultimately mounted, in the early sixties all three objectives of the postwar environmental movement worked in nuclear power's favor. As we have seen, the 1962 Report to the President argued that nuclear power would preserve scarce natural resources. Proponents pointed out that nuclear plants were far more pleasing aesthetically than fossil fuel plants with their ash-spewing smokestacks. Utilities cited restrictions on coal-burning plants – levied for both aesthetic and health reasons – as one reason they were turning to nuclear power. As a rare editorial in *Nucleonics Week* crowed, "Let the Public Choose the Air It Breathes."[147] Utilities concurred. Connecticut Light and Power's chair, for instance, stated in 1965 that "atomic power is bound to be increasingly attractive to communities as concern over air pollution intensifies."[148] A number of important conservation groups either supported, or were ambivalent about, nuclear power.[149] The coal lobby, which was nuclear power's only organized national opponent until the late sixties, felt it had more to lose from a battle over potential health risks than the nuclear community. When the Atomic Industrial Forum arranged a December 1963 meeting with the National Coal Association, the coal lobby suggested that both sides cease discussing health hazards. The Coal Association was "willing not to bring up the health and safety problem in the future" if the Forum would agree to do the same.[150]

HEATED WATER EXPANDS

As environmentalism became a national force, the biologists and ecologists at the forefront of the movement began to examine the effects of nuclear plants more closely. They concentrated on the thermal effects of the superheated waste water released by nuclear power plants. As Samuel Walker, official historian at the Nuclear Regulatory Commission, has noted, conventional power plants also generated waste heat. The problem at nuclear plants, however, was more acute.

---

147 Quoted in J. Samuel Walker, "Nuclear Power and the Environment: The Atomic Energy Commission and Thermal Pollution, 1965–1971," *Technology and Culture* (October, 1989): 1.
148 Ibid., p. 2.   149 Ibid.
150 Luedecke to Tremmel, December 27, 1963 in AEC 152/168, January 7, 1964, AEC/DOE, p. 2.

Comparably sized nuclear plants generated 40 percent to 50 percent more waste heat than conventional plants.[151] Aligned with the scientists were traditional conservation constituencies, such as fishing and wildlife organizations, mobilized by the more visible impacts resulting from operating nuclear plants.

The first challenge on thermal pollution came from a federal bureau with, as in the case of the U.S. Geological Survey, a mandate that conflicted with the AEC's promotional agenda and with an independent base of expertise drawn from disciplines outside physics, chemistry, and engineering. The Fish and Wildlife Service, a division of the Department of Interior, began in the mid-1960s to suggest that the Atomic Energy Commission consider nonradiological effects of reactors on the environment, particularly the impact of thermal pollution. This the AEC staunchly resisted, claiming it lacked the authority to rule on non-radiological effects.[152]

In March 1966, the head of the Fish and Wildlife Service increased his pressure on the AEC to consider the thermal effects of reactors. What disturbed the AEC's director of regulation, Harold Price, the most was not the opposition itself, but another federal agency's effort to expand the scope of debate. Fish and Wildlife had not only sent its criticism to several state agencies, it had "openly and publicly challenge[d] the position of the Commission with respect to authority over thermal effects."[153] There was more bad news on this front for Price. John D. Dingell, chair of the Fish and Wildlife's oversight committee, and staunch protector, also joined the fray. This further broadened the debate, particularly when Dingell held hearings in May 1966 and grilled AEC officials.[154]

The Atomic Energy Commission fought a two-front war against this assault. The most visible one marshaled legal arguments that continued to eschew AEC jurisdiction. Less visibly, but crucial to the defense of nuclear power should the first front fail, the AEC scrambled to marshal expertise on the effects of thermal pollution. Its efforts were halting and defensive at first. For instance, in response to the president's Advisor on Science and Technology Donald Horning's inquiry about thermal pollution, the AEC stated in September 1968 that "the AEC competence in directing pertinent research on such subjects as heat transfer and fluid flow, indicates that increased AEC research activity might be desirable."[155] By 1969, the AEC could tell the Office of Science and Technology, "The AEC has for a number of years been deeply involved in environmental engineering research and development." The commission not only cited its work on the dissipation of waste heat, but pointed to its investigations into seismology as well.[156]

The most visible public challenges on thermal pollution did not arise until 1968 and 1969. As with Bodega Head, experts in the biological sciences led the way. The best documented case is the Bell Station on Cayuga Lake in upstate

151 Ibid., p. 3.   152 Ibid., p. 3.   153 Quoted in ibid., p. 6.   154 Ibid, pp. 6–7.
155 Draft response to Horning, attached to AEC 152/238, September 30, 1968, AEC/DOE.
156 Draft letter to David Freeman, attached to AEC 152/240, April 3, 1969, AEC 152/240, p. 17.

New York. New York State Gas and Electric (NYSGE) filed an application for the Bell Station in 1967. Public opposition developed in 1968. Dorothy Nelkin's case study of the dynamics of that opposition points out that the initial source was a group of scientists affiliated with the Water Resources Center at Cornell. Far from being adamantly opposed to the plant's construction, the center assumed it would be built and in fact applied for a research grant – which was turned down – from the AEC to study the plant's environmental impact. Several scientists pursued the question of environmental impact anyway, and issued the first critical report on the plant and its potential impact on the lake's ecology. Nelkin concluded that the voluntary action of these scientists was crucial. Without it a controversy would not have developed.[157]

The scientists were soon joined by the Cayuga Lake Association – an organization formed in 1956 to deal with sewage problems. That organization was supplanted by the Citizens Committee to Save Cayuga Lake (CCSCL). CCSCL was dominated by professionals associated with Cornell. Its nerve center was its scientific advisory body. The disagreement between scientists was initially subtle and complex, according to Nelkin. Like the question of radiation release and its health impact, which was also raised by opponents, the major problem was that both sides had difficulty acquiring the data from which to draw definitive conclusions. As the debate became increasingly strident New York State Gas and Electric concluded that whatever impacts there might be would be negligible; the opposition focused selectively on those data most supportive of its dire predictions.[158]

Owing to growing concerns on the national level about the environment, the opposition found little difficulty in expanding the scope of debate. In fact, Senator Edmund Muskie's Subcommittee on Air and Water Pollution discussed the Cayuga site before formal hearings on the state level were held.[159] Undoubtedly the opposition sought out these forums because of the AEC's refusal to consider a reactor's impact on the environment except in terms of safety. If grass-roots groups felt snubbed by the AEC, the commission's actions both humiliated and enraged powerful congressional leaders. According to one staffer, Muskie felt that the AEC was "thumbing its nose at the intent of Congress."[160]

Under mounting public pressure NYSGE officially postponed its request, even though AEC approval appeared likely. Until the Bell withdrawal (and another at Easton at about the same time), reactor cancellations had always been the product of imminent or actual AEC regulatory opposition. Although public opposition had created unfavorable publicity and expanded the scope of debate,

157 Nelkin, *Nuclear Power and Its Critics*, pp. 37, 38.
158 Ibid., pp. 57–61, 76, 98; see also *Nuclear Safety* 10, no. 1 (January–February 1969): 108–9.
159 Senate Subcommittee on Air and Water Pollution of the Committee on Public Works, "Hearings on the Extent to Which Environmental Factors Are Considered in Selecting Power Plant Sites," 90th Cong., 2d sess., April 17–19, 1968 [hereafter Muskie Hearings], pp. 941–61. Nelkin, *Nuclear Power and Its Critics*, p. 73.
160 Walker, "Nuclear Power and the Environment," p. 9.

it had never actually stopped a project. That public opposition in and of itself was the major cause for withdrawal signaled a new phase of the commercialization process. As late as 1967, cooling systems for waste water were rare. In less than three years, public scrutiny, or even its threat, made them so common that they quickly became the visual shorthand for nuclear power. The large majority of reactors located on smaller bodies of water under construction in 1970 included cooling systems.[161]

The nuclear industry had always feared that local site opposition would arouse national interest. Finally, it looked as though that concern was justified. Writing in *Nuclear Safety* – a journal published by the Oak Ridge National Laboratory – the chair of the Atomic Industrial Forum's Public Affairs and Information Committee cited an article published in *Sports Illustrated* in 1969 entitled "The Nukes Are in Hot Water" as the first indication of a change in the peaceful situation that had existed heretofore. The author was concerned about the issue getting more national press, which it soon did. In fact, a survey of the site approval process offered plenty of warning signs long before the *Sports Illustrated* article.[162]

Some of the earliest signs that environmentalism was beginning to find more permanent political and institutional support appeared at the state level. The New Jersey Department of Health, for instance, considered the limited information regarding thermal pollution submitted by Jersey Power on behalf of its Oyster Creek plant to be "a major problem." The Department recommended that the issue be resolved, before the AEC granted a construction permit.[163] Even the Atomic Energy Commission, which eschewed responsibility for regulating thermal pollution, acknowledged that states were encouraged to establish water-quality standards under the Federal Water Quality Act of 1965. The state of New Jersey held hearings the same year the act was passed to determine the impact of released waste water at the Oyster Creek plant. Jersey Power eventually satisfied the concerns of the state agencies. Other states, however, proved less cooperative as concerns about thermal pollution mounted.[164]

When Vermont Yankee applied for a construction permit in November 1966, Vermont's attorney general insisted that the plant include cooling towers. Because of the plant's location, New Hampshire and Massachusetts officials also expressed concern. "Vermont will receive a million-dollar injection into its economy," Bay State Attorney General Elliot Richardson quipped, "Massachusetts will receive hot water."[165] Public officials in all three states denounced the AEC at Muskie's

161 Walker, "Nuclear Power and the Environment," p. 6.
162 H. G. Slater, "Public Opposition to Nuclear Power: An Industry Overview," *Nuclear Safety* 12, no. 5 (September–October, 1971): 448–56. R. H. Boyle, "The Nukes Are in Hot Water," *Sports Illustrated*, January 20, 1969, pp. 24–8.
163 Weisburd to Jensch, November 1964, "Jersey Central Power and Light" file, Box 578, JCAE.
164 *Nucleonics* 23, no. 1 (January 1965): 26; *Nucleonics* 23, no. 2 (February 1965): 20.
165 Walker, "Nuclear Power and the Environment," p. 8.

hearings held in Montpelier, Vermont.[166] New Hampshire ended up suing the AEC over the matter, and losing. Similar disputes broke out in Florida and Michigan. "Like the controversy over the Vermont Yankee reactor," Walker concluded,

they usually started when state officials or conservationists raised questions about the thermal effects of a plan, became matters of dispute when a utility refused to build cooling towers or take other action ... and ended only after considerable acrimony and/or concessions by the power company. In several cases, what began as local issues received widespread attention as a part of growing national concern over environmental quality in general and thermal pollution in particular.[167]

STATING THE CHALLENGE

State governments that had challenged the Atomic Energy Commission's jurisdiction over safety determinations in the 1950s but failed, found a new framework and with it new sources of expertise. They began to adopt, at least in part, an environmental perspective. Minnesota, which had led the way in the fifties, remained on the cutting edge of the challenge. In January 1969, the Northern States Power Company applied to the Minnesota Pollution Control Agency (MPCA) for a permit to discharge industrial waste from its Monticello reactor.[168] After public hearings during which several scientists, environmentalists, and the mayor of St. Paul urged the agency not to grant a permit, the agency decided to select a consultant to offer an expert opinion on the matter. It chose Dr. Ernest C. Tsviglov, professor of engineering at the Georgia Institute of Technology, to conduct a study of the reactor and recommend standards that Minnesota should apply. At a subsequent meeting, the director of the agency recommended that a discharge permit be approved because Dr. Tsviglov "feels this permit is okay." At the same meeting, however, the Minnesota Committee for Environmental Information, a group made up primarily of scientists and professionals on the faculty of the University of Minnesota, issued a statement expressing concern that issuance of a permit was being considered before the study was completed. Despite the director's recommendation, the agency voted to delay issuance of the permit.[169]

In early 1969, Dr. Tsviglov issued his report. He recommended that operating approval for Monticello be limited to one year, that radioactive releases be monitored carefully, and, most significantly, that radiation release limits set by the state be considerably more restrictive than those set by the AEC. Regarding the Northern States Power Company's contention that the MPCA "prove the

---

166 Ibid.   167 Ibid., p. 11.
168 That reactor had begun construction in 1967, having already obtained a construction permit from the AEC.
169 *Scientist and Citizen* 10, no. 6 (August 1968): 154–5; quotation from p. 155.

need for more restrictive standards," Tsviglov sounded much like Vischer testifying before the Joint Committee on Atomic Energy a decade before, responding: "It is not the proper or desirable function of the MPCA to defend its desire to protect the public and the environment to the full extent that is feasible and reasonable."[170]

The Atomic Energy Commission refused to back down. It firmly believed, as the commission told the Office of Science and Technology in April 1969, that "regulatory control over radiological effluents from nuclear plants should continue to be exercised exclusively by AEC."[171] When the Pollution Control Agency accepted Tsviglov's recommendations, the issue wound up in the courts. In March 1971, a federal district court ruled against the State of Minnesota, determining that Congress had preempted the field of radiation regulation by nuclear power plants. The decision was upheld on appeal.[172]

Although a defeat for state prerogatives, the Monticello controversy helped nationalize opposition to nuclear power in several respects. First, the legal battle attracted a great deal of press, publicizing a problem that previously had not been associated with reactors. The Monticello controversy also led to commercial nuclear power's first rebuke by a national scientists' organization. The Scientists Institute for Public Information was an outgrowth of the fallout controversy that had prompted the organization of the Minnesota Committee for Environmental Information. Local branches had been active in other reactor controversies, such as Oyster Creek, New Jersey.[173] Those challenges, however, had remained localized. In the case of Monticello, the Scientists Institute for Public Information used its national forum, *Scientist and Citizen*, to alert a national audience to a new set of concerns. As the August 1968 issue put it:

Public concern about radioactive fallout has abated since the advent of the ban on atmospheric weapon testing, but there remain many other possible sources of radiation contamination, and the need continues for careful assessment of the benefits and risks associated with nuclear technology. This need will intensify as large scale nuclear installations spread through the land.[174]

The Monticello case was also significant because Minnesota captured markedly more public support from other states than it had in its earlier effort (in the 1950s) to carve out a larger state role. Twelve states joined Minnesota in its court battle. Three other states lowered their standards to match Minnesota's. The National Conference of Governors passed a unanimous resolution in September

170 "Cooling It in Minnesota," in *Environment* 11, no. 2 (March 1969): 21.
171 AEC 152/240, p. 8.
172 Del Sesto, *Science, Politics, and Controversy*, p. 150.
173 See, for instance, Charles Roberts, co-chair New Jersey Scientists' Committee for Public Information, to Samuel Jensch, October 28, 1964, "Jersey Central Power and Light" file, Box 578, JCAE.
174 *Scientist and Citizen*, August 1968.

1969 endorsing the right of states to set standards more restrictive than those of the AEC.[175]

There were other indicators that a number of states were beginning to balance the developmental perspective that dominated their early entry into nuclear affairs with the kind of environmental sensitivity that Wolman quietly had been urging public health officials to promote for almost twenty years. By 1970, for instance, New York's Atomic Development Authority, which under the guidance of Oliver Townsend in the 1950s had been one of the most aggressive promoters of nuclear power, took seriously environmental concerns. As Development Authority director James Cline told an Atomic Industrial Forum conference, New York would most likely consider in the next year a new regulatory procedure that would provide "a balanced and comprehensive review of environmental and other factors prior to construction of any new power generating facility."[176] In June 1971, Harold Price briefed the executive director of the Joint Committee on Atomic Energy about the avalanche of state activity in this field. Florida, Maryland, New York, New Hampshire, Oregon, and Vermont were proposing legislation that would subject nuclear sites to significant levels of environmental review. Four other states contemplated similar actions.[177]

It is easier to catalogue the tangible results of environmentalism – particularly the new level of scrutiny applied by states in siting decisions – than to capture the variety of ways it emerged in fifty states and hundreds of controversies. It is useful to return to Massachusetts to examine in detail at least one state's transition from vigorous nuclear promoter to leading critic. The emerging environmental perspective played a crucial role in that transformation.

No state supported commercial nuclear power more vigorously than Massachusetts in the fifties and early sixties. That support was most visible at the

---

175 Council of State Governments, *Environmental Quality and State Government* (Lexingtom, Ky.: Council of State Governments, 1970), pp. 11–12; and *Science* 172 (June 18, 1971): 1215. *Environment* 11, no. 7 (September 1969): S–2. While Minnesota pursued its court case, Dr. Ernest Sternglass embarked on a crusade to publicize the dangers of even extremely low levels of radiation. Sternglass, a professor of radiation physics at the University of Pittsburgh, contended that low doses of fallout from nuclear weapons tests may have caused more than 400,000 infant deaths since the early 1950s. Although Sternglass could find virtually no support for his conclusions – and despite the fact that they were attacked by many prominent scientists – his publicity campaign, launched in 1969, included articles in the *Bulletin of the Atomic Scientists* and *Esquire* (Philip Boffey, "Ernest J. Sternglass: Controversial Prophet of Doom," in *Science* 166 [October 10, 1969]: 195–200). Ironically, the controversy widened when two AEC-funded researchers – Arthur Tamplin and John Goffman – in rebutting some of Sternglass's more extravagant claims, conceded that fallout might have accounted for more than 8,000 fetal deaths in 1963 (ibid., p. 200). Efforts by the AEC to suppress Goffman and Tamplin's research further exacerbated the debate (Philip Boffey, "Goffman and Tamplin: Harassment Charges Against AEC, Livermore," in *Science* 169 [August 28, 1970]: 838–43).
176 "The Development of Environmental Policy Action at the State Level," November 16, 1970, "New York State–Atomic Development" folder, Box 266, JCAE, p. 7.
177 Price to Bauser, June 15, 1971, "Cooperation with States–New Hampshire" folder, Box 266, JCAE.

top, with leading politicians, science-based institutions such as the Massachusetts Institute of Technology, and large businesses calling for local efforts to aid development. Characteristic of the optimistic spirit that anticipated great things from commercial nuclear power was a story told by Rowe, Massachusetts, chairman of the board of selectmen and real estate entrepreneur Jack Williams. Williams recalled that within weeks of the 1955 announcement naming Rowe as the site for the Commonwealth's first commercial reactor, he received a phone call from frozen food mogul Clarence Birdseye. The future, Birdseye told Williams, lay in irradiated foods. Birdseye was considering building a huge processing plant to take advantage of the new reactor's surplus radiation. There was even speculation about Rowe's becoming a city.[178]

Support for nuclear power was not limited to a political and economic elite. Periodical accounts of the average Rowe citizen's response to nuclear power's arrival consistently portrayed reactions as highly favorable.[179] The only concerns expressed had to do with fears of unruly development, and in the case of some fishing enthusiasts, rumors that the reactor might heat up the Deerfield River. That such concerns were not taken terribly seriously was perhaps best demonstrated by Howard J. Cadwell, president of Western Massachusetts Electric (a member of the Yankee Rowe consortium). Cadwell told the Associated Press, "I'm a trout fisherman myself. I think the plant won't do a bit of harm and might actually improve the fishing."[180] Twenty years later, in 1980, Rowe still liked its reactor, the *Boston Globe* reported.[181]

The initial reaction to the Pilgrim reactor, announced by the Boston Edison Company in January 1967, was similar.[182] "If this does for Plymouth what Yankee Atomic did for Rowe the taxpayers will be very happy indeed," the chair of the Plymouth Home National Bank told one reporter.[183] The state supported nuclear power actively throughout the sixties. The attorney general of Massachusetts, Elliot Richardson, intervened in the Atomic Energy Commission's licensing hearing as a supporter of the plant. Robert Ward of the *Boston Globe* captured the pioneering spirit that many claimed nuclear power brought back to the Bay State. In an article headlined "New Pilgrim Coming to Plymouth," replete with the subhead "Log of the New Pilgrim," Ward reported that "like the sturdy 180-ton ship which brought colonists to these windswept shores, the newcomer will signal the beginning of a new era."[184]

Massachusetts, of course, was not immune to the national debates that unsettled

178 Barry Werth, "Windfall: Or How Rowe, Massachusetts, Learned to Stop Worrying and Love Its Nuclear Power Plant," *New England Monthly* (November, 1989): 40.
179 See, for instance, *Boston Herald*, April 10, 1955; *Boston Globe*, April 24, 1955; *Christian Science Monitor*, February 16, 1956; ibid., April 2, 1959.
180 *Christian Science Monitor*, April 2, 1959.
181 *Boston Globe*, May 1, 1980, p. 15.
182 See, for example, *Boston Traveler*, January 4, 1967; *Boston Globe*, January 4, 1967; *Boston Herald American*, January 4, 1967.
183 *Boston Globe*, January 4, 1967.   184 *Boston Globe*, January 9, 1970.

the nuclear community. The fallout scare was widely publicized, but with the exception of Paul Benzequin's warning, it did not seem to affect attitudes toward commercial nuclear power. Nor did the release of WASH-740 leading to Price–Anderson indemnification legislation in 1957 stir much of a reaction in the Commonwealth.[185] The nuclear-related controversy that struck closest to home was a flap over the AEC's ocean disposal of low-level wastes. In the summer of 1959 a group of Cape Codders protested a proposal that would have allowed dumping two hundred miles due east of the Cape. The dispute seems to have received as much national as local attention.[186] In the end, the AEC backed off. Had the AEC stated it planned to permit dumping "west of the Azores," one local selectman quipped, rather than "east of Massachusetts," there would have been no controversy.[187] Support for commercial nuclear power appears to have remained steady throughout the sixties, even as the plans to increase the percentage of nuclear-generated power grew. By the spring of 1967, a blueprint for New England's power needs called for eleven operating nuclear reactors by 1972.[188]

There were isolated instances of doubt expressed by 1967. A *Boston Herald* reporter checked out the reaction to the Pilgrim announcement at Currier's, a downtown Plymouth restaurant. One patron expressed characteristic enthusiasm. "I think that's wonderful," she replied upon hearing the news. Then she added a troubling aside: "I hope it isn't dangerous."[189] A review of correspondence docketed as part of Pilgrim's licensing reveals similar scattered instances of concern. Mrs. T. J. McTaggart's challenge was the most sophisticated. The letter was directed to Secretary of the Interior Udall, as McTaggart explained, because of his "militant interest in preserving the natural beauty and the natural resources of the country." Before beauty, however, McTaggart had safety on her mind. She noted that the plant was originally slated to be built in metropolitan Boston. Although the reasons for the site change had not been made public, McTaggart told Udall, she concluded "that they wished to put in [sic] in a thinly populated area in case of accident." The Cape was not thinly populated enough, as far as this nearby cottage owner was concerned.[190] McTaggart had certainly not been privy to the internal safety debate that in fact had led to moving the reactor to a more sparsely settled area. Nor had that debate been made public. She correctly surmised, however, that some doubt existed, otherwise the reactor would not have been moved.

Another indication that concern was mounting came from the ranking Republican member of the Joint Committee on Atomic Energy, Vermont Senator George Aiken. In August 1968, an aide to the senator told an Associated Press

185 See, for instance, May 21, 1957, from unnamed Boston newspaper in Boston University, School of Public Communications, "atomic power" file, Boston, Mass.

186 See summary in Mazuzan and Walker, *Controlling the Atom*, p. 361.

187 Quoted in ibid.   188 *Boston Globe*, May 21, 1967.

189 *Boston Herald*, January 8, 1967.

190 McTaggart to Udall, June 9, 1967, general correspondence (category H) Pilgrim Reactor, Public Document Room, Nuclear Regulatory Commission, Washington, D.C.

reporter that the JCAE had "been faced this year with growing resentment toward nuclear plants by a public afraid of nuclear accidents." Aiken, who continued to support development of nuclear power – the headline covering this press release stated "Aiken Sees Need of Hike in N-Power" – straddled the safety question by reiterating that he would oppose authorizations for "unsafe" reactors.[191] Whether Aiken was projecting the fears of a nuclear insider or whether he had a particularly keen feel for shifting public sentiment is unclear. His comments were the first made public in Boston, however, that suggested there might be serious public concerns about the safety of nuclear reactors.

In August 1970, the first explicit public doubt about reactor safety was aired. Under the headline "Warning Issued on Use of Nuclear Plants," the *Boston Globe* covered a forum on power needs and their relation to the environment held at the State House. David R. Inglis, a University of Massachusetts physics professor, drew most of the press. Nuclear power plants were hazardous in general, Inglis warned. "There is so much radioactive stuff there, it is a hazard we must face," he told the audience. Inglis also cited the Fermi reactor accident, claiming that disaster had been only narrowly avoided.[192]

Approximately one month later, the national debate between experts about radioactive emissions from nuclear reactors was brought home to Massachusetts residents. The catalyst was an all-day teach-in sponsored by the League of Women Voters to discuss the merits and liabilities associated with a 1,100-megawatt nuclear plant proposed for Ipswich. Dr. Manson Benedict, head of MIT's Department of Nuclear Engineering, defended nuclear power. Dr. Ernest Sternglass, professor of radiation physics at the University of Pittsburgh and a vocal national critic of the low-level emissions released by nuclear reactors, charged that the plant would be dangerous. Citing "an extensive study" that Sternglass claimed linked low-dose radiation to leukemia, he warned, "If we are going to have nuclear energy in the future, we must tighten up our standards now."[193]

According to the *Globe* article, consideration of the reactor proposal came down to three factors: "public health, ecology, and taxes." The prospect of lowering taxes had always been an important appeal to residents of a prospective nuclear site. Safety, however, had for the most part been taken for granted, and ecology largely ignored. It was the latter two issues that by 1970 had begun to dominate the debate. As Larry Bogart, executive director of the Citizens Committee for the Protection of the Environment, told the audience, "There are too many nuclear plants too large, too soon, too little research and too much risk."[194]

Whether at forums to consider the relationship between power needs and the environment, or at teach-ins that explored the ramifications of nuclear power for

191 *Boston Record*, August 2, 1968.
192 *Boston Globe*, August 29, 1970.
193 *Boston Sunday Globe*, September 27, 1970, p. 56. For Sternglass role and point of view, see Ernest J. Sternglass, *Secret Fallout: Low Level Radiation from Hiroshima to Three-Mile Island* (New York: McGraw-Hill, 1981).
194 Ibid.

a specific community, scientific disputes were broadcast to far larger audiences in the new context of environmental concerns. Although the *Globe* article did not mention it along with public health, ecology, and taxes, there was an even more pervasive issue that cut across all of the substantive issues discussed. That issue was implicit in the very structure of both meetings where questions were raised, a forum and a teach-in. The issue was participation.

After 1970 the state moved rapidly and played a far more active role in protecting the environment. Politicians moved equally quickly to capture at least the rhetoric of participation. In January 1971, Massachusetts governor Francis Sargent announced that ecology problems would receive top priority in the Commonwealth's planning. Sargent stated that he would draft an environmental bill to enhance the state's role in protecting open space and wetlands and would establish an "environmental strike force" of full-time government lawyers to prosecute polluters.[195] Little more than a year later, the Massachusetts legislature groaned under the weight of more than three hundred bills on environmental issues.[196] Sargent signed the most important of these into law on July 18, 1972. The Environmental Policy Act, modeled on similar federal legislation, required that public agencies consider the environmental impacts of their projects. It also gave the Massachusetts secretary of environmental affairs the authority to reject projects or suggest environmentally sound amendments.[197]

Demands for greater environmental sensitivity seemed inextricably linked to demands for greater citizen participation in decision making. Governor Sargent hoped to capitalize politically on both. In announcing the 1971 environmental policy legislation, Sargent insisted that his first priority was to "demonstrate that government is open to its people, that effective avenues exist to permit people to challenge their government out of deeply felt convictions."[198] Sargent may have had in mind a recent Op. Ed. piece written by Robert Healy. Entitled "In the 1970s, People Count," the essay labeled the 1960s the "decade of the alienated voter." The 1970s, Healy continued, "may well be the decade of people's power." Environmentalism may have been regarded by many as a fad, Healy argued, but it soon demonstrated that it had real clout in Washington. Healy pointed to national controversies ranging from pesticides to the SST. There were even more local battles, the extension of a runway at Logan Airport, for instance, that demonstrated the effectiveness of "people's power."

Although local politicians scrambled to capitalize on the sentiment for greater environmental protection and enhanced citizen participation, state government was implicated in past practices that now came in for closer scrutiny. It was just such an attack that put one local group of scientists on the state's political map. That group was the Union of Concerned Scientists (UCS). The UCS was formed in 1969 in response to student activists who challenged scientists to take public

---

195 *Boston Globe*, January 21, 1971, p.1.   196 *Boston Globe*, March 19, 1972, p. 2.
197 *Boston Globe*, July 19, 1972, p. 20.   198 *Boston Globe*, January 21, 1971.

positions on the issue of technology. Faculty members at MIT who were
sympathetic to some of the students' concerns – but who disagreed with strident
tactics that might embarrass MIT – formed a working group to plan a grass-
roots action to demonstrate these concerns.[199] They agreed on a research stoppage
planned for March 4, 1969. Rather than conduct business as usual, the day
would be devoted to examining the "misue of scientific and technical knowledge,"
which, in the words of the faculty group, "presents a major threat to the existence
of mankind."[200] These concerns were explicitly linked to the federal government's
use of science and technology in Vietnam. The style of the work stoppage was
clearly modeled after the antiwar teach-ins that spread through campuses in the
sixties. The work stoppage came off as planned, drew prominent speakers,
including antiwar activists Noam Chomsky, Howard Zinn, and George Wald,
and received national publicity.[201] As campus politics veered toward more radical
alternatives, the Union of Concerned Scientists faded into the background on
campus.[202] It did not retire from politics, however.

A position paper entitled "Beyond March 4" distilled into several pages the
UCS critique of the application of science and technology.[203] Although the
statement quoted Winston Churchill's warning that "the Stone Age may return
on the gleaming wings of Science," this was no antiscientific tract. "We are
immersed in one of the most significant revolutions in man's history. The force
that drives this revolution is [the] relentless exploitation of scientific knowledge,"
the statement proclaimed. "That many of these transformations have been
immeasurably beneficial goes without saying," the report continued. As with all
revolutions, the technological revolution unleashed destructive forces that society
had to grapple with. Among these, the UCS cited the "irrational and perilous"
arms race and man's inability to protect the environment. Nor was the statement
politically radical. "Our proposal," UCS insisted, "is based on the conservative
working hypothesis that the existing political system and legal tradition have
sufficient powers of adaptation to eventually contain and undo the destruction
and peril generated by technological change."

The key to achieving this, the UCS argued, was access and participation by
a broader range of scientists and experts. They, in turn, had an obligation to
educate and inform the public. As the UCS put it, "The government that we

199 Jonathan Allen, ed., *March 4: Scientists, Students and Society* (Cambridge, Mass.: MIT Press, 1970),
    pp. xv–xxii. The best secondary source on subsequent UCS activities is Joel Primack and Frank
    Von Hippel, *Advice and Dissent: Scientists in the Political Arena* (New York: Basic Books, 1974).
200 Faculty statement, "Union of Concerned Scientists" in Allen, *March 4*, pp. xxii.
201 Dorothy Nelkin, *The University and Military Research: Moral Politics at MIT* (Ithaca, N.Y.: Cornell
    University Press, 1972), ch. 3. For national coverage, see *New York Times*, February 9, 1969, and
    March 5, 1969; Bryce Nelson, "MIT's March 4: Scientists Discuss Renouncing Military Re-
    search," and Elinor Langer, "A West Coast Version of the March 4 Protest" *Science* 163 (March
    14, 1969): 1175–8.
202 Nelkin, *University and Military*, p. 104.
203 "Beyond March 4," Union of Concerned Scientists (UCS) files, Cambridge, Mass. For a chro-
    nology of events, see "15 Years of UCS," *Nucleus* [UCS quarterly report] (Spring 1984).

see today is in many respects no longer democratic because the vast bulk of its constituency cannot begin to scrutinize some of the gravest issues." The UCS attributed this not only to the complexity of scientific issues, but to "the shroud of secrecy that enfolds so much of the government's operations" as well. Congress, in the eyes of the UCS, had "largely surrendered its constitutional duties." Only the scientific community, according to the UCS, could provide comprehensive and searching evaluations of advanced military technologies, environmental impacts of industrialization, and the applications of biological research. "The scientific community must meet the great challenges implied by its unique capacity to provide these insights and predictions. It must engage effectively in planning for the future of mankind ... " the statement concluded.[204]

The Union of Concerned Scientists soon emerged as the self-proclaimed representative of the portion of the scientific community that purported to harbor concerns about the social consequences of technology's application. Organized into committees under the direction of MIT physicist Henry Kendall, UCS relied on the volunteer efforts of mainly local scientists for its technical support. The UCS Committee on Environmental Pollution, for instance, included approximately 125 members drawn primarily from the Boston area. The membership consisted, according to a March 4, 1970, update on its activities, of "lawyers, economists, geophysicists, physicians, physicists, biologists, biochemists, engineers from several disciplines, meteorologists, and representatives from a number of other disciplines."[205] The UCS issued its first report in April 1969. It criticized President Nixon's proposed antiballistic missile system. The UCS also published reports on arms control and biological warfare.[206] On the environmental front, James MacKenzie, the only full-time scientist employed by UCS, prepared a study of air pollution. He challenged existing standards in a petition sent to the Secretary of Health, Education and Welfare.[207]

At the same time that UCS pursued this national if not international agenda, it found that demands for participation and access might meet with less resistance at the local level. Its first full-scale report on the environment was distinctly local in nature. Released in October 1970, "The Massachusetts Department of Public Health – A Critical Study of Failure in Government" charged that the DPH had failed to meet its responsibilities in the areas of air pollution control, pesticide regulation, and meat inspection.[208] The UCS was as blunt about the reasons for this failure as the report's title. It was a classic critique of iron-triangle politics. The public's interest, according to the UCS, had been subordinated to corporate interests. "In the development of pesticide and air pollution regulations the

204 Ibid.
205 "UCS Committee on Environmental Pollution: An Overview," March 4, 1970, UCS, p. 1.
206 "15 Years of UCS."   207 "An Overview," pp. 3–4.
208 Committee on Environmental Pollution, Union of Concerned Scientists, "The Massachusetts Department of Public Health – A Critical Study of Failure in Government," October 21, 1970, UCS.

Department often allows economic factors to outweigh public health considerations," UCS charged.[209] Even after regulations were in place, large economic interests continued to exercise a disproportionate amount of influence by hammering out compromises about how the already watered down regulations would be enforced. "Our study suggests that narrow economic interests are often the primary consideration when these compromises are made," the report concluded.[210]

The solution? Enforcement should be separated from promulgation of the rules. Follow-up studies should be conducted to determine the effectiveness of enforcement. Most significantly, the UCS called for procedures that would "encourage active participation in environmental decisions" by the public and by staff of the agency's own subdivisons.[211] This last recommendation was designed to address what the UCS considered to be a major problem. As the report concluded, "It is unfortunate that regulatory agencies within the Department feel that they function most effectively when there is a minimum of public involvement." That was because the Department, according the the UCS, viewed comments from citizens groups as "emotional outbursts from technically unqualified people." The UCS warned that "citizens have a legitimate role to play in making decisions that will affect the quality of their environment."[212]

The UCS report assaulted a classic iron triangle. As part of a newly emerging issue network, the UCS drew upon the authority of some of the top experts in the state to support its positions. It brought expertise from a variety of disciplines to bear on what it claimed were interrelated problems handled by an overly specialized approach. Symbolically, the Union of Concerned Scientists sought to make the most of its scientific training. Its report on the Department of Public Health, for instance, was released at the MIT Faculty Club.[213]

The UCS membership soon expanded to include a far broader segment of the public and, more significantly, financial support from that public and a number of foundations. By 1974, UCS operated on an annual budget of $175,000 and could afford to obtain the services of environmental lawyer Anthony Roisman, who served as the organization's Washington counsel.[214] By 1979, the UCS staff had grown to thirteen people spread between two offices in Cambridge and Washington, D.C. At least half of its funding came from more than four thousand supporters who made monthly pledges to support UCS.[215]

Despite the frontal assault that the UCS launched at the Massachusetts Department of Public Health, UCS scientists maintained crucial links to agency technical personnel through shared scientific disciplines as well as social concerns. That is why the UCS recommended that staff of the Health Department's subdivisions be actively encouraged to participate in the decision-making process.

209 Ibid., p. 52.   210 Ibid.   211 Ibid., p. 54.   212 Ibid., p. 53.
213 *Boston Globe*, October 22, 1970.
214 Undated [probably 1973], "The UCS Fund: Description, Budget and Funding Proposal," UCS.
215 Union of Concerned Scientists, "The First Fifteen Years," UCS.

In the future, the UCS would come to rely on information "leaked" by agency personnel and UCS staffers would be recruited from disgruntled technically trained bureaucrats frustrated by agency efforts to "contain" internal debates. The UCS would even advocate by 1975 expanding the Public Health Department's role in nuclear safety regulation.

In Massachusetts the UCS was only one of many contributors to parallel trends toward greater environmental concern, challenges to authority, and demands for greater citizen participation. Its contribution, however, was distinctive. It lent expert analysis and credentials to a battle that continued to be cast by defenders of the status quo and a number of environmentalists as well as one that pitted the people versus the experts. The latest to cast the debate in these terms was John Collins, director of environmental health for the Commonwealth of Massachusetts. At March 1970 hearings on air pollution standards, Collins questioned public involvement. "[T]here are many people who don't understand the problem," Collins testified. "I am not deprecating them, but they just technically do not understand the problem, but they want to have input into it. They are not gainfully helping the situation," Collins concluded as delicately as he could.[216] The Health Department's expert consultant, Dr. Melvin First, professor of applied industrial hygiene at Harvard, was no more receptive to public participation, according to eight Harvard graduate students in environmental sciences and engineering who also testified at the hearings. According to them, Dr. First believed that the public could not make intelligent choices because it did not know enough about air pollution.[217] When these specialists asked Dr. First about their own testimony on behalf of the UCS, and that of other expert witnesses for groups such as the Massachusetts Audubon Society and the Sierra Club, First charged that these groups were not representative of the general public.[218]

Though the Commonwealth of Massachusetts had not moved as quickly as the Union of Concerned Scientists would have liked, it in fact had made great strides since the mid-fifties toward becoming an independent actor in the nuclear regulatory process. Federal preemption still precluded any direct role for states in the reactor licensing process. But the Commonwealth did begin to play a more active role in other aspects of nuclear power. The requirement that all reactors develop emergency evacuation plans required state cooperative efforts. In Massachusetts, that led to agreements between Yankee Rowe, the Massachusetts State Police, and the North Adams Hospital. Radiation emergency drills were conducted annually.[219] With the implementation of the National Environmental Policy Act of 1969, information about reactor licensing was disseminated far more widely to agencies in states and at the federal level.[220] In the fall of 1970,

216 Collins quoted in *Boston Sunday Globe*, October 25, 1970, p. A4.
217 Ibid.   218 Ibid.
219 Johnson to Morris, February 26, 1968, in Yankee Rowe file, PDR/NRC.
220 See, for instance, Price to Orlebeke, October 12, 1970, "Pilgrim" file, PDR/NRC.

the Atomic Energy Commission sent out information about Pilgrim's environmental impact to Arthur Brownell, the Massachusetts commissioner of natural
resources.[221] As pressure from citizens concerned about the environment mounted
after 1970, the Commonwealth moved from neutral observer to active critic. In
the spring of 1971, the Atomic Energy Commission received thirty requests for
a public hearing on the commission's consideration of an operating license for
Pilgrim and several petitions to intervene in the proceedings. Among those
requesting to intervene were not only the Sierra Club and the Union of Concerned
Scientists, but the Massachusetts attorney general as well.[222]

Letters from citizens to their legislators and the AEC almost all reflected the
integrating force of the environmental perspective. Reactors could not be considered separately from their impact on the water and air surrounding them,
many of the critics argued. Whether a handwritten note from a Kennedy
constituent, which called the reactor a "risk on people's lives and also the fish in
the water thereabout," or a far more formal request for a hearing by a resident
who signed his name "Chemist; A.B., M.A." and worried about "the probability
of nuclear pollution to the area through operation or accident, [and] the potential
extent and results of thermo-pollution," critics sought broader review and greater
sensitivity to environmental and health concerns.[223]

As was the case in the Bodega Head controversy, health and safety issues
seemed to grow in importance as the approval process dragged on. In the case
of the Pilgrim reactor during the summer of 1971, the health issue most often
cited was the threat of low-level pollution publicized by Ernest Sternglass.[224] Of
all the letters docketed for Pilgrim's operating license, the most tortured came
from Lawrence Laskey, president of the the Merchants Bank of Cape Cod. Laskey
described himself as "a little country banker" who was certainly not a scientist.
Although Laskey readily admitted that he had "practically zero" knowledge of
nuclear power, he was concerned about charges by eminent scientists like Sternglass
– who Laskey noted had impeccable credentials – that operating the Pilgrim
reactor might be ill-advised. Laskey submitted a list of seven questions that
probed into the charges and countercharges made during the Sternglass controversy
on the effects of low-level emissions. Laskey's tone was almost apologetic, but
his experience had taught him that "many men and women in government have

221 Price to Brownell, October 12, 1970, "Pilgrim" file, PDR, NRC.
222 Johnson to Murray, June 17, 1971, "Pilgrim" file, PDR, NRC.
223 Mr. and Mrs. Barry Thomas to Kennedy, April 6, 1971; John McCue to Secretary of the AEC,
    May 18, 1971, "Pilgrim" file, PDR, NRC.
224 As Mark Weissman, project editor at Houghton Mifflin's School Science series, wrote to Senator
    Kennedy, the Atomic Energy Commission had fired its own scientists and suppressed findings
    that correlated infant mortality to radioactive releases. "Dr Ernest Sternglass, the great biologist
    from the University of Pittsburgh," Weismann noted, "recently showed that during the time
    of our above-ground testing in Nevada the number of miscarriages in pregnant women ...
    increased by one quarter of a million" (Weissman to Kennedy, May 18, 1971, "Pilgrim" file,
    PDR, NRC).

proven to be wrong ... despite the fact that they talk with a tremendous amount of authority and know-how." Laskey insisted that "I am not a troublemaker, but I will be damned if I can sit on my fanny and be satisfied with the reports from the Atomic Energy Commission while there are many experts in the field who do not agree with the findings and general thinking of the Commission."[225]

Massachusetts residents, along with the rest of the nation, soon had a new, even more unnerving, dispute between experts to weigh in making their calculations about nuclear safety. At a July 26, 1971, press conference, the Union of Concerned Scientists released a thirty-page report that called for a moratorium on reactor licensing. None should be licensed, the UCS argued, until "their safety systems can be shown to be reliable and adequate to prevent a major accident." Safety system failure, the UCS warned, and the resulting core meltdown could lead to an accident "more disastrous than any accident the nation has ever experienced." In its press release, the UCS stressed that it had relied on official AEC documents and reviews. The report's conclusions, therefore, were "based in major part on official sources."[226] As we have already seen from Clifford Beck's private warning that even routine testing of safeguards was not yet well developed, the UCS assertion that "experiments should have been performed years ago, before the program to construct power plants was begun," was essentially correct.[227] As we shall soon see, so was the UCS claim that its concerns were based on official sources and reports. The major difference between the UCS position and that of many within the AEC's safety community was not technical or scientific. The largest divide was over access, publicity, and open discussion. In short, the issue was political "containment."

Far more powerfully than the runaway chain reaction that had led to reactor meltdown and killed three workers ten years earlier, the UCS efforts actively to publicize internal AEC doubts and its demands for public access to these previously insulated decision-making forums – when framed in the context of a far more differentiated and specialized safety community and a public sensitized through the environmental perspective to the interrelated nature of health and safety issues – vastly broadened the debate over reactor safety. Despite the elaborate series of events, both inside and outside the nuclear community, that ultimately closed the "safety gap" carved by nuclear enthusiasts long ago, there was hardly a spontaneous reaction to the UCS release. It was only after a determined, well-organized effort that the UCS and other nuclear opponents achieved a critical mass of public concern. The UCS issued another report in October, blasting the Atomic Energy Commission's stopgap measures on the emergency core cooling system (ECCS). The authors of the first two reports published an article in

225 Laskey to Wallig, August 10, 1971, "Pilgrim" file, PDR, NRC.
226 Quotations from UCS press release, July, 26, 1971, UCS. See also Ian A. Forbes, Daniel Ford, Henry Kendall, and James Mackenzie, "Nuclear Reactor Safety: An Evaluation of New Evidence," UCS.
227 Quotation from July 26 press release.

Environment that spread the word to one of their key constituencies. Another press release, issued in March 1972, maintained the opponents' momentum. This was only the beginning.

It was this sustained effort – carried out by a burgeoning issue network, backed by expertise, plugged into official channels of communication, connected to the grass roots, committed to "going public,"[228] and received by a public rapidly embracing the environmental perspective – that activated what otherwise may have remained merely an unstable set of conditions. As with the production of nuclear chain reactions, generating a public counterreaction was a complex and sophisticated business that took both resources and an organizational base to sustain the effort over time. Ultimately, the UCS helped change the framework for the debate from that of science versus emotion to one that increasingly concentrated on technical disputes between the experts themselves. As with its response to public health officials earlier in the year, the UCS – relying on highly credentialed volunteers and, more significantly, using the arguments generated by debate within the Atomic Energy Commission – challenged decades of "official" assurances. These assurances consistently insisted that qualms raised publicly were simply the product of emotional fears linked to the bomb. The UCS substituted the nuclear community's own doubts about safety for vague public unease, by connecting expertise to "the people."

In Massachusetts, the Union of Concerned Scientists connected broader safety concerns to a specific reactor by intervening in the operating licenses hearings for Pilgrim.[229] The story was no longer just that reactors might be subject to accidents. It was, as the *Boston Globe* reported on October 14, 1971, "Chance of radiation accident at Plymouth."[230] That charge was made, according to the *Globe*, by the Sierra Club and the UCS, "a coalition of Boston area scientists and engineers."[231] The UCS intervention in the Pilgrim proceedings received ongoing local coverage. UCS members also addressed rallies and meetings designed to tackle broader environmental concerns. UCS chair James MacKenzie, of the Audubon Society and lecturer in the environmental studies program at Brandeis University, brought the findings of the UCS physicists to a rally at Toms River. He told the crowd that "if 'outsiders' such as the physicist-author Ralph Lapp and the Union of Concerned Scientists had not exposed the AEC's failure to perform the needed safety research it might not have happened until a serious accident occurred."[232] The public, MacKenzie told the rally, had only recently become aware "of what the AEC has known for some time."[233] By linking expert analysis of the AEC

228 On "going public" see Sam Kernell, *Going Public: New Strategies of Presidential Leadership* (Washington, D.C.: CQ Press, 1986).
229 See, for instance, *Boston Globe*, October 10, 1971, p. 49, two articles under the headline "Nuclear Plant Foes Aim at Licenses."
230 *Boston Globe*, October 14, 1971; see also *Boston Herald*, October 14, 1971.
231 Ibid.
232 Quotation from UCS press release, March 25, 1972, UCS.
233 Ibid.

position to the reactor approval process in specific sites, the UCS helped fuel the growing "not in my backyard" (NIMBY) phenomenon.

As the hearings on the safety of the emergency core cooling system dragged on through 1972, the Union of Concerned Scientists consistently stressed its local implications. When the UCS released internal AEC documents obtained through the group's intervention, Henry Kendall emphasized that this meant that operating power levels at Pilgrim should be reduced and the operating license granted to Pilgrim questioned.[234] The UCS effort to sustain doubts about reactors in general, and Pilgrim specifically, was particularly urgent at this time, coming during an energy crunch. A front-page headline from the same *Boston Globe* that covered the UCS release, screamed, "No Light ... No Heat ... No Industry – It Could Happen in US."[235] Several months later, in February 1973, the UCS put safety on the *Globe*'s front page. "Safety Defects Reported at Plymouth Nuclear Plant,"the headline blared.[236] Again using the AEC's own internal documents, the UCS reported that Pilgrim had recorded twenty-six "unusual or abnormal occurrences" since it had opened in June 1972. Besides documenting incidents that seemed to be serious, Daniel Ford captured quirky incidents that in and of themselves did not necessarily threaten safety, but added up to a portrait of bumbling incompetence, or outright disregard for safety. In this case, it was the story of officials peering into the reactor to inspect it. "While looking through a pair of binoculars its plastic fitting fell off and was lost in the core. It was never recovered," Ford reported, "and is presumed to have disintegrated."[237] In August, when the Atomic Energy Commission's revised regulations on emergency core cooling systems did not go far enough to satisfy the UCS, the group held a press conference, resulting in a *Globe* article headlined, "New A-plant Rules Still Worry Experts."[238] Ford was again back on page one of the *Globe* in March 1974. Presented in a David-versus-Goliath context, the headline shouted "Student's Challenge Idles Pilgrim Nuclear Plant." Intervening as a public citizen, Ford convinced a three-person AEC licensing and safety board to prolong what had started as a routine maintenance shutdown in order to examine Ford's contention that cracked fuel housings might be hazardous.[239] While the board looked into the matter other problems were revealed, disabling the plant for months.

In each of its charges, the UCS had used internal documents to illustrate that there was serious, informed debate about the safety of some aspects of nuclear power. In the wake of Watergate, access to these documents became as big a story as the technical problems they revealed. *Globe* coverage of the Pilgrim hearings reflected this shift. Headlines like "Board Denies Request for Pilgrim Plant Data," and "How Secret? New Debate on N-plant," undoubtedly caught the eye of yet another segment of the public previously indifferent to the technical

---

234 *Boston Sunday Globe*, October 29, 1972, p. 79.
235 Ibid., p. 1.   236 *Boston Globe*, February 6, 1973, p. 1.   237 Ibid.
238 *Boston Globe*, August 26, 1973, p. 23.   239 Ibid., p. 12.

skirmishes over the emergency core cooling system.[240] By the end of the year, the UCS was charging the AEC with a "massive cover-up" of reactor safety problems. The problems identified were old; in fact, even the UCS charge of cover-up was old. Only the headline – "Scientists Accuse AEC of Cover-up" – was new.[241] As Paul Langner, who wrote the story, put it, what was new was that "yesterday's action by the Union of Concerned Scientists was a gathering together of their case for presentation to the public and the President." Langner's story focused on how the UCS had obtained its documents. Some had been pried loose from the AEC via Freedom of Information request. Others, according to Kendall, "came to us in brown envelopes."[242]

Ford and Kendall elaborated on these charges in a presentation to the American Association for the Advancement of Science in January 1975, entitled, "Nuclear Safety: Misrepresentation and Manipulation of Information."[243] The presentation concluded with a ringing attack on the Atomic Energy Commission for betraying the public trust. "However well-intentioned the managers of the nuclear program may have been," the UCS representatives told the AAAS, "We cannot escape the conclusion that much of what they have effected or condoned has been profoundly undemocratic."[244] Quoting another critic of the AEC's practices – Harold Green – Ford and Kendall sought to explain these actions. The "establishment" genuinely believed, Green had argued, that

it knows best for the American public and that the public, if told candidly about the risks, would irrationally oppose nuclear power. The public is asked, therefore, to accept uncritically the conclusions of the experts that nuclear power is safe, but it is never told about the gaps in actual experience and knowledge or the extent to which the expert judgements are based on theoretical analysis or sheer prediction.[245]

The UCS was hardly the only visible opponent of nuclear power in Massachusetts. It would have been hard to miss Sam Lovejoy of Montague, Massachusetts. Located ninety miles west of Boston in the Connecticut River Valley, Montague was targeted by Northeast Utilities in 1973 as the site for two nuclear reactors.[246] Although the town of Montague voted three to one in favor of the plants in 1974, Northeast Utilities had hoped that the promise of jobs, economic development, and lower taxes would yield a victory by a ratio of ten to one. Even more distressing was the reaction of nearby towns. Apparently these towns feared the negative consequences of a nuclear plant, and saw little hope for offsetting economic benefits. Wendell, directly to the east, voted ten to one against the

240 *Boston Globe*, April 23, 1974, p. 5; *Boston Globe*, April 26, 1974, p. 41.
241 *Boston Globe*, November 13, 1974, p. 24.
242 Ibid.
243 Daniel F. Ford and Henry W. Kendall, "Nuclear Safety: Misrepresentation and Manipulation of Information," January 31, 1975, Union of Concerned Scientists, UCS.
244 Ibid., p. 30.   245 Ibid., p 29.
246 Stephen Davis, "Lovejoy's Nuclear War," *Boston Sunday Globe Magazine*, December 1, 1974, p. 74.

reactors in its town meeting. Residents of Leverett and Shutesbury also opposed the reactors by smaller margins. One resident of Leverett felt that "the closer you get to the university the more they are opposed." Langner, however, noted that the divisions were not as simple as that.[247]

Everybody knew where Sam Lovejoy stood on the matter. That is because on Washington's Birthday, 1974, dressed commando style in dark clothing, Lovejoy cut through a fence and toppled a meteorological tower that stood several hundred feet above the proposed reactor site. The tower had been erected by Northeast Utilities to measure wind patterns at the site. Lovejoy then turned himself in and was charged with willful and malicious destruction of property. Lovejoy defended himself, and protested when the judge dismissed the case on a technicality.[248]

Lovejoy's attitude towards nuclear power was not unlike that of a number of other local residents who had embraced many of the ideas of the "counterculture." Langner identified commune farmers and other members of the counterculture as the only group that could be labeled "unreservedly opposed" to nuclear power. They constituted the most radical opponents to the reactors. Organized into a group called Nuclear Objectors for a Pure Environment (NOPE), radical opponents fielded a candidate who ran unsuccessfully for the board of selectmen. Lovejoy joined the Alternative Energy Coalition of Massachusetts, which may have been the only organization to address the Nuclear Regulatory Commission as "Good People" and close "with love."[249] Lovejoy and his supporters undoubtedly drew attention to the nuclear debate. Their countercultural life-styles also lent a touch of credibility to the nuclear community's charges that opponents' criticisms were emotional and based on deeply held antitechnological beliefs.

Ultimately, however, it was undoubtedly sustained challenges to the scientific basis of the AEC's authority, and charges that those outside the nuclear community had been excluded from the discussions on which AEC rulings were based, that began to undermine for record numbers of Massachussetts residents not only confidence in the safety of nuclear power, but confidence in expertise as well. The *Boston Globe* reflected this on its editorial page and in the way it framed its reports on nuclear power. A lead editorial on November 17, 1974, charged the Atomic Energy Commission with deception. The editorial was drawn in large part from material presented by the Union of Concerned Scientists. The *Globe* expressed concern about the substance of some AEC decisions. The editorial, however, was even more disturbed by "concealments and deceptions [that recalled] efforts to cloak Vietnam failures and Watergate crimes." Unaware of the historical irony, the *Globe* concluded that the AEC "can take no refuge in even the pretext of national security."[250]

More subtle than editorials blasting the AEC were indications that the framework in which the nuclear debate was presented was beginning to shift. An

247 *Boston Globe*, June 19, 1974, p. 3.   248 Davis, "Lovejoy's."
249 Alternative Energy Coalition to NRC, July 23, 1975, "Yankee Rowe" file, PDR, NRC.
250 *Boston Globe*, November 17, 1974, p. A6.

article by Paul Langner in December 1973, entitled "Safety of Growing Nuclear Power Industry Stirs Controversy," stated that "Until a few years ago, the battles were fought among the members of a priesthood as arcane as any in history – the nuclear scientists."[251] The article was accurate as far as it went, but that a reporter was aware of battles within that priesthood was the real news. Until the mid-seventies there had been so few breaches in political "containment" that only the elect were aware that such ecclesiastic debates took place at all. A subtle change transformed the tone of the "boilerplate," or standard background, that was repeated in each article. Langner, in March 1974 for instance, referred to the spotty history of nuclear safety, whereas only a few years earlier, even when covering nuclear mishaps, reporters had reminded readers that nuclear power had a far better safety record than most industries.[252] In May 1974, the *Christian Science Monitor* ran a series of articles asking whether New England had overcommitted itself to nuclear power and emphasizing that nuclear power was, at best, a mixed blessing.[253]

The framework changed as stories critical of nuclear power seemed to multiply geometrically. There had been no major accident. Nuclear power was certainly not a major campaign issue, even at the local level. Public opinion polls were just beginning to be conducted on the issue. Yet there seemed to be a steady stream of bad news that triggered more bad news. On September 22, 1974, the *Boston Globe* ran four negative stories about nuclear power.

The antinuclear-issue network that sprang to life in the early seventies proved every bit as capable of generating news about nuclear power – independently of what actually happened to be going on inside reactors – as the nuclear community had been for the preceding twenty-five years. In fact, the issue network had tapped into enough suppressed bad news to keep it humming all the way to Three Mile Island. No wonder that the AEC's director of regulation, L. Manning Muntzing, wrote to the president of Yankee Atomic Electric Company in March 1974 about "a matter of growing concern – the credibility, from the public viewpoint, of the nuclear industry regarding the availability and reliability of nuclear power plants."[254] Fifteen years after Yankee Rowe's startup, Muntzing informed the Yankee president that the AEC's regional offices would begin issuing "prompt and factual public announcements concerning incidents."[255]

A *Boston Globe* article in March 1975 entitled, "Costs, Safety Cloud Future of Nuclear Power Production," summed up the situation. "It may be hard to admit, but when it comes to actually solving this nation's growing energy woes, the experts turn out to be as confused as the rest of us."[256] Irwin Bupp, author of one

---

251 *Boston Globe*, December 26, 1973, p. 3.
252 For spotty history, see *Boston Globe*, March 31, 1974.
253 *Christian Science Monitor*, May 24, 1974, and May 31, 1974.
254 Muntzing to Allen, March 12, 1974, "Yankee Rowe" file, PDR, NRC.
255 Ibid.   256 *Boston Globe*, March 23, 1975.

of the first scholarly accounts of the rise of light-water reactors and the source of their economic demise, drew a parallel to Vietnam. "The industry is talking now like our government officials were talking in the 1960s, about being over the hump, seeing the light at the end of the tunnel. Personally, I think the war is far from over."[257] The conditions that nurtured the antinuclear-issue network did not create debate between experts. They merely injected far more experts into the debate and then projected that debate – sometimes distorting and intentionally magnifying it – onto as large a screen as possible. In Massachusetts, and nationally, the Union of Concerned Scientists was the central organizational amplifier in that network. As a *Boston Sunday Globe Magazine* feature on the Union of Concerned Scientists put it, "the UCS's strategy differed in a significant way from that of the countless public-interest groups that have attacked nuclear power over the years. Instead of issuing a generalized press release ... UCS launched a 14-month investigation ... using the technical expertise of the group's handful of scientists and engineers."[258] It was by translating its experts' findings and its own political point of view into clear language accessible to informed non-scientists, and by selling these distillations to the press, that the UCS helped bring nuclear power to a standstill.

State government provided a crucial forum that opponents hoped to use to expand the debate. It was also an important potential ally. In Massachusetts, the state continued to increase its role and stress the environmental implications, broadly defined, of nuclear power in direct response to growing public concern. As already noted, the Commonwealth's attorney general had intervened in the spring of 1971. The attorney general's office, however, sought to draw a distinction between environmental and safety issues. "[T]he state agencies that have worked with Boston Edison are satisfied that they meet all regulations and the plant is safe," Assistant Attorney General Roger Tippy told the *Boston Herald* that fall. The attorney general had intervened on environmental, not safety, grounds.[259]

Public concern, however, soon pushed the state into safety matters. The Union of Concerned Scientists testified early in 1973 before the Social Welfare Committee of the Massachusetts legislature that incidents at Pilgrim raised safety concerns and that the state should oversee operations there directly, rather than leave safety entirely up to the AEC.[260] The UCS used this forum to release their report on the twenty-six "incidents" at Pilgrim – a strategy that captured front-page *Boston Globe* coverage.[261] The UCS supported a bill introduced by committee chair Chet Atkins. The bill would have given the Public Health Department the authority, funding, and staffing to monitor reactors in the state. The debate

257 Ibid.
258 Peter Cowan, "Questioning the Safety of Nuclear Power Plants," *Boston Sunday Globe Magazine*, August 22, 1976.
259 *Boston Herald*, October 14, 1971.   260 *Christian Science Monitor*, February 7, 1973.
261 *Boston Globe*, February 6, 1973, p. 1.

emphasized the nature of federal preemption in the field of reactor safety. "Massachusetts has more regulation of beauticians and barber shops than of nuclear power plants," Atkins chided his colleagues.[262]

By 1974, the state took explicit steps to address the growing safety concern. The legislature passed the Electric Power Plant Council Siting Act, which established a state unit with the authority to approve siting of all new reactors.[263] When Boston Edison announced that it planned to construct two new power reactors at its Plymouth site, the attorney general joined the UCS and the Massachusetts Wildlife Association in challenging the plants. The intervenors challenged not only the thermal impact of the plants but also Boston Edison's competence to construct and manage the plants, as well as the Atomic Energy Commission's ability to monitor safety at the plants.[264] In August 1975, Massachusetts Environmental Affairs Secretary Evelyne Murphy used the powers granted to her under the Environmental Policy Act to reject Boston Edison's environmental impact report on its proposed reactor at Plymouth. The report, according to Murphy, failed to address the plant's potential for "catastrophic human and environmental damage," and was "seriously deficient" in its treatment of safety issues.[265] At the same time a Special Massachusetts Legislative Commission on Nuclear Power Plant Safety prepared to conduct hearings on nuclear power, and the Massachusetts Commission on Nuclear Safety, appointed by Governor Sargent, examined the adequacy of the state's nuclear accident evacuation plans.[266] *Christian Science Monitor* staff writer Lance Carden summed up the trajectory of the fast-moving chain reaction: "Debate between some members of the academic-scientific community and the utility companies over the safety and environmental aspects of nuclear power-plant construction in Massachusetts moved into the public spotlight a few months ago. Now it's getting into the political arena."[267]

Massachusetts was not alone among New England states in actively intervening in both environmental and safety issues. Four states intervened in hearings held to consider Vermont Yankee's request for an operating license in October 1971. The plant, located on the Connecticut River (the border between Vermont and New Hampshire), also affected Massachusetts, which had planned to divert water not far downstream to a state-owned reservoir. Not only did all three contiguous states intervene, the state of Kansas also intervened to challenge plans for transporting waste from the reactor to a proposed waste site in Kansas.[268] By 1975, a number of New England states had moved from intervention in federal hearings to tough siting laws of their own. Vermont passed the toughest law. It gave the legislature the final decision on whether to construct any proposed atomic power plant in Vermont. According to the *New York Times*, anger at the cost of

262 *Christian Science Monitor*, February 7, 1973.   263 *Boston Globe*, May 2, 1974, p. 25.
264 *Christian Science Monitor*, December 5, 1974.   265 Ibid., August 6, 1975.
266 Ibid., September 25, 1975.   267 Ibid.
268 *Boston Globe*, October 10, 1971, p. 49.

Vermont Yankee power – originally estimated to cost 4 mills per kwh in 1966 but actually costing 30 mills per kwh in 1974 – was an even stronger source of public pressure than safety concerns.[269] In Maine, the legislature was considering a virtually identical bill, while the New Hampshire legislature reviewed a bill calling for "perpetual continual surveillance" of atomic plants built in the state. Connecticut, meanwhile, was considering a bill that called for an independent council to evaluate the safety and environmental effects of nuclear power.[270]

HOT AIR RISES: NATIONALIZING THE DEBATE

By the late sixties, environmental concerns as they touched upon nuclear power had gained a toehold in Congress. The Joint Committee on Atomic Energy had carved out a sizable chunk of congressional turf over the years. A large part of the deference shown to the JCAE derived from the relative monopoly on expertise the agency commanded through its alliance with the Atomic Energy Commission. Even in the fallout controversy – where Senator Hubert Humphrey was able to use his position on the Foreign Relations Committee to conduct hearings – the JCAE upstaged policy competitors, conducting far more comprehensive hearings, relying largely on research conducted by, or funded in part by, the AEC. Environmental concerns, precisely because of their broad-ranging scope, because they engaged the skills of experts outside the AEC's reach, and because their political appeal developed so rapidly, outstripped the JCAE's – or any other single congressional committee's – ability to adapt.

As the issue caught on, it intruded farther into the JCAE's bailiwick. The congressional assault was led by Senator Edmund Muskie, the first to hold hearings on problems related to thermal pollution. Pursuing Congressman Dingell's earlier contention that the AEC already had authority to rule on thermal pollution before licensing reactors, Muskie urged the commission to do so. The AEC, on the other hand, contended that although it had the technical expertise to police thermal pollution, it lacked the legislative authority.[271] The question of expertise in this field was an important one. Despite AEC contentions, that agency's position was far from secure. The field of ecology was dominated by a small, highly select body of experts, as one newsletter informed the nuclear community.[272] As with seismological expertise, the Atomic Energy Commission did not enjoy its usual preponderance of expertise in this field.

Another Capitol Hill offensive was spearheaded by Senator George Aiken in 1968. The Aiken–Kennedy bill (S. 2564), entitled "A Bill to Insure a Reasonable Opportunity for an Electrical Utility to Participate in the Benefits of Nuclear

269 *New York Times*, April 1 and 3, 1975.   270 *Boston Globe*, April 27, 1975.
271 Muskie Hearings, parts 1–4. *The Olwell Report* 1, no. 1 (March 1, 1968): 1 in Box 849, Anderson Papers. Congressman Dingell to Senator Anderson, March 18, 1966, in Box 845, Anderson Papers.
272 *Olwell Report*, March 1, 1968, p. 3.

Power," was intended to make nuclear power more accessible to municipal or cooperative systems. Aiken made it clear that safety was not his primary concern. But the bill required the Water Resources Council and the Federal Power Commission to rule on the thermal pollution question – a provision the JCAE effectively resisted.[273] According to Aiken, one objective of the legislation was "to protect and conserve natural resources." Still trying to push some of the components of that unsuccessful legislation in 1970, Aiken noted that S. 2564 was "welcomed by responsible citizens who were becoming increasingly concerned about protecting our environment from air and thermal pollution."[274]

In defining the JCAE's position, Chairman Holifield – a longtime supporter of public power – criticized his congressional colleagues for singling out nuclear power. According to Holifield, the legislation would stop the development of this new source of power. Amendments to the administration's Electric Reliability Act introduced by Congressman Richard Ottinger – a member of the House Communications and Power Subcommittee – also would have made the Federal Power Commission responsible for licensing nuclear facilities after determining environmental impacts. The most broad-ranging legislative proposals continued to come from congressmen representing the fossil fuel industry. Congressman John Saylor of Pennsylvania, who at the time of the Ravenswood controversy had called for a moratorium on nuclear development, now asked for the creation of a federal committee on nuclear development to reevaluate the existing civilian nuclear program.[275]

Although initially unsuccessful, the congressional assault on the AEC's jurisdiction – and implicitly, the JCAE's as well – reflected structural changes in the Congress that went beyond the emergence of environmental concerns. Congress was moving from the golden age of committee government toward a proliferation of subcommittees and dispersed power that undermined neat jurisdictional boundaries. Congress was also racing to recapture the expert advice that increasingly had been lodged in administrative agencies.[276] Emilio Q. Daddario, a key figure on the House Science and Technology Committee and architect of the legislation that eventually established the Office of Technology Assessment, pushed for an enhanced congressional role in the review of technology. As a report issued by his Science, Research and Development Committee noted as early as 1966, there were "dangerous side effects which applied technology is creating, or is likely to create for all humanity."[277] It would be several more years

---

273 Senator Aiken press release, January 15, 1968, Box 45, Holifield Papers.
274 Senator George Aiken press release, March 4, 1970, in AEC 152/261, March 12, 1970, AEC/ DOE, pp. 9–10.
275 *Sacramento Bee*, January 3, 1968, in Box 45, Holifield Papers. Ottinger to colleagues, January 24, 1968, in Box 4, Holifield Papers; *Olwell Report*, March 1, 1968, pp. 6–10.
276 For a good summary of the political science literature on this trend, see Dodd and Schott, *Congress and the Administrative State*, pp. 58–154. For an excellent analysis on congressional efforts to professionalize in fields related to nuclear power, see Susan Elizabeth Fallows, "Technical Staffing for Congress: The Myth of Expertise," Ph.D. diss., Cornell University, 1980.
277 Quoted in Dickson, *The New Politics of Science*, p. 236.

before the legislature would establish the Congressional Budget Office and the Office of Technology Assessment, but in the interim, less formal mechanisms for ensuring congressional access to expertise sprang up. In the environmental field, one such effort was a congressional conference on the environment, sponsored by twenty-four liberal congressmen and senators. As the announcement put it, "It is time that the decision-makers in government and the general public should listen closely to what the scientists and planners have to say about this vital area."[278]

From the nuclear community's perspective, the most important result of congressional activism on the environment was the Water Quality Standards Act of 1970 and National Environmental Policy Act (NEPA), which established a permanent executive organization – the Environmental Protection Agency – for its enforcement. Although the U.S. Public Health Service and the Department of the Interior had challenged the AEC's jurisdiction in isolated instances, the EPA presented a powerful new bureaucratic force. It was not hamstrung by having to juggle both developmental and regulatory responsibilities – as was the case with Interior – nor submerged in a larger agency with concerns far afield from nuclear power – as with the Public Health Service.[279]

The Environmental Protection Agency soon made its presence felt. Responding to the AEC's sizable June 1970 reduction in permittable public exposure to radiation from reactors, the EPA assistant general counsel commented that "it is not yet necessary" for the EPA to take over emissions control on radioactive wastes. "But if we decide we need the authority we'll damn well get it." At a press conference announcing the AEC's reductions, Regulatory Staff director Harold Price denied that the AEC was responding to critics or potential EPA pressure. "Although we are taking action in a controversial area, as far as I am concerned," stated Price, "we are not reacting to anybody." But clearly a struggle was under way. Via the environmental impact statement – the EPA's most important means of penetrating the AEC's jurisdiction – the new agency intervened in several AEC hearings to press for a change in the longstanding policy of refusing to consider nonradiological issues such as thermal pollution. Behind the EPA's intervention lay pressure from grass-roots environmental groups.[280]

At Calvert Cliffs, for instance, citizens groups filed suit against the AEC for failure to apply an environmental impact statement adequately. The 1971 D.C. Court of Appeals decision was a landmark: "We believe that the Commission's crabbed interpretation of NEPA makes a mockery of the Act," the court stated. This case led to a full airing of the thermal pollution question at AEC proceedings. It also opened the door to considerations of broader safety-related issues. As

278 Senator Hughes to colleagues, October 15, 1969, Box 46, Holifield Papers.
279 For a firsthand account of the possibilities envisioned by the first generation of EPA staffers, see John Quarles, *Cleaning Up America: An Insiders View of the Environmental Protection Agency* (Boston: Houghton Mifflin, 1976).
280 Quoted in Robert Gillette, "Reactor Emissions: AEC Guidelines Move Towards Critics Position," in *Science* 172 (June 18, 1971): 1215.

Harold Green, head of George Washington University's law, science, and technology program, pointed out: "Now it becomes clearly possible to talk [in hearings] about exactly what a reactor means for people of a given city. Its safety and economics are going to have to be considered."[281]

The environmental perspective served to unite a number of disparate elements that had sporadically sought entrance into the nuclear debate. It helped nationalize the ramifications of grass-roots citizens organizations. It provided more administrative forums – on both the state and federal level – for challenging AEC hegemony over nuclear matters, and it provided the ideology that congressional committees could use in attacking the Joint Committee's jurisdictional control over all aspects of nuclear power. Access to independent expertise was crucial to the success of each challenge. Expertise, in fact, was prerequisite to entering the debate.

As what had once been highly insulated debate within the AEC's safety community continued to expand into more public forums, it eventually caught the attention of those recently sensitized to health and safety issues through the environmental perspective. Longstanding safety questions within the nuclear community – such as concerns about the adequacy of emergency core cooling systems, a renewed debate about the probability and severity of a nuclear accident, and the still unresolved problem of high level radioactive wastes – were now played out in crowded and politically charged arenas. On a number of issues, organized grass-roots opposition and scientists working through independent organizations began to overtake administrative agencies – even those that had previously critiqued various aspects of nuclear power – in their concern about nuclear safety. The intensity of the debate had increased as it spread beyond the AEC's political "containment" to state and federal agencies. It grew even greater as participants with organizational bases entirely independent of the government – such as the Union of Concerned Scientists – entered the discourse. In many instances, these groups merely echoed longstanding internal AEC debates, or debates pursued between the AEC and institutional critics such as state health agencies, the Department of Interior, or the EPA. The new groups, however, greatly increased the visibility of the struggle.

These new independent organizations, rather than seek bureaucratic consensus through quiet dialogue, took every opportunity to publicize their differences. They sought governmental allies at both the state and federal level; they also

---

281 The court ordered the AEC to include independent assessment of water quality and other environmental factors at every important decision-making stage, including a case-by-case cost-benefit assessment of environmental and nonenvironmental factors. These assessments were to be applied retroactively to all licensing proceedings begun after January 1, 1971 (Calvert Cliffs Coordinating Committee v. United States Atomic Energy Commission, D.C. Circuit, No. 24, 839 [July 23, 1971], in JCAE Subcommittee on Legislation, "Hearings on AEC Licensing Procedure," 92d Cong., 1st sess., 1971, pp. 573–612). Robert Gillette, "AEC's New Environmental Rules for Nuclear Plant May Open New Debate, Extend Delays, Raise Plant Costs," *Science* 173 (September 17, 1971): 1113.

pushed those allies toward more intransigent positions. The independent organizations, led by a vanguard of their own experts, turned each of these debates into a caricature of the scientific and bureaucratic consensus-seeking process. Along the way, these groups exposed a substantive issue far easier to grasp by the American people than the technical details of eutraphication, radiation standards, or accident probabilities: the issue of participation and access, an issue that one did not have to be an expert to grasp.[282]

A look at the origins of the emergency core cooling system debate within the Atomic Energy Commission illustrates how an internal discussion evolved as the forum broadened. Although most Americans did not hear of the problem until the Union of Concerned Scientists entered the picture, the debate began far earlier than 1971. The ECCS was the subject of heated dispute within the AEC's safety community. Unlike the pressure vessel issue, or metropolitan siting, no member of that community sought to broaden the scope of debate beyond the safety community. Rather, doubts about the adequacy of proposed ECCS systems were explored internally by a special task force, which reported in 1967 that the ECCS was adequate.[283] However, the "Ergen Report" (labeled for its chair, William K. Ergen) called for further research and testing of its findings. While the Ergen task force was preparing its report, a parallel study into ECCS safety had been undertaken by the Nuclear Safety Information Center (NSIC) at the request of the Division of Reactor Development and Technology. The NSIC's report, issued in October 1968, concluded that the overall design used by manufacturers was conservative, but that in some areas there was an absence of experimental data to support the theoretically derived calculations. The task force's and NSIC's call for further research plus continued expressions of concern from the Advisory Committee on Reactor Safeguards prompted a series of experiments designed to confirm the computer codes used to calculate ECCS safety. Small-scale models would be used to simulate conditions that to date had been projected but never actually tested.[284] It was the results of precisely these kinds of tests that had to be made public regardless of the results, Ergen had argued several years earlier, to avoid any suspicion that the AEC was not carrying out fully its safety mission.

Internal debate in the AEC's safety community had always been carefully followed by the industry. Industry newsletters such as *Nucleonics* and *Nuclear News* were an important source of such information. But as safety analysis became more specialized and as organized opposition to nuclear power began to develop, that opposition began to monitor the increasingly formalized communications that flowed within the safety community. In the January–February 1969 issue of

282 Campbell, in *Collapse of an Industry*, captures this evolution in his discussion of the AEC's "internal legitimation" and "external legitimation" crises (see ch. 4).
283 "Emergency Core Cooling," Report of Advisory Task Force, US AEC Report TID-24226, 1967.
284 Ibid. Joseph Hendrie to Robert Hollingsworth, March 20, 1969, and David Okrent to Glenn Seaborg, October 12, 1966, in JCAE Licensing Hearings, 1971, pp. 1111–20. See also ACRS testimony at Licensing Hearings, June 22, 1971, p. 99.

Environment an article appeared entitled "A Mile from Times Square," written by Sheldon Novick. In it, Novick outlined the dangers posed by urban siting and larger reactors. The article quoted extensively from the NSIC report – describing in graphic detail what happens when a reactor core melts. It also quoted one of the Ergen task force's more pessimistic conclusions. Current knowledge, Ergen reported, was insufficient to assure that containment would be maintained should emergency core cooling systems fail and a substantial portion of the core melt. The Novick article went on to list the various problems that might afflict emergency core cooling systems.[285]

The issue was picked up in a broader forum in Ralph Lapp's article entitled "Safety" in a January 1971 issue of *New Republic*. Lapp noted that very little attention had been paid to the problem of nuclear accidents. Relying on data from the Ergen Report, Lapp sketched out a meltdown scenario and concluded that the emergency core cooling system would have to work within five to ten seconds. Noting that phrases like "a matter of speculation," or "it is not known" recurred throughout the report, Lapp urged that "we, not just a few experts in a closed community, audit the nuclear books and lay the basis for public confidence in our nuclear future." He proposed the establishment of a permanent Nuclear Safety Board, and several other steps including separation of the regulatory function from the AEC and its assignment to the EPA. With the formalization of safety review, the necessity for regularized communications between the growing number of bureaus responsible for safety within the AEC, and the far more organized and aggressive array of actors concerned with the public health aspects of reactors, internal discussion could quickly become national news.[286]

In the meantime, experimental data from tests planned during the years of quiet consensus produced serious political repercussions when introduced in this new, charged environment. In semi-scale tests conducted during the winter of 1970–71 at Arco, Idaho, on a teakettle-size model of a pressurized water reactor and emergency cooling system, water was injected into a deliberately ruptured cooling system. The ECCS was activated, but instead of refilling the core with cooling water, the newly injected water was blown out the same break by rising steam that prevented the water from ever getting to the core. This raised a number of embarrassing questions about the reliability of the calculations that such experimental data were intended to confirm. Inside the AEC, a regulatory staff task force had been working on detailed internal guidelines for ECCS performance criteria since issuance of the Ergen Report in 1967. After the semi-scale tests were received, and after further evaluation, the task force issued Interim Acceptance Criteria including some additional conservative assumptions for the evaluative models it proposed. It rejected calls for resistance to any changes,

---

285 Sheldon Novick, "A Mile from Times Square," *Environment* 11, no. 1 (January–February 1969): 10–15.
286 Ralph Lapp, "Safety," *New Republic* 164 (January 23, 1971): 18–21; quotations from pp. 20, 21.

as well as arbitrary limits on reactor power. The AEC accepted the Regulatory Staff's recommendation, and in June 1971 the criteria became effective immediately.[287]

Outside the AEC, the issue was also heating up. Robert Gillette, who covered nuclear power for *Science*, wrote an article entitled "Nuclear Reactor Safety: A Skeleton at the Feast?" published on May 28, 1971. Gillette noted that while new orders were pouring in for reactors, important technical issues remained to be settled. Although Milton Shaw, AEC's director of reactor development, told Gillette that the issue had been exaggerated by "some people who have taken a little data and made a big thing out of it," Gillette noted that the agency had already held up license hearings because of the new information.[288] At the same time that Gillette's article appeared, the *Washington Post* ran a front-page story on the test failures. The *Post* estimated that five reactors were currently facing delays because of the problem, and up to fifty might soon face delays. The *Post* article quoted only AEC spokespersons, who understandably sought to play down the significance of the tests. "These tests were not fair models of real reactors," stated one AEC official. Even he, however, acknowledged the "outside possibility" that there might be a fundamental design problem.[289]

Even before the *Post* article, Businessmen for the Public Interest – which was supporting a number of interventions around Lake Michigan – learned of the problem. That group informed the Union of Concerned Scientists, which at the time was preparing to intervene in the Pilgrim reactor operating license hearings. The UCS – using AEC research reports and the expertise of MIT physics professor Henry Kendall – prepared a report on possible implications of the semi-scale test failures and held a press conference in July 1971. The UCS was the first party explicitly to seek as large an audience as possible for the issue. It found one. The press conference was covered by both NBC and CBS and aired the same evening.[290]

This was only the beginning of UCS involvement in the issue at the national level. Daniel Ford, who had been the coordinator of research at the Harvard Economic Research Project, joined the UCS and participated in its Pilgrim reactor intervention. Ford also participated as an intervenor in the Indian Point 2 hearings. Ford and other UCS members also embarked on a fact-finding mission to Oak Ridge, Battelle Memorial Institute – an AEC reactor safety contractor – and the National Reactor Testing Station in Idaho, the site of the semi-scale tests. There, they began to establish contacts with scientists who were dissatisfied

287 Ford, *Cult of the Atom*, p. 102; William Cottrell, "The ECCS Rule–Making Hearing," in *Nuclear Safety* 15, no. 1 (January–February 1974): 37–8; James Schlesinger testimony and submissions to the JCAE Reactor Safety Hearings, January 23, 1973, pp. 6–63; and Seaborg testimony in Licensing Hearings, June 22, 1971, pp. 1–22.
288 Robert Gillette, "Nuclear Reactor Safety: A Skeleton at the Feast?" in *Science* 172 (May 28, 1971): 918–19; quotation from p. 918. Glenn Seaborg to John Pastore, April 27, 1971; press release "Announcing Regulatory Hearings," May 27, 1971; in Licensing Hearings, pp. 486, 493–4.
289 *Washington Post*, May 26, 1971, p. 1.
290 Primack and Von Hippel, *Advice and Dissent*, pp. 210–11.

with the way safety matters were being treated. In January 1972, the UCS published articles in *Nuclear News* and *Environment* that directly challenged the adequacy of the AEC interim criteria. These articles incorporated many of the criticisms gleaned from Oak Ridge and AEC scientists during the UCS visits to these facilities.[291]

In an effort to ward off delays at individual site hearings, the AEC announced that it would hold rule-making hearings aimed exclusively at reviewing ECCS interim acceptance criteria (referred to as "generic hearings" because they considered a generally applicable system, as opposed to site-specific reactors) in January 1972.[292] This was a direct response to the growing number of challenges to the ECCS at individual site hearings. By the end of 1971, emergency core cooling systems were being, or were expected to be, contested at fourteen separate reactor hearings. Holding generic hearings was an important strategic move by the AEC – allowing individual site hearings to progress, despite the ECCS controversy. But it also offered an opportunity to the opposition. Intervenors at various individual reactor hearings joined to form the Consolidated National Intervenors (CNI). The CNI included the Union of Concerned Scientists. The AEC's efforts to isolate the problem had inadvertently resulted in the national organization of its opposition.[293]

The hearings dragged on for almost two years, during which more than 50,000 pages of evidence were accumulated. By cross-examination, by using newly released internal documents, and by gaining access to a steady stream of leaked documents, the Consolidated National Intervenors exposed deep disagreements between AEC-funded scientists. The testimony of an Oak Ridge expert on fuel assembly, Phillip Rittenhouse, was devastating. He stated that in areas where he was personally expert, there was insufficient evidence to confirm emergency core cooling system performance during an accident. "Beyond that," Rittenhouse told the safety board, "I can only say that I have talked to a number of people who work in the area of ECCS, both the materials people, people who work primarily with codes, people who are experts, if you will, in heat transfer, fluid flow. And I get the genuine feeling from all of these people that they believe there are things we just do not know well enough yet." The following day, Rittenhouse read into the record a list of thirty scientists and engineers at Oak Ridge National Laboratory, the Reactor Testing Station in Idaho, and the Atomic Energy Commission central staff who had doubts about the emergency core cooling

---

291 Ibid., p. 217; Ford, *Cult*, pp. 109–15; I. A. Forbes et al., "Nuclear Reactor Safety: An Evaluation of New Evidence," *Nuclear News* 14, no. 9 (September 1971): 32–40; and "Cooling Water," in *Environment* 14, no. 1 (January–February 1972): 40–7.

292 "Generic hearings" explore a feature found in a number of reactors, as opposed to "site specific hearings", which concentrate on a particular reactor and its relationship to its environment.

293 Cottrell, "ECCS Rule-Making," p. 39. Individual site approval was halted temporarily. But late in 1972, with the ECCS hearings dragging on, the AEC instructed its local hearings boards to disregard the ECCS issue (Primack and Von Hippel, *Advice*, p. 229).

system. Many of them subsequently expressed in their own words reservations about the Interim Acceptance Criteria.[294]

In its concluding statement, the Regulatory Staff proposed several amendments to the Interim Acceptance Criteria – making them more conservative. The AEC estimated that implementation would result in a 5 percent power loss for all nuclear plants operating through mid-1976, or equipment changes amounting to approximately $10 million for a 1,000 MW(e) reactor. William Cottrell, director of the nuclear safety program at Oak Ridge and vocal critic of the ECCS criteria, termed these modifications in the criteria "significantly more conservative." In Cottrell's opinion the AEC accommodated "most of those in the nuclear community who previously expressed concerns about the adequacy of the 1971 criteria." In December 1973, the commission approved the Regulatory Staff's modified criteria.[295]

The Union of Concerned Scientists was not satisfied with the results. After hearing the Regulatory Staff's final statement, Consolidated National Intervenors joined with Friends of the Earth and Ralph Nader in filing suit against the AEC.[296] In retrospect the opposition had every reason to be pleased with the results of their efforts. Besides forcing the Regulatory Staff to modify its original Interim Acceptance Criteria, the generic hearings had several broad ramifications for the scrutiny of reactor safety by those outside the AEC. Most significantly, they brought together disparate and localized elements of the opposition and provided a forum that virtually forced on them a national organization. This allowed critics to pool their resources and consolidate their access to expertise. The revelations about the management of safety programs were also partially responsible for a shake-up in the safety program. Implemented by AEC chair Dixie Lee Ray in 1973, the reorganization stripped Milton Shaw of all responsibility for safety. This change was undoubtedly prompted as well by a series of articles by Robert Gillette sharply criticizing the safety program. The intervention also garnered ongoing publicity for the critics of nuclear power. The *New York Times* began to follow the issue regularly, and in May 1973, ABC television ran a documentary on nuclear reactor safety.[297]

The sequence of events that led to broad-ranging discussion of the emergency core cooling system is central to understanding the more general evolution of

294 Ford, *Cult of the Atom*, pp. 122–8 (quotation from p. 126); *New York Times*, March 12, 1972; Cottrell, "ECCS Rule-Making," p. 140.

295 *Nuclear Safety* 14, no. 5 (September–October 1973): 528; quotation from Cottrell, "ECCS Rule-Making," p. 57; *New York Times*, December 28, 1973.

296 *Nuclear Safety* 14, no. 6 (November–December 1973): 679.

297 *New York Times*, January 4, 1973, May 22, 1973, May 30, 1973, June 15, 1973. *Science* 172 (May 28, 1971): 918–19; 173 (July 9, 1971); 177 (May 5, 1972); 179 (January 26, 1973): 360–3; 180 (June 1, 1973): 934–5. ABC documentary cited in Primack and Von Hippel, *Advice*, p. 230. Besides its generic hearings on emergency core cooling systems, the AEC also held hearings to consider permissible levels of radioactive releases and matters relating to the nuclear fuel cycle (Ebbin and Kasper, *Citizen Groups*, p. 57).

public opposition to nuclear power. Research pursued at Brookhaven National Laboratory in regard to the probability and consequences of a nuclear accident raised new questions in the Advisory Committee on Reactor Safeguards about the adequacy of emergency core cooling systems. Stopping short of expressing its concerns outside the safety community, the ACRS settled for a task force to look into the matter in greater depth. At the same time, a study at Oak Ridge National Laboratory raised serious questions. Although these doubts and the results of the experimentation recommended by the task force were not widely publicized, they were communicated formally to the growing number of bureaus and committees within the AEC concerned with the matter. Outside the AEC, where a growing number of groups were organizing around a set of issues illuminated by the environmental perspective, an organizer for the scientists' Institute for Public Information, Sheldon Novick, had begun monitoring the AEC's findings. It was his article in *Environment* that made concerns expressed within the AEC public knowledge. Faced with increasing national press on the issue and local interventions citing ECCS safety at individual site hearings, the AEC moved to consolidate and isolate the problem through the generic rule-making hearings. In doing so, it inadvertently helped organize the opposition nationally and attracted even more national press to the dispute.

The increasingly public debate among an expanding array of experts left the public confused. The vice-chair of the JCAE, Senator John Pastore, summed up this reaction best when, kicking off the JCAE's 1973 safety hearings, he complained, "the public today is absolutely confused. The public is not being told in categorical terms 'yes' and 'no'.... We are receiving remonstrances from the public everytime we suggest the building of a reactor at a particular site. And what is even worse is the fact that right within the agency, itself, you haven't made up your minds whether it is or isn't. You have people who are being paid to do a job who are saying that it is good and other people who are being paid to do the same job who say that it isn't good."[298]

THE AEC/JCAE UNDER FIRE: CIRCLING THE WAGONS

While relative newcomers to Joint Committee on Atomic Energy leadership like Senator Pastore hedged their bets on nuclear power, the bond between the Atomic Energy Commission and the JCAE's oldest and most powerful member was strengthened as criticism mounted. When Minnesota proposed lowering its allowable radiation standards, JCAE chair Chet Holifield intervened personally. In a letter to the chair of the Minnesota Pollution Control Authority shortly before it was to render a decision, Holifield expressed "very deep concern" that the MPCA was attempting to regulate in an area specifically preempted by the federal government. Federal preemption, Holifield continued, "was taken out

298 Reactor Safety Hearings, January 23, 1973, p. 6.

of recognition of the Atomic Energy Commission's and the Federal Radiation Council's vastly greater expertise respecting the potential hazards of radiological effects, and the control thereof, than any other single state reasonably could be expected to have or acquire."[299]

In a letter to Representative Clark R. MacGregor of Minnesota, Holifield attributed opposition to the public's ignorance. Because "the public had obviously not been apprised of the significant expenditure of federal research funds on the effects of radiation and the relatively low risk from radiation that may be released during customary operation at nuclear power reactors," Holifield wrote MacGregor, "I would be happy to arrange a briefing on technical points involved for the benefit of interested Minnesota senators and congressmen." Holifield promised to have plenty of "recognized experts from within and without the Government" on hand. At least one member of the Minnesota congressional delegation did not applaud Holifield's concern. Referring to Holifield's letter to Robert Tuveson – which was read before the MPCA vote – Joseph E. Karth wrote, "As far as I'm concerned, I want you to know I'm most unappreciative of your precipitous interference."[300]

The Joint Committee also intervened in state politics to promote the development of nuclear power. Chet Holifield and Craig Hosmer wrote jointly to Governor Ronald Reagan of California in 1969. The letter began by citing the critical need for thermal power plant sites – particularly in California – noting among other problems the decreasing availability of waterfront property, the shortage of cooling water, and seismic problems. It congratulated the state on establishing a power plant siting committee and adopting a policy on siting. The letter objected, however, to the slant of that policy. The focus of state policy and the committee, Holifield and Hosmer protested, was on aesthetics and the environment. "Clearly," the veteran JCAE members wrote, "what is called for is a statewide program aggressively to assure that sufficient numbers of suitable power plant sites actually become available to meet the state's requirements through the end of the century."[301]

The letter proposed beefing up the siting authority to assure its capacity for fulfilling the long-range need for sites, giving it the authority to purchase and set aside suitable sites, as well as the authority to "resolve any conflicts which may arise between the public's need for electric power and other public values." The letter suggested, as one example of meeting the urgent need for sites, that in instances where seismic conditions required engineered safeguards, it might be desirable for the authority to bear the cost of these safeguards rather than individual utilities.[302]

299 Holifield to Tuveson, May 3, 1969, Box 10, Holifield Papers.
300 Holifield to Representative MacGregor, May 8, 1969, Box 10, Holifield Papers; Karth to Holifield, May 15, 1969, Box 13, Holifield Papers.
301 Hosmer/Holifield to Reagan August 5, 1969, Box 10, Holifield Papers.
302 Ibid.

At this time the Joint Committee on Atomic Energy's attention was also devoted to fending off congressional encroachment on issues related to environmental concerns – a battle it ultimately lost. The committee again and again sounded two central themes. The growing demand for electricity required the development of nuclear power; and nuclear power was unfairly being singled out for closer environmental scrutiny than fossil fuel plants. In 1969, after a series of congressional hearings on pollution, the JCAE decided to hold its own hearings on the environment. In a letter to Senator Lee Metcalf, Holifield first spelled out the basics. "[C]oal and oil fueled power plants contaminate the environment far more than nuclear fueled power plants. [E]lectric power will be needed from all fuel sources in the future." Holifield then made the intent of his hearings quite clear. "I intend to hold hearings this year and place upon the public record both sides of this issue," he wrote Metcalf. "Then the public can decide, not between nuclear and fossil fuel plants, but how much additional cost they wish to pay for their electricity in the interest of attaining zero or near zero environmental pollution." Besides public hearings, the JCAE sought to educate its colleagues in the Congress. It countered the seminar held by liberal congressional environmentalists with its own congressional seminar on "Nuclear Power and the Environment," where Milton Shaw discussed "America's Electrical Crisis."[303]

The committee also entertained proposals for improving nuclear power's public image. An October 1969 memo from former staffer Charles Hamilton warned that growing criticism by organized citizens groups and states "could result in a Naderization of nuclear energy that would tragically delay the benefits of the peaceful atom." To counter these growing concerns, Hamilton proposed establishing an "independently supported organization" that would enlist the cooperation of "professionally sound and knowledgeable men with impeccable credentials who can present the facts from a technical and historical point of view." This environmental information center would be headed by a small board, consisting of some of the top leaders in atomic energy.[304]

The JCAE also advised the Atomic Energy Commission as to how it should deal with the wave of environmental legislation it faced. In July 1969, Jim Murray of the AEC asked Bill England of the JCAE's staff whether the commission should comment on Representative Henry S. Reuss's bill to transfer additional functions to the Environmental Quality Council. England told Murray that Holifield could probably judge better than the committee staff. His own advice was that since one version of the bill would probably pass, it would be unwise for the AEC to testify critically. "Not only would the AEC's testimony probably be in vain, but might only serve to provide additional ammunition to certain

---

303 Holifield to Metcalf, Box 9, Holifield Papers. Agenda for Congressional Seminar on "Nuclear Power and the Environment," Box 9, Holifield Papers.
304 Charles Hamilton memorandum: "An Approach Towards Assuring Wider Public Acceptance of Atomic Energy," October 13, 1969, Box 9, Holifield Papers.

Congressmen (e.g., Reuss and Dingell) and others who have been very critical of what they consider to be AEC's reluctance to exercise responsibility in connection with protecting the environment."[305]

The JCAE also tried to limit the impact of the National Environmental Act. An amendment to the Atomic Energy Act sponsored by Chet Holifield and Craig Hosmer that passed the House 345 to 0 in October 1970 would have required the EPA to consult with the National Committee on Radiation Protection, the National Academy of Sciences and the JCAE before acting in the field of radiation safety. But even this mild imposition by the JCAE met stiff resistance. A *Washington Post* article published the day after House passage quoted one source as stating, "Let's face it, this is just a device to keep the joint committee's stake in the field of radiation safety and to make sure that the friends of the Atomic Energy Commission keep the upper hand in the policing of atomic energy matters." Holifield lashed out at the article, and another in the *Times*, arguing that the principal effect of his provision would be to assure that radiation standards would be studied by the "foremost scientific talent" and that "their expert advice will be made known to the Congress and the public at large." But the *Post's* prediction that the amendment would meet stiff resistance in the Senate proved true. John Pastore, JCAE vice-chair and sponsor in the Senate, withdrew the amendment after the White House and Senators Philip A. Hart, Gaylord A. Nelson, and Muskie announced their opposition.[306]

Opposition on environmental grounds was particularly distressing to Chet Holifield. Earlier in his career, he had been a staunch advocate of conservation measures. He had been a leader in demanding greater public access to information and ensuring the integrity of the Advisory Committee on Reactor Safeguards' review on safety matters. But as increasingly assertive demands were articulated in all three of these areas by well-organized interests capable of sustaining them in spite of JCAE opposition, Holifield grew increasingly embittered. In June 1971, following President Nixon's speech announcing a scaled-up program for achieving a successful breeder reactor, Holifield wrote to former AEC chair Admiral Strauss, expressing his approval of the speech and his satisfaction with the public's response. Holifield added that "as a result of the initial approach, stressing anti-pollution, a great many environmental extremists have become disarmed or confused."[307]

A more serious indication of the JCAE's hardening position was Holifield's reaction to Oak Ridge National Laboratory's entry into the field of environmental research. An amendment to the Atomic Energy Act sponsored by the JCAE in 1967 provided authority for the AEC to support research on environmental problems. Upon passage of "Sec. 33," as the amendment was called, the AEC

305 Bill England note to Ed Bauser, July 25, 1969, Box 8, Holifield Papers.
306 *Washington Post*, October 3, 1970, and December 7, 1970; JCAE press release, October 3, 1970, Box 9, Holifield Papers.
307 Holifield to Strauss, June 11, 1971, Box 11, Holifield Papers.

encouraged its national laboratories to provide assistance in environmental research. At the time, Holifield was a strong supporter of the legislation.[308]

As we have already seen, Alvin Weinberg – Oak Ridge National Laboratory's director – was a resourceful operator, hardly oblivious to changes in potential funding sources. It is not surprising that Weinberg sought to make the most of the authority provided him in Sec. 33, particularly in light of the increasing number of questions raised about the environmental impact of nuclear power. By 1970, faced with cutbacks in AEC funding, Weinberg was casting a wider net in search of support for these efforts. "I cannot really say what the future will bring," he told his laboratory. "As far as AEC is concerned, we must remember that the AEC's total budget is falling, not rising. It therefore seems to me unlikely that our budget will increase during the next couple of years....On the other hand, our involvement with other matters of national concern, particularly the environment," Weinberg speculated, "could well provide new, substantial sources of additional support."[309]

Weinberg sought a silver lining in the laboratory's fiscal cloud – just as he had a decade earlier in his "State of the Laboratory" speech that anticipated the AEC's push for breeder reactors. This time, however, the alternatives he proposed did not fall neatly within the AEC's primary functions or range of technical expertise. Following words with action, Oak Ridge National Laboratory successfully applied for a National Science Foundation grant for a comprehensive research program on "man and the environment." As Colonel Seymour Shwiller of the JCAE staff wrote to the JCAE's executive director, Captain Edward Bauser, the $1.5 million one-year grant was a relatively small amount. Shwiller also pointed out that about 20 percent of ORNL's $95 million budget came from federal agencies other than the AEC. But, Shwiller concluded, "considering the wording of the initial proposal and the scope of the initial NSF program (including political science and sociology)...this could constitute the foot in the door for going beyond the limits of Section 33 into new competence in new areas."[310]

Holifield needed no additional warnings. He had already fired off a letter to Weinberg charging him with empire building. It wasn't the $1.5 million that angered Holifield. Rather, it was that Weinberg had promoted the idea behind the National Environmental Laboratory Act of 1970 (S. 3410) – sponsored by Senators Howard Baker and Muskie and Congressman Sam Ervin in the House.[311]

---

308 Amendment to Section 33 (AEC Omnibus Legislation – 1967), in JCAE, Atomic Energy Legislation Through 90th Congress, 2d Session, p. 17. John Reich to Edward Bauser, February 16, 1970, Box 9, Holifield Papers.

309 *Oak Ridge National Laboratory Review* 3, no. 3, cited in Shwiller to Bauser, June 17, 1970, Box 9, Holifield Papers.

310 Ibid.

311 While introducing the legislation, Baker credited Weinberg with mentioning to him "the germ of an idea for a network of national laboratories that would bring together first-rate minds from many different disciplines." The National Laboratory Act, Baker continued, was the direct outgrowth of an Oak Ridge Task Force report, presented to Baker in December 1969 (*Congressional Record*, February 6, 1970, p. 2703).

Holifield resented Weinberg's effort to build a political base outside the nuclear community, not to mention outside the JCAE. That Muskie was one of nuclear power's toughest critics could not have softened the blow.

In his response to Holifield, Weinberg defended his position. Citing the JCAE's sponsorship of the 1967 amendment as evidence, Weinberg suggested that the JCAE agreed that much talent existed at the national laboratories that could be brought to bear on the environment. He also reminded Holifield that despite AEC and JCAE support, Oak Ridge had reduced its staff by 630 people in twenty-nine months. "The issue is not that of building empires," Weinberg responded, "it is rather preventing the existing structure from crumbling." Weinberg distinguished the lab's plans to expand its environmental mission via NSF funding from the National Environmental Laboratory legislation, defending his own move as a small step – merely an attempt to impose some coherence on "rather separate pieces of environmental research." But he also reiterated his support for the national lab legislation, stating it was probably a good idea, although he couldn't "prove" this.[312]

Although the bill did not pass, the exchange between Weinberg and Holifield illustrates the variation within the nuclear community's response to increasing environmental concern. Just as the JCAE had hoped to take the lead in space, when it faced tough competition for high-tech funding from the space program in the late fifties, it sought to capture a piece of the environmental action when that perspective emerged as a permanent threat to nuclear power in the sixties. Holifield, like many conservationists, viewed nuclear power as less of a threat to the environment than fossil fuel plants. But environmentalists began to challenge nuclear power over the thermal pollution issue, and within Congress, the JCAE had no special claim to expertise in this area. By 1970, its position had hardened: it concentrated on defending its remaining turf and ensuring that the National Laboratories remained a part of that turf. As Congressman Bingham, after charging the AEC with neglecting its regulatory responsibilities and calling for an end to its dual responsibility of promoting and regulating, put it, "My colleagues on the Joint Committee have themselves over-reacted to criticism. It appears they have harassed witnesses who have appeared before them and they have become defenders of the AEC."[313]

Weinberg, on the other hand, sought to expand the efforts of his laboratory into areas where national concern and political demands promised a steady source of funding. Rather than circling the wagons, he sought to establish new contacts in the Congress to assure ongoing support. He understood that the old days of iron-triangle politics – the kind of narrow policymaking the AEC had worked so hard to build – were coming to a close. The relatively narrow objective of commercializing nuclear power at all costs was now honeycombed with a variety

---

312 Weinberg to Holifield, May 28, 1970, Box 10, Holifield Papers.
313 Quoted in Graham to Holifield, February 12, 1970, Box 7, Holifield Papers.

of considerations, many of them loosely connected by the environmental-issue network. Weinberg believed that national laboratories offered a unique opportunity for "reintegration at the working level" of problems that become fragmented at the bureaucratic level. Citing past work in desalting and civil defense, Weinberg argued that similarly, with the environment, "it is important that we view the problem broadly, that we integrate economic, demographic, and other social factors with technical factors."[314]

Weinberg might well have added to his charge of bureaucratic fragmentation a note on its congressional counterpart. As other congressional committees asserted themselves on the environmental issue, the JCAE's undisputed margin of discretion – once virtually unlimited – was reduced. If an interdisciplinary approach in the labs meant sharing that discretion with other committees and administrative agencies other than the AEC, the JCAE wanted no part of it.

The Atomic Energy Commission joined the Joint Committee on Atomic Energy in its defensive posture. Along with the nuclear industry, the AEC embarked on a public relations campaign to counteract nuclear power's growing negative image. But the campaign, begun in 1969, soon veered from education – the solution that Chairman Seaborg had so eloquently espoused in the early sixties. It was beginning to look as if attaining a greater degree of literacy – as Seaborg had called for in 1961, and which Americans in fact had done in increased numbers – had not solved atomic energy's problems. Howard Brown, AEC assistant general manager for administration, summarized the new position: "We find that Mr. Average Citizen needs something in addition to the facts – he wants to be reassured that his government is doing the best it can."[315]

There is no better way to conclude this final, almost pathetic chapter in the once proud and powerful nuclear community's history than by returning to the public statements of Glenn Seaborg. By the end of his tenure in the early 1970s, Seaborg, the aggressive, optimistic chair who had called for more creative participation in the early 1960s, had little confidence left. At a December 1970 symposium entitled "Science and the Federal Government" sponsored by the Association for the Advancement of Science, Seaborg acknowledged that science was temporarily in a doldrum. "Science no longer has the carte blanche it had after Sputnik."[316] Part of the problem, according to Seaborg, was that scientists may have oversold their physical powers. Another part of the problem – and one Seaborg emphasized more heavily – could be blamed on a small but "vehement group of critics [of the atom] who traded on public apprehension and misunderstanding."

314 Weinberg to Holifield, May 28, 1970, Box 10, Holifield Papers.
315 For a summary of the AEC's public relations campaign, see Howard Brown, "AEC Goes Public – A Case History," in *Nuclear Safety* 11, no. 5 (September–October 1970): 365–9; quotation from p. 368.
316 Seaborg, December 30, 1970, Box 78, JCAE Files.

For Seaborg, however, the solution remained the same. "We must create today," Seborg insisted, "a 'participating citizen' who will have to become something of a 'super-citizen,' one who has been educated, in the fullest sense of the word, to understand and evaluate today's facts of life and the web of complexities into which they have been woven."[317] Seaborg's plea was easier to understand a decade earlier. As he aptly pointed out, science was then enjoying a golden age. An important concomitant of that confidence was public deference to a small group of experts. What Seaborg and many others missed in the seventies, however, was that the educated participating citizen had arrived, although not in the form the AEC would have preferred. It was not an attack on expertise that finally tapped underlying fears. Nor was it catastrophic accidents. It was disputes among experts with different institutional affiliations looking at the question from the perspective of different disciplines.

Rather than the sporadic debates resolved consensually in the insulated environment of the Reactor Safeguards Committee, the disputes of the seventies were publicly joined by a host of institutions with the resources to engage expertise of their own and to fight a war of attrition. These disputes were perceived by many of the parties as zero-sum games. Most significantly, as the debate persisted, questions of participation and access often overshadowed the technical issues initially debated. Internal debate deliberately concealed, and would-be participants deliberately excluded, challenged unencumbered intellectual discourse – crucial to the public's notion of scientific process – and access to a fair hearing for all affected interests, crucial to the ideals of pluralist democracy. With these assaults on Kuhn and Madison, is it any wonder that the politics of expertise lost some of its glamor in the seventies?

317 Seaborg speeches, June 4, 1970; March 22, 1971, Box 78, JCAE Files; quotation from June 4 speech.

9

Conclusion:
harnessing political chain reactions

The powerful chain reaction triggered by unprecedented reliance on expert guidance housed in large-scale bureaucratic structures funded at public expense is perhaps the most significant political development in America in the past fifty years. Plotting the course of this chain reaction revises our understanding of the rise and fall of commercial nuclear power in the United States. It redirects the emphasis away from external pressures, whether World War II or growing environmental consciousness, toward the professionals and administrators who staked proprietary claims to this new federal policy. Nuclear power's trajectory, I argue, is best explained by an internal dynamic fueled by the constantly shifting elaboration of expertise and administrative capacity within the nuclear community and the juxtaposition of the prominstrative state and an older, less professional, and far more decentralized American political landscape. This perspective in turn shifts the chronological focus of my study away from the more dramatic beginning or demise of nuclear power to the crucial years between 1945 and 1975. That story, and the techniques I have used to construct it, however, have broader implications.

For historians, my approach provides a methodology and narrative form that capture the high degree of contingency and political choice embedded in what more commonly have been presented as the overly determined and numbing forces of professionalization, bureaucratization, science and technology. My account repoliticizes large segments of social and public activities often abandoned to the supposedly predictable forces of modernization. At the same time, it captures the incremental compromises negotiated with older political traditions, reminding social scientists of the powerful political legacies of the past. Newly empowered with the authority of experts and the resources of the federal government, the prominstrative state ultimately had to accommodate the pluralist, porous, and often parochial landscape that continues to shape American politics even today.

My study also begins to lay a new foundation for understanding post–World War II American politics. The politics of commercial nuclear power shared much in common with other post–World War II programs. World War II and the infusion of expertise it inspired was a watershed as monumental as the turn-of-the-century denouement described by Stephen Skowronek. There, the administrative state replaced a nineteenth-century state of "courts and parties." Following World War II it was the fusion of professional and administrative capacity that again reshaped the political system.

Federally funded experts designed policy agendas with little reference to public demand, yet these agendas often found the inside track to policy implementation. There was plenty of debate among the experts in the proministrative state, but initially it took place in highly insulated forums. There was a dynamic to the politics of expertise, however, that virtually ensured a reaction to its protagonists' preference for autonomous policymaking. Faced with flagging demand when it came time to implement their programs, experts and bureaucrats were forced to go public, fighting for shallow but broad public support with promises increasingly difficult to fulfill. Contradictory missions spawned by the experts' organizational loyalties also exposed experts to broader public scrutiny. Specialization, inherent to expertise and administrative control, proved to be the agent most destructive to expert insulation. Specialization spread the debate and destroyed internal consensus on agendas. Ultimately the experts' promotional activities, their pursuit of contradictory organizational missions and the never-ending process of organizational and intellectual specialization undermined the authority of experts.

Though often insulated from public participation, proministrators were never isolated from political competition. As they translated research agendas into new programs they gobbled up turf once controlled by other political participants and created new policy vistas that eclipsed once powerful political actors. Often it was only access to expertise and administrative capacity that stood between traditional participants and the political clout that had formerly been theirs. Ironically, one of the most distinctive aspects of the proministrative state – the prolific production of more experts – ensured that access to expertise would not remain a barrier for long. Congressional entrepreneurs, competing federal agencies, state and local government, and, ultimately, issue networks that reached down to mass-based constituencies, all acquired access to expertise. As each new participant entered the fray equipped with its own "independent" experts, the once exalted political clout of expertise waned. Expertise remained a necessary but hardly sufficient component of the proministrative state.

LINKING THE BEGINNING TO THE END

The first Hollywood account of the bomb, released by MGM in 1947, was entitled *The Beginning or the End*. The film's title reflected the ambivalence of most contemporary accounts about America's powerful weapon. "The beginning or the end" also neatly describes the two periods in nuclear power's history featured in virtually all of today's scholarly literature.[1] I have concentrated instead

---

1 For instance, see Rhodes, *The Making of the Atomic Bomb*; Campbell, *The Collapse of an Industry*; and Marone and Woodhouse, *The Demise of Nuclear Energy?*

on nuclear power's crucial developmental years between 1945 and 1975 in order to dramatize two submerged but ongoing challenges that link nuclear power's grandiose origins to its ignominious decline. One plot traces the quest by pro-ministrators to create organized demand for nuclear power. The second is built around the inherently contradictory viewpoints embedded in the nuclear community that ultimately undermined the consensus so crucial to the experts' political authority. As demand increased, these rifts were nurtured by the very growth and development that the nuclear community struggled so hard to sustain.

Commercial nuclear power moved directly from the laboratory to the nation's policy agenda: its producers – not its would-be consumers – were its best-organized constituency. Even while the frenzied race for the bomb raged toward its conclusion, administrators and professional researchers planned atomic power's introduction into the civilian world. The political planning was initiated by the nation's first prominstrators – men like Vannevar Bush and James Conant in the Office of Scientific Research and Development. The OSRD, in turn, was responding to scientific developments at the Manhattan Project's University of Chicago laboratory. Having completed their weapons-related work, scientists there pursued new self-directed research agendas, one of the most attractive of which was producing power from the now-tamed atom. These agendas were rubber-stamped by the new civilian agency, the Atomic Energy Commission. Farrington Daniels was the most aggressive though least successful proponent of civilian nuclear power. Daniels argued that reactors were the most important product of the Manhattan Project. Though Daniels failed to obtain funding for his demonstration reactor, Eugene Wigner, Alvin Weinberg, and Walter Zinn quietly advanced a number of different reactor concepts during the AEC's first five years.

While the laboratory scientists and engineers cleared the way scientifically and technologically, and were a crucial catalyst in placing commercial nuclear power on the policy agenda in the first place, they could not provide the political muscle required to implement a major federal program. That kind of support required public promotion, a task that congressional entrepreneur Brien McMahon and AEC chair David Lilienthal were delighted to pursue. McMahon, no less than the laboratory scientists eager to continue their research, was a producer when it came to nuclear power. Controlling this newly created political turf, though fraught with risks that warned off more seasoned politicians, offered the freshman senator an opportunity to chair a committee rich in foreign and domestic policy possibilities. McMahon jumped at the chance to parlay his position on the Joint Committee on Atomic Energy into a powerful platform. Nor was the senator bashful about public promotion. His effort to establish civilian (which also meant JCAE) control depended in part on convincing his colleagues that nuclear power had significant civilian implications. McMahon pointed to com-mercial nuclear power as the primary example of such a civilian use.

As chair of the newly created Atomic Energy Commission, David Lilienthal's stake in producing commercial nuclear power was equally high. He understood the need for an organized constituency and initially hoped to create the same kind of favorable public attitude that had breathed life into the Tennessee Valley Authority. There were, however, no nuclear equivalents of the gadgets that Lilienthal had once piled in front of farmers to demonstrate the practical benefits of TVA power. Instead, Lilienthal faced a series of gloomy predictions by the nation's leading experts about the timing and cost of nuclear power. There was some popular support for an administrative apparatus distinct from the military, but the public soon lost interest in this esoteric field and its administrative details.

By 1950, all of nuclear power's producers – laboratory scientists, Atomic Energy Commission, and Joint Committee on Atomic Energy – recognized that if there was to be civilian demand for nuclear power, it would have to be manufactured by the state. Particularly as the cold war heated up, the military proved to be one open-ended source of demand for reactor development. Admiral Hyman Rickover summed up the situation best when he noted that since there was no economic incentive for the electric industry to invest in nuclear power's development, "the program and the drive must come from the navy itself."[2] Indeed, the first power-producing demonstration reactor constructed by the Atomic Energy Commission was a direct outgrowth of Rickover's naval propulsion program. Particularly during the AEC's first five years, the military's insistence that scarce enriched fuel and even scarcer expertise be directed toward weapons programs was a barrier, not a boon, to civilian development. Even military projects turned civilian, such as Rickover's Shippingport reactor, were doomed by their disregard for cost – a crucial consideration if nuclear power was going to compete commercially.

Potential nuclear manufacturers were another logical source of demand. Westinghouse and General Electric were aware of the dismal prospects for economically competitive nuclear power, and the great complexities that remained to be resolved. As the Joint Committee on Atomic Energy's most influential member, Chet Holifield, pointed out (in the safety of executive hearings) in 1953, it was odd that the public hue and cry for private ownership of reactors was not coming from the big companies "that really know what this thing is about."[3] Starting with David Lilienthal's effort to create a "home" for the nuclear industry in the Atomic Energy Commission, the AEC and the JCAE actively sought to stimulate interest among large manufacturers. That task proved difficult even for as skilled a financier and salesman as Lewis Strauss. Strauss argued that it was up to GE and Westinghouse to help maintain America's status as the world's technological leader. At the same time, Strauss struggled valiantly to ensure that nuclear development heeded the dictates, not just the rhetoric, of free market capitalism. His insistence that manufacturers take risks undoubtedly delayed nuclear development. But it cemented the powerful economic interest group that

2 See Chapter 3.   3 See Chapter 4.

the AEC had long sought to develop. By the early 1960s, nuclear manufacturers were themselves seeking to create demand among that industry's most crucial consumers – the nation's utilities.

The problem remained one of basic economics. Because of America's abundant supply of low-cost fuel, nuclear power couldn't compete in the early sixties, even in the best of circumstances. Furthermore, nobody was willing to vouch that such favorable circumstances would prevail for this untested technology. The AEC's reactor development chief summed up the problem in December 1961. "The glamor of being 'first' is no longer there, but the cost of being 'among the first' is."[4]

The Atomic Energy Commission's Report to the President, issued in November 1962, rekindled some of the glamor and substituted rhetoric and optimism for concrete technological developments to surmount the utilities' fears that nuclear power was not economically viable. Seaborg's bold new vision for breeder reactors sought to capture some of the technological glitter recently refracted toward space, while expert assurances that complex technical problems had been resolved and that costs for nuclear power were on the threshold of competitiveness addressed the utilities' questions about the bottom line. Although the report was only one of several factors that helped launch the reactor bandwagon of the mid-1960s, it left its mark, coming at the zenith of America's faith in expertise and confidence in science and technology. That the AEC remained the only authoritative "independent" source of operational and cost data undoubtedly enhanced the report's influence. Given the lengthy period between design and operation of nuclear reactors, the bandwagon gathered a great deal of momentum before utilities acquired some firsthand experience with the true costs and the questionable reliability of nuclear power. Once they did, not even a dramatic rise in fuel costs triggered by OPEC's actions in 1973 could entice utilities to scramble aboard again.

By the mid-1970s, the nuclear community's greatest political asset – its experts' authority – was shattered. A brief headline in the *Washington Post* the morning of October 30, 1984, captured succinctly just how far the dynamic inherent in proministrative politics had carried the nuclear community's reliance on "expertise." "DOE Hires Phobia Expert to Examine Public Concerns," the highly skeptical piece proclaimed. According to the DOE's latest hired gun – Robert L. Du Pont, the president of the Phobia Society of America – the 40 percent to 50 percent of the American public that feared nuclear power suffered from irrational fears. Soothing these fears, of course, was crucial to developing nuclear power. For decades, the public consensus maintained by experts within the nuclear community had kept such fears at bay. In fact, until the late sixties, the classified documents within the nuclear community chronicled a litany of concerns that made public opinion on safety issues appear to be naively optimistic about such

4 See Chapter 7.

matters. This "safety gap," however, evaporated as disputes between those experts seeped into the political mainstream. The Department of Energy could no longer rely on physicists, chemists, and engineers alone to sell nuclear power. Experts from fields as questionable as "phobiology" were now crucial. Despite the cavernous gap in public authority that yawned between an Oppenheimer in 1947 and *Good Morning America* guest DuPont in 1984, the DOE stuck to its prominiﬆrative formulas. DuPont's was "a legitimate scientific inquiry," a DOE spokesperson told the *Post*. "This is high-quality research being done by someone who is a pioneer in this field. We don't apologize for the study."[5]

The story, however, did not end there. Sure enough, DuPont's techniques were disputed by fellow "phobiologists." Addressing the Phobia Society, Robert Ackerman attacked DuPont's techniques, stating they were "a misuse of psychiatric labeling." This interdisciplinary dispute did not take long to reach a mass audience, appearing on the front page of the *Post* days later. The quick transmission and high profile of the dispute had been helped along by a powerful issue network. The *Post* article also quoted Representative Richard Ottinger, who labeled DuPont's contract a part of the DOE's pronuclear propaganda campaign. A spokesperson for the Safe Energy Communication Council claimed that the DOE was "trying to obscure the problems of nuclear power by shifting the blame to those who have legitimate concerns." The people-versus-the experts framework, as Glenn Seaborg had recognized decades earlier, was not the exclusive property of scholars.

My account suggests that the nuclear experts' authority was shattered by projecting internal debate beyond the insulated forums that had once contained that debate. There was a clash of "values," as Seaborg and most scholars were quick to point out, but it was influenced as much by competing disciplinary perspectives and organizational missions as by anti-intellectualism or an attack on science or authority. Many of the accusations hurled by protesters in the late seventies owed their history to questions raised by leading experts working within the nuclear community decades earlier. Those doubts, however, had been confined to insulated forums and initially reached limited audiences. It is not surprising that the challenge to siting nuclear plants, for instance, should be associated with protestors like Samuel Lovejoy, who in the mid-seventies toppled a meteorological tower to stop a proposed plant in Montague, Massachusetts. No doubt Lovejoy epitomized all of the "anti" trends associated with the counterculture of the late sixties. But Lovejoy acted precisely to publicize his concerns, and he realized his objectives. A long line of experts working within the nuclear community, starting with Edward Teller and including some of the AEC's leading authorities on safety, also raised serious questions about reactor siting and containment. Virtually all of them, however, were committed to the political strategy of "containment." Publicity was the last thing they wanted. They

5 *Washington Post*, October 30, 1984, p. A1.

understood only too well that their broader political clout rested on public unanimity.

Initially, these experts and the administrators that supported them had a number of advantages when it came to debating differences among themselves and then closing ranks publicly. Following the war there were just a handful of experts familiar with nuclear reactors. Their field was highly esoteric. Development and safety concerns were undifferentiated – the same scientists were responsible for both. Two disciplines – physics and chemistry – dominated the field. There was only one agency (the AEC), one oversight committee (the JCAE), and one political jurisdiction (the federal government) involved in development and safety. Perhaps most significantly, nuclear power was in the early planning stages. Full-scale implementation was decades away.

Like the components of a nuclear chain reaction, however, each element crucial to nuclear power's inception proved highly unstable. The mixture of expertise, administration, and politics that initiated this chain reaction was constantly reconfigured, initially because of internal, not external, pressures. This in turn took its toll on the insulation and the experts' public unanimity so crucial to triggering the program in the first place. Two responsibilities inherent to successful policymaking in the prominiistrative state stretched "political containment" to the breaking point. One was the promotional obligations imposed upon advocates of programs lacking organized political support. The second was a product of the discretion nuclear proponents carved out after World War II, the luxury of self-regulation. Once it became exposed to the elaboration of professional agendas and the articulation of organizational structures, bombarded by conflicting disciplinary perspectives, and threatened externally by a host of interests – ranging from federal agencies to independent citizens groups newly armed with expertise of their own – even nuclear policymaking mutated toward a more democratic and participatory style of politics. Whether its advocates could then sell a program for which there was still no organized consumer demand remains an open question even today.

Promotion sometimes led to inherently contradictory objectives. For instance, the most important promotional function provided by nuclear experts during the program's early days was assuring the public that this energy source was safe. Yet utilities would not consider investing in nuclear power unless the federal government provided indemnification against accidents. The Price–Anderson Act, passed in 1957 to do just that and still in effect today, owed its political life to expert demonstration that a nuclear accident, should one occur, would be so catastrophic that private insurers could not possibly be expected to take such a risk. The experts delivered, producing the WASH-740 report that projected deaths in the thousands and property loss in the billions should a catastrophe occur. Try as it might, the AEC could not make WASH-740 disappear after it had served its political purpose. It became a rallying point for those who doubted the safety of nuclear power plants.

To stimulate demand for nuclear power, the AEC also embarked on a high visibility campaign to market a technology that it argued was on the threshold of competitiveness. Increased visibility and marketing responsibilities for experts were not coincidental: they were integrally related to sparse demand for expert-generated agendas. In the case of nuclear power, promotion paid off; it contributed to the reactor "bandwagon" of the mid-sixties. Here, too, however, contradictory demands were placed on experts by the nature of their mission. Since nuclear power was so competitive, some politicians asked, why were large federal subsidies required? Why, asked the coal lobby, did the Atomic Energy Commission, despite billions of private investment dollars flowing into commercial nuclear power, insist that nuclear power had no "practical value"?

Self-regulation, when exposed to the dynamic developments within the professions and organizations on which it relied, proved even more corrosive to the strategy of "political containment" than the experts' contradictory promotional responsibilities. Specialization – both bureaucratic and professional – was one significant contributor to the decline of expert authority. Here, the experts' tendency to narrow and deepen their inquiry fit readily into the bureaucracy's organizational preferences. The question of nuclear safety was addressed by just such a division of labor, a division that was institutionalized as a series of specialized organizational units within the Atomic Energy Commission. This had the effect of formalizing debate between the experts. Although each unit was committed to sustaining vigorous internal debate and to reaching a consensus for public consumption, each was also tempted to breach that consensus when its organizational mission was threatened. This destabilized the process by which public consensus was achieved within the expert community. Over time it broadcast what had been an internal debate to wider audiences.[6] The pressure was so great that even as staunch an advocate of secrecy as Harold Price, the AEC's director of licensing, was forced in the early sixties to call for a public moratorium on urban siting.

The scope of debate was also inexorably broadened as the experts developing nuclear power tackled complex agendas that cut across professional disciplines.[7] In organizational terms, these issues spanned agency and federal jurisdictions as

6 There is a significant literature discussing the breakdown of public consensus within the professions. See, for instance, Haskell, *Authority*, pp. xiii–iv; Aaron, *Politics*, pp. 155–9; Heller, "What's Right with Economics," pp. 1–3; Culliton, "Science's Restive Public," p. 149; Hohenemser et al., "The Distrust of Nuclear Power," pp. 25–34.

7 The concept of expanding the scope of debate is based on E. E. Schattschneider's discussion of "managing the scope of conflict" (*The Semisovereign People*, especially ch. 1, "The Contagiousness of Conflict").

   John L. Campbell adds a valuable refinement to Schattschneider, noting that the policymaking stage in America tends to be far more insulated than the implementation stage (*Collapse*, ch. 5). This certainly applies in the case of nuclear power, as Campbell deftly demonstrates. Like other scholars, however, Campbell frames his analysis in terms of the people versus the experts, thus missing the structural reasons behind the expanded scope of debate and conflict within the expert community.

well. As each new discipline was drawn into the discussion, its representatives – now fused to an institutional base – brought with them their agency and state, local, or federal allegiance. They also brought the varied perspectives that such amalgams of discipline and political institutions tended to produce. Committed to particular issues, these networks of officials and consultants further broadened the debate, and extended it to the full range of forums available in America's decentralized, pluralist political system.[8] Unlike the iron triangle bounded by AEC, Joint Committee on Atomic Energy, and nuclear industry, these networks did not necessarily consider the development of a standardized national program for nuclear power to be their top priority. Depending on the makeup of each issue network, states rights, public health, and water standards, for example, might be far higher priorities. Abel Wolman stood at the center of just such a network, urging his environmental engineering and public health colleagues to look into the safety of nuclear power. The U.S. Geological Survey and Secretary Stewart Udall engaged yet another. By the 1970s, an increasing number of crosscutting concerns challenged the production-oriented approach implicit in the nuclear iron triangle, and reflected quality-of-life issues that lay at the heart of the growing environmental movement.[9]

Experts were everywhere by 1975. At its inception, the Atomic Energy Commission so dominated the handful of experts in the field that the Joint Committee on Atomic Energy found it difficult to oversee that agency's activities. The nation's only congressional oversight body itself had limited access to expertise. But the proliferation of experts – a direct product of the federal government's new role in the production of expertise – made them available to a whole range of political actors who originally had been left out of the esoteric world of nuclear power.

By the mid-1970s, policymakers were deadlocked, with experts seemingly on all sides of every issue. Yet no combination was able to deliver a knockout punch. By this time, national citizens groups, citing experts of their own, had weighed in. This dramatized the situation. Public disputes between experts left the program adrift, and the public confused. Experts were doing what they had always done – expressing a range of viewpoints while debating their colleagues – but now they were doing it publicly, and in increasingly political forums. In 1973, Senator Pastore summed up the results of this phenomenon, noting that the public was confused because it was not being told in categorical terms "yes and no." And what is even worse, the Joint Committee on Atomic Energy chair continued, "is the fact that right within the agency, [Atomic Energy Commission] itself, you haven't made up your minds.... You have people who are being paid

8 On issue networks, see Heclo, "Issue Networks and the Executive Establishment," in King, ed., *The New American Political System.*

9 On the transition from production-oriented to quality-of-life related federal policies, see Hays, *Beauty, Health, and Permanence,* introduction and ch. 1.

to do a job who are saying that it is good and other people who are being paid to do the same job who say that it isn't good."[10]

The elaboration of expertise into specialized units, the inevitable differences in perspective that arose when a number of disciplines examined the same problem, and the political agendas engaged by various host institutions brought disputes between experts to the public's attention. Despite a natural inclination toward a more insulated policymaking style, experts were also forced to enter highly visible arenas in order to garner necessary public support. At times, it was the experts who aroused the people. At a minimum, these visible fissures between experts undermined public confidence in expertise. It was the elaboration of expertise within the decentralized and specialized American political system, not an attack on Galileo, that waylaid the nuclear bandwagon.

REPLICATING THE RESULTS

The patterns I have discerned in the history of commercial nuclear power have both molded and buffeted a broad cross section of federal policies since World War II. The dynamic forces set in motion by prominstrative politics are not readily apparent, however. I have been able to reconstruct some of them by drawing on and synthesizing important new currents in historiography and the social sciences. If similar "chain reactions" have swept through other policy areas, we will not be able to replicate them without a clear idea of what to look for and how to look.

As historians began to recognize several decades ago, the key to understanding twentieth-century American politics lay in shifting political history's concern with electoral politics and presidential leadership toward the study of bureaucratization – its causes and its impacts. An approach to history labeled "the organizational synthesis" did just that.[11] While Parsonian and Weberian sociology steered the original organizational scholars toward rigid functionalist portrayals of bureaucracy, subsequent work in the social sciences shifted the debate from description to an exploration of the sources of power.

At the same time that social scientists reexamined organizational theory, they revised their views on the sources of professional autonomy. Increasingly, sociologists rejected structure-functionalist descriptions of professionals that assumed experts existed and were granted authority because a complex and interdependent society required it.[12] Instead, scholars explored how these experts

---

10 U.S. Congress, Joint Committee on Atomic Energy, "Nuclear Reactor Safety Hearings," 93rd Cong, 1st sess., 1973 (published 7/75), January, 23, 1973, p. 6.
11 For an extended discussion of the organizational synthesis, see Balogh, "Reorganizing the Organizational Synthesis."
12 Peter B. Evans, Dietrich Rueschemeyer, and Theda Skocpol, *Bringing the State Back In* (New York: Cambridge University Press, 1985), p. 4.

achieved their power.[13] Historians, employing this new sense of skepticism, have begun to analyze professional development in the same terms used to analyze other groups, asking how professionals shaped social, political, and cultural relations.[14] Much as particle physics shattered scientists' notions of what constituted the building blocks of nature, the revolt against Parsonian structure-functionalism in the social sciences challenged historians' lock-step conceptualization of bureaucratization, professionalization, technological advance, and evolution from entrepreneurial to corporate capitalism.

Historians, on the other hand, have some valuable perspectives of their own to contribute to the sensitive technology required to follow politics into its administrative and professional forums. As Stephen Skowronek demonstrated, American would-be state builders were forced to come to grips with the antistatist culture in which they operated. Cultural historians recently have stressed the interplay between values championed by the rising professions and more deeply rooted cultural beliefs.[15] It was not promotional obligations, specialization, and interdisciplinary approaches alone that broadened the scope of debate in nuclear power. Rather, it was the interaction between these tendencies and America's strong commitment to free market capitalism on the one hand, and multiple points of political access on the other, that shattered the experts' public consensus. The prominodernative tendencies in post–World War II America permanently altered the political system but hardly eradicated powerful traces of its antecedents – administrative state or, for that matter, "courts and parties." In fact, the long-term success of expert policy agendas often turned on their advocates' ability to translate the policy's benefits into a currency more familiar to traditional political actors – whether contracts for a key legislator's district, the promise of more resources for an agency head, or economic protection for a tightly organized interest group.

Each element that contributed to nuclear power's political chain reaction shaped – often decisively – a wide variety of post–World War II programs. Before illustrating that point, let me reiterate the prominodernative pattern I have discerned

13 The best summary of this literature is Friedson, "Are Professions Necessary?" in Haskell, *The Authority of Experts*.
14 The best example of this is Haskell, *The Emergence of Professional Social Science*.
15 John Higham, "Hanging Together: Divergent Unities in American History," *Journal of American History* 61 (June 1974); Richard L. McCormick, "The Discovery that Business Corrupts Politics: A Reappraisal of the Origins of Progressivism," *American Historical Review* 86 (1981): 247–74. In both the Higham and McCormick models, the technical–professional interests outlasted the wave of broader public concern and involvement. Both Thomas McCraw, in *Prophets of Regulation*, and Tomlins, in *The State and the Unions*, have shown in their studies of regulators that the professionals were hardly insensitive to popular beliefs. Perhaps the best example, however, of how cultural values influenced professional perspectives is Allan M. Brandt's study of venereal disease – *No Magic Bullet: A Social History of Venereal Disease in the United States Since 1880* (New York: Oxford University Press, 1987). Prevailing attitudes toward sexual practices and class bias, Brandt argues, helped shape the way doctors and scientists defined and treated venereal diseases. This proved to be a major factor in explaining why venereal disease failed to join the list of infectious diseases eradicated by medicine in the early twentieth century.

in nuclear power in its simplest form. The process began with agenda setting. Expert-driven policy agendas sought to extend the boundaries of professional research agendas or to apply new research findings on a large-scale basis in order to achieve social benefits. In either case, the need for funding, comprehensive organizational capacity, and control over complex policies that ultimately would touch the lives of millions drove professionals into an alliance with the federal government, and prompted professionals to trade a portion of their autonomy for an alliance with federal administrators. While policy planning occurred in highly insulated forums, the resources required and, in some instances, the political impact anticipated required that prominstrators sell their programs, often in dramatic fashion. From the start, the major developmental obstacle for prominstrative policies was demand from would-be consumers. Prominstrative policies were producer- not consumer-driven, and producers labored to create the kind of lasting interest group support required to institutionalize their programs. At the outset, the prominstrator's greatest asset – the authority of the programs' experts – was also its most vulnerable commodity. The need to market programs' services clashed with the desire for political insulation. More significantly, expert authority was highly dependent on the maintenance of public consensus among experts. As policies moved from planning to implementation, a number of tendencies pushed expert debate out of insulated forums into more public arenas. The most important of these was specialization – both professional and organizational. Another was the interdisciplinary approach required to implement complex programs. A final tendency was the proliferation of experts in large part generated by the prominstrators themselves. The proliferation of experts allowed political actors previously excluded from the program to join the debate and introduce a variety of political perspectives into it. Issue networks linked to competing agencies at all three levels of America's federal system were often the critical agents. They spread debate from the institutional arena to broader audiences. The cacophony of competing expert opinion eroded the special advantages held by prominstrators at the outset, and forced programs toward more traditional bases of interest group or partisan support. The simultaneous erosion of expert public consensus across a spectrum of policy areas undermined the general confidence in experts – regardless of their particular field. Thus experts today begin from a very different base of support than they did thirty years ago. Nevertheless, America remains more dependent on its experts than ever.

In turning to the first phase in the prominstrative pattern – expert agenda setting – America's post–World War II history is indeed ironic. Many of the agendas adopted by critics of existing scientific or technological practices – whether efforts to eliminate pollution, or challenges to radiation standards – originated with new and improved scientific or technological abilities to measure and test. Though often seen by their adversaries and scholars alike as antiscientific or antitechnological, critics were highly dependent on the latest scientific thinking and technology for measuring effects on the body or the environment. "As science advances and our ability to detect risks improves, our opportunities for influencing risk have proliferated," a *Science* article evaluating epidemiology stated.[16]

Greater capacity to detect, however, led directly to new policy agendas. The decision to undertake the study in the first place was often the result of a policy decision made by prominstrators at the federal level. More often than not, such decisions were guided by professional and agency research agendas, not organized political demand from the population targeted for services or treatment. As Alvan Feinstein, director of Yale's clinical epidemiology unit, put it, "The episodes have now developed a familiar pattern. A report appears in a prominent medical journal; the conclusions receive wide publicity ... and another common entity of daily life becomes 'indicted' as a 'menace' to health – possibly causing strokes, heart attacks, birth defects, cancer." The reported evidence, he continued, "is almost always a statistical analysis of epidemiologic data, and the scientific tactics that produce the evidence are almost always difficult to understand and evaluate."[17] Tens of thousands of such decisions to embark on new federally funded research were made each year, flooding the nation's research agendas.

For example, in 1984 the *New England Journal of Medicine* published a pair of scientific studies that implicated even moderate use of alcohol in the development of breast cancer.[18] Other recent examples of expert agenda-setting included charges that silicone implant surgery was linked to cancer in rats, that depression was far more debilitating than previously recognized, that the Environmental Protection Agency had neglected some of the most hazardous risks – such as indoor pollution – or that the ozone layer was disappearing.[19] In each instance, experts played a leading role in determining what would be studied and how. The answers to this first set of questions in turn led to a new set of questions and policy recommendations, ranging from warnings printed on beer bottles to calls for international mechanisms to deal with the effects of global warming.

In all of these instances, federal experts were integrally involved, either in granting funds for the original research or requiring and analyzing the data submitted. One-third of the funding for the depression study, for instance, came from the federal government, the EPA sponsored the study of its own priorities, NASA flew into the ozone hole, and the silicone discovery was based on data submitted by a drug-manufacturing company to get Food and Drug Administration (FDA) approval.[20] The relationship between these scientifically

16 Richard J. Zeckhauser and W. Kip Viscusi, "Risk Within Reason," *Science* 248 (May 4, 1990): 563.

17 Alvan R. Feinstein, "Scientific Standards in Epidemiologic Studies of the Menace of Daily Life," *Science* 242 (December 2, 1988): 1257.

18 Stephen Lyons, "'Crying Wolf' About Risks?," *Boston Globe*, October 16, 1989, p. 29.

19 On silicone, see *Boston Globe*, November 11, 1988, p. 3; on depression, see *Boston Globe*, August 18, 1989, p. 1; on revised risk priorities, see *Boston Globe*, September 12, 1988, p. 1; on ozone hole, see *Boston Globe*, September 1, 1987, p. 2.

20 *Boston Globe*, September 1, 1987, p. 2.

driven agendas and the federal government, however, did not end there. The National Center for Health Services Research and Health Care Technology Assessment, the National Institute on Aging, and the National Institute of Mental Health will no doubt play a crucial role in implementing studies' recommendations. Their fate ultimately will turn on federal agencies' ability to incorporate findings into the massive federal grant programs, particularly Medicaid and Medicare, administered by the federal government. EPA and its congressional oversight committees ultimately will determine whether past environmental priorities are revised. The FDA must make the final decision on the safety of implants. The very timing of NASA's exploration of the ozone hole was the product of prominstrative decision making. "If this were for pure science," commented chief program scientist Robert Watson, "we would have waited another year, but because of the importance to society and policy, we felt it would be wrong to wait."[21]

Administrative and expert agendas seemed inextricably linked. This was one of the cornerstones of the prominstrative alliance. When completed, expert research spawned a number of new policy recommendations. To be sure, more research was one of the most common recommendations. Often, however, the study also called for additional federal action. Economists, for instance, could theorize at universities to their heart's content, but as Leonard Silk reported in his *New York Times* column, the "Economic Scene," "truth is of scant value unless the public and government respect it and use it."[22] It was no different when it came to the nation's nutrition. Improving nutrition could best be addressed through federal aid to school lunch programs, hospitals, nursing homes, and food stamps. The most direct existing path cut through the armed forces and subsidiary veterans' benefits. Both bureaucracies had long served as models for social programs ranging from universal access to prenatal care to tuition support. In other words, not just federal dollars were required – although that always seemed to be an important prerequisite – federal administrative capacity and authority were required as well if experts hoped to have a social impact.

That impact, however, was often sought more fervently by providers than by would-be consumers. Organized consumer demand turned out to be harder to create than pathbreaking science or technology. An article in the *New York Times* neatly captured this dilemma.[23] A gripping photograph showed a premature infant (born to a drug-addicted mother) and its teddy bear engulfed by a contraption that was obviously keeping the child alive. Graphs told the bleak story. The decline in America's infant mortality had leveled off; the gap between blacks and whites had actually increased in percentage terms. The central theme of studies released by the Bush administration, the National Academy of Sciences, and the National Commission to Prevent Infant Mortality, was that the United States

21 *Boston Globe*, September 1, 1987.
22 Leonard Silk, *New York Times*, January 3, 1986, p. D2.
23 *New York Times*, August 12, 1990.

had the techniques available to reduce infant mortality substantially without spending billions. There appeared to be only one hitch, summed up by Dr. Anthony Robbins, former president of the American Public Health Association: "You have to create a demand for services, for prenatal care, through education and marketing."[24]

Some of America's more abstract promindistrative programs so outpaced demand that it was not even clear what services they might provide. The space station was a classic example. In FY 1989, NASA earmarked $900 million to begin work on a space station expected to cost more than $30 billion by its completion before the end of the century. But as Albert Wheelon, a retired aerospace executive and member of the president's commission to investigate the *Challenger*, put it, "One problem with the space station is that we are not sure why we are building it."[25] Some speculated that without the space station, the multibillion-dollar space shuttle might have little else to do once the backlog from the *Challenger* accident was completed. The nub of NASA's problem, wrote one *Times* reporter, was "that it finds itself with the means, an advanced transportation system, and no ends."[26] Or, as the Congressional Budget Office cautioned, "the future NASA program will become increasingly determined by choices made years earlier: Strong incentives will exist to fund new missions in order to rationalize the use of the infrastructure already provided."[27] Public opinion polls showed that there was broad, albeit shallow, support for the program. Given the competition for funds in the 1990s, that may not prove sufficient for implementation. NASA's ability to sustain organized support will ultimately determine that agency's fate.

Enter the prominidistrative state. In ways remarkably similar to those documented in the history of commercial nuclear power, the Reagan administration launched a program of "privatization," designed to create demand for space services such as rocketry, space stations, and space cameras. The race for international prestige was now against the Japanese, not the Soviets. "It's worth noting," a Commerce Department official warned in a recent speech, "that the two largest Japanese construction companies... both have space project offices which are today designing spaceports and moon bases."[28] There have also been calls for the federal government to limit liability in case of accident. Experts, the *Times* reported, "say similar limits helped spur the founding of the nuclear-power industry."[29]

At the same time that some space prominidistrators labored mightily to create demand, others struggled to keep their heads above an avalanche of unused data already beamed back from space. Known by insiders as the "Black Hole," a huge federal warehouse at the Washington, D.C., National Records Center housed more than 70,000 tapes in an area the size of eighteen football fields (nicknamed by some scientists "tape landfills"). Hundreds of thousands more tapes were

24 Ibid.   25 *New York Times*, October 9, 1988, p. E2.   26 Ibid.   27 Ibid.
28 *New York Times*, January 24, 1988, p. 1; quotation from p. 28.
29 Ibid., p. 28.

scattered around NASA labs throughout the country. NASA scientists, it turned out, had only been able to analyze closely one percent of the tapes. Yet nobody at NASA suggested that perhaps we did not need all of this data. Rather, as William J. Campbell, who heads an artificial intelligence program at Goddard, sees it, NASA had to spend more on its software so that it could analyze all of this data. Nor was NASA unique. As one computer scientist at Goddard Space Center emphasized, the mismatch between raw data and real analysis plagued policies ranging from energy exploration to medical imaging. "Everybody has the same problem, NASA's just ahead of the rest."[30]

Given the Herculean task of organizing demand that it faced, it is not surprising that NASA, like the Atomic Energy Commission, actively cultivated a powerful lobby of suppliers. By 1987, however, the agency had overstepped even its own sense of propriety. This occurred when NASA's Office of Industry Affairs sent a memo to contractors asking for their help in lobbying Congress at budget hearings. The memo requested that contractors report back within nine days on the results of their efforts.[31] Although NASA eventually apologized to Congress for this violation of the law, NASA spokesperson David Garrett sounded less than chastened. "I think every agency in town certainly does things to protect their turf," Garrett insisted.[32]

Although NASA symbolized the extremes of the prominstrative struggle to generate organized demand and sell its program in the absence of demand for its services, its tendency to create new demands in order fully to utilize past resources was typical of patterns in other policy areas. The study of depression cited earlier underscored the need to diagnose depression properly, particularly for patients who are not complaining of depression. The discovery of such high rates of depression – including sizable numbers of patients who did not know they were depressed – obviously called for major changes in the way patients were screened.[33] The EPA study concluded that the agency "desperately needs a long-term research and development program" and called for the creation of an environmental research institute.[34]

Experts turned to the federal government in part because it assured them an environment relatively insulated from other political interests in which to carry out research and promised them a relatively large degree of control over the implementation of their research findings. In return, experts promised federal administrators a reliable, often predictable, and most of all respected source of meeting poorly articulated political demands. The requirement that experts engage in highly visible public salesmanship, however, clashed with the desire of all prominstrators to maintain the insulated environment crucial to the control they sought.

Experts often regretted the publicity that they eventually received. Few,

30 Ibid.   31 *Boston Globe*, September 1, 1987.   32 Ibid.
33 *Boston Globe*, August 18, 1989, pp. 1, 6.
34 *Boston Globe*, September 12, 1988, p. 1.

however, understood the degree to which that publicity ensured their very professional survival. Brian MacMahon, chair of epidemiology at Harvard's School of Public Health, for instance, wrote a 1981 study that linked coffee to pancreatic cancer. When attacked, he acknowledged that it would have been better to publish his work in a more obscure journal. As he told a *Boston Globe* reporter, had he done that, "It would have come to the attention of the scientists, but there would not have been the hoopla that there was about it."[35] Had MacMahon published only in obscure journals, however, it is not clear that he would have obtained crucial funding for the project in the first place.

Prominstrators were well aware that insulation was crucial to success. Nor did they need the shield of national security to establish initially what they considered to be the requisite degree of insulation (although it undoubtedly helped). Whether researching the ozone hole, lobbying for the privatization of space, reducing infant mortality, or setting dietary standards, prominstrators managed to keep their reports private – at least for a while. In the case of ozone, the official justification cited international treaties about reducing industrial chemicals. The press was denied access to Reagan's privatization plan, the White House task force on infant mortality's report, and the National Academy of Sciences report on dietary standards.[36]

Regardless of the conscious efforts made to preserve prominstrative insulation, the processes of specialization, the need for multidisciplinary approaches to complex problems, and the proliferation of experts and agencies continued to erode any such efforts. Specialization seemed to hit economists the hardest. The *Journal of Economic Perspectives* was established in 1987 specifically to get economists to communicate across their subspecialties and start talking to each other again.[37] Specialization was at work in epidemiology as well where the emergence of the subdiscipline directed toward noninfectious disease expanded the reach of the discipline but also led it toward more tenuous conclusions. As Alvan Feinstein put it, "statistics are like a bikini bathing suit: what is revealed is interesting; what is concealed is crucial."[38] Feinstein's was hardly the first critique penned by an epidemiologist directed toward his colleagues. Most of its antecedents, however, had appeared in specialized journals, fueling the intraprofessional debate, but at the same time containing it within the boundaries of the profession.[39] It is not surprising that given the nature of scientific debate, the openness of American communciations, and the public interest in health, the epidemiological debate spread to more public forums. As Noel Weiss of the University of Washington noted, "the problem is that other people are listening to us communicating, and they're making more out of it than we are."[40]

35 *Boston Globe*, October 16, 1989, p. 30.
36 On ozone, see *Boston Globe*, September 1, 1987; on infant mortality, see *New York Times*, August 12, 1990; on dietary standards, see *New York Times*, October 8, 1985.
37 *New York Times*, September 27, 1987.   38 Feinstein, "Scientific Standards," p. 1263.
39 *Boston Globe*, October 16, 1989.   40 Ibid.

Even in America's most highly insulated forums, however, the specialized debates that began behind barriers of self-consciously maintained and legally protected secrecy, relentlessly crept toward more public forums. No debate was more closely guarded than that surrounding America's production of nuclear weapons. Although it took decades, even in this field, internal criticism ultimately made its way into the headlines. Specialization, the division between those charged with producing weapons and those charged with ensuring their safety, was again the driving force behind these revelations. A *Boston Globe* editorial sized up the situation at America's first weapons production plant in Hanford, Washington, labeling the latest revelations of nuclear waste dumping "numbing proof that excessive secrecy, single-mindedness and fear can add up to shockingly irresponsible behavior." The source of the revelations publicized by the *Globe* was "specialist" Cleve Anderson, who ran a committee studying long-range nuclear waste management at Hanford in the 1950s. Anderson ultimately had to file a Freedom of Information request to obtain and make public his own report. While it took Anderson decades to substantiate his charges of cover-up, an intense debate between development and safety interests in weapons production was covered by the *New York Times* at virtually the same time that the debate unfolded within the Department of Energy in late 1988. As the lead paragraph in the *Times* coverage put it, "A dispute between safety specialists and production managers in the Energy Department erupted in public today over when it will be safe to reopen a nuclear reactor that makes bomb fuel."[41]

As policymakers sought to tackle increasingly difficult problems, they employed a variety of professional disciplines. These disciplines and their organizational hosts often worked at cross-purposes. It was NASA database expert Barry Jacobs and/or "frustrated" James Green, head of NASA's National Space Science Data Center, who fueled the flap over unused space data. And artificial-intelligence director William Campbell told the *Journal* reporter that the $2 million (out of a $7.8 billion budget) earmarked by NASA for new software techniques "is peanuts."[42] In the battle against AIDS, presidential commissions and popular accounts stressed the problem of feuding disciplines and jealous agencies. Conflicts between the National Institutes of Health and the Centers for Disease Control, for instance, and scientists competing for limited grant money retarded the effort.[43] Even the seemingly innocuous campaign for "designated drivers" came under fire, once it was closely scrutinized by psychologists employed by market research agencies. Because teens viewed the designated driver as a parent, the deputy director of research for the firm Saatchi and Saatchi told a *Boston Globe* reporter, "they feel compelled to overturn the parent and in many cases, they try to get him drunker than they are."[44] Major accounting firms discovered they could make

41 *New York Times*, December 10, 1988, p. 10.   42 *Wall Street Journal*, January 12, 1988.
43 Diane Johnson and John F. Murray, "AIDS Without End," *New York Review of Books* (August 18, 1988), p. 58.
44 *Boston Globe*, September 29, 1988, p. 7.

more money and argued that they could do a more comprehensive job if they supplemented traditional financial audits with management-consulting and data systems analysis. Improving a company's bottom line, however, was not necessarily compatible with certifying the accuracy of that bottom line.[45] The interdisciplinary conflict was far more intense in more esoteric programs such as genetic engineering, where ecologists charged that a National Academy of Sciences panel appointed to study the safety of these new techniques was biased toward genetic engineering because molecular biologists outnumbered ecologists in the NAS.[46]

From the public's point of view, experts seemed to be everywhere by the 1980s. "Television and radio are chockablock with experts, seething with specialists prepared to advise on matters that span the range of human interests," wrote one *New York Times* columnist.[47] "You need only pick up a telephone to ask for expert help," she continued. In the nation's courts, trials increasingly became duels of expert witnesses.[48] The talking heads – "a legion of consultants and analysts" – as one journalist put it, dominated election campaign coverage in 1988. As another columnist concluded, they also dominate parenthood. "Experts tell parents how to have sex, how many children to have and how to raise them. They tell women how to behave during pregnancy and make many feel guilty if they decline natural childbirth or cannot breast-feed adequately. Experts coach parents on turning little tots into serious academics or 'superbabies.'"[49] So that they could be the experts of the future?

Specialization, multidisciplinary approaches to problem solving, and the sheer proliferation of expertise combined to make experts more accessible to political actors who had at one time been excluded from esoteric policy debates. Issue networks were the primary agent connecting a growing community of experts to these would-be policymakers. As the number of participants involved in policy debate increased, the means of communicating became more public and more accessible. Where internal memoranda once laid out options to a handful of decision makers, speciality book publishers mass-produced monographs designed to influence other experts, lobbyists, consultants, and politicians. With presses and bookstores for every political persuasion, what had been fierce scholarly debate was translated into strident public battle.[50]

Coursing through federal agencies, congressional committees, state and local governments, even citizens groups, experts linked issue networks to mobilized constituencies. Sometimes citizens groups led elected officials and state and local bureaucrats into the fray. When Public Citizen, a Washington-based Nader-founded group, released the results of a study charging that radon tests were

45 *New York Times*, May 13, 1984, 3, p. 1.
46 *Boston Globe*, August 24, 1987, p. 44.
47 Ellen Currie, "Hers," *New York Times*, July 31, 1986, p. C2.
48 *Boston Globe*, August 6, 1989, p. 25.
49 Mary Meehan, "Experts and Wisdom," *Baltimore Sun*, p. A11.
50 Suzanne Gordon, "Public Policy Publishing: Lobbying in Print," *Washington Post*, "Book World," July 28, 1985, p. 5.

inaccurate, federal and state regulatory bodies quickly began testing of their own. Although these agencies argued that the net effect of the Public Citizen group's report was to increase the risk of radon poisoning (because even inaccurate tests warned the public of dangerous radon levels), the same agencies increased their overall level of scrutiny.[51] Relying on their own experts, close readings of publicly available reports and correspondence, and informal contacts with disgruntled agency experts, citizens groups were instrumental in publicizing internal prominstrative debates. The flap over the safety of reopening weapons fuel production reactors, for instance, was made public at a forum sponsored by the Nuclear Control Institute, a group opposed to the proliferation of nuclear weapons.[52] The Public Citizens Health Research Group set off the uproar about the safety of silicone implants by publicizing the data submitted to the FDA by Dow and calling for a ban on the procedure.[53] The Southern California Federation of Scientists, a group opposed to the Strategic Defense Initiative, was instrumental in publicizing internal criticism of SDI's technical progress.[54]

More often than not, however, the scope of debate was broadened by elected officials, or political jurisdictions that had previously been excluded from the discussion. For instance, by the spring of 1984 two House committees were looking into the issue of auditor independence and the Securities and Exchange Commission's oversight of that process, at the same time that the Federal Deposit Insurance Corporation prepared to bring suit against five accounting firms.[55] The most devastating attack on SDI's technical capabilities came from scientists at the forefront of the government's efforts. "I'm very alarmed at the degree of hype, promises and a failure to focus on what this national program really is – a research program with lots of unanswered questions," Dr. George Miller, director of defense programs at Livermore National Laboratories told the *New York Times* in December 1985.[56] It was congressional opponents, however, who relied on internal disputes to break the story open. As the *Boston Globe's* lead headline shouted on Sunday, June 12, 1988, "Senate Study Contributes to Avalanche of Doubts on SDI." The article chronicled the negative findings of the Senate staff report, which was based on "classified briefings and interviews with 120 scientists and technicians involved in SDI." Another Senate probe led by John Glenn, directed at the nation's weapons production plants, exposed rifts between the experts. It was Glenn's hearings in September 1988 that revealed a memo written by a Department of Energy contractor chronicling a history of serious accidents at the South Carolina weapons production reactor. The accidents had never been publicly exposed before.[57]

Whatever the sequence, access to expertise appeared to be one ingredient crucial to joining the debate. The newer political actors had less incentive to

51 *New York Times*, April 8, 1989, p. 11.   52 *New York Times*, December 10, 1988.
53 *Boston Globe*, November 11, 1983.   54 *Boston Globe*, October 21, 1987.
55 *New York Times*, May 13, 1984.   56 *New York Times*, December 16, 1985, p. 1.
57 *New York Times*, November 28, 1988, p. 1.

maintain any semblance of public consensus. In the case of America's nuclear weapons production plants, members of the congressional committee directly responsible for overseeing these operations – the Armed Services Committees – missed the problem because they had relied too heavily on DOE expertise.[58] Other committees were hesitant to take on the issue for lack of independent expertise. As one congressional staffer put it, "there is an intimidation factor at work. Because there are no scientists on their payroll the committees have to accept the word of the Department of Energy and its contractors."[59] Once it became apparent, however, that experts within the DOE were sharply divided, politicians and agencies previously excluded scrambled to gain access to experts who were in turn eager to spread the debate into a political forum more conducive to their point of view.

The political actors that moved most effectively to do that were the states. By the end of 1988, even in South Carolina, which had a long history of state-promoted nuclear industry and housed 1,700 DOE workers, some local and state officials led the charge against building a new weapons production reactor.[60] The state's top environmental official acknowledged that the government "got caught up in a time warp in the late 1970s" and "found itself 20 years behind."[61] Environmental officials in Ohio were more aggressive. Originally told by the federal government that the state "did not have the expertise to regulate Portsmouth weapons facilities," Ohio's governor, Richard Celeste, in December 1988 obtained the nation's first legally enforceable court order mandating that the DOE clean up the Portsmouth site. "Their initial position was nobody including the U.S. Environmental Protection Agency, had any authority over them," gloated Ohio's attorney general, Anthony Celebrezze, Jr. "They've come a long way in their attitude."[62] When the issue was civilian nuclear reactors or radioactive waste dumping, a number of state governors ventured farther. Mario Cuomo and Michael Dukakis actively opposed opening reactors that billions of dollars had already been poured into. In Colorado and Idaho, governors pressured the federal government to find other sites for nuclear waste.

Disputes that started in insulated forums made their way into the public domain. Feuding epidemiologists, conflicting kremlinologists, ebullient and doom-predicting economists, squabbling nutritional experts, warring weapons production experts, and frustrated SDI scientists, all took their case to the public, despite the clear understanding that it might undermine the general standing of their professions and programs. Flora Lewis summed up the SDI physicists' dilemma, pointing out that "they have been arguing inside the program for some time, to

58 Ibid.　　59 Ibid.
60 *New York Times*, December 8, 1988, p. B18.
61 "Federal A-Plants Used Dumping Practices Banned for Others," *New York Times*, December 8, 1988, p. B18.
62 *New York Times*, November 23, 1988, p. 1.

no avail.... That is why their concerns are seeping into public print, despite the gag rule."[63]

These disputes, combined with scandals that shook a number of the professions, and some spectacular failures – ranging from the *Challenger* disaster to the 1983 consensus forecast by economists that underestimated growth in the GNP by almost 100 percent – undermined the authority of experts.[64] The signs of this erosion in confidence were everywhere. In 1988 a governor's task force in New York agreed unanimously that physicians should be reexamined periodically as part of the state's licensing procedure.[65] Epidemiologists complained of the disastrous consequences that the "cry wolf" phenomenon might produce. Alvan Feinstein concluded his article with the warning that

epidemiologic studies of noninfectious disease have produced their own adverse side effect: an 'epidemic of apprehension.' The epidemic grows with each new alarm about a new menace in daily life. Uncertain about how to distinguish the many false alarms from the few that might be true, the public and nonepidemiologic scientists are confronted by evidence that is peer group-approved but scientifically inadequate.[66]

Following the warnings about radon testing kits, there was a marked decline in test sales; in just one month consumers demanded 30,000 refunds. As the president of a leading test-kit maker told a *Times* reporter, "There has been a tremendous loss of public confidence."[67] A retired bank clerk in Mayfield, Ohio, who had used a test kit, read Public Citizen's warning, and demanded a refund, captured the public mood best: "I'm still very confused about it."[68]

Analysts pointed to a decline in the public's faith in auditors, once thought of as the watchdogs capable of detecting fraud or anticipating a business's financial collapse.[69] Economists fared even more poorly in the 1980s. Once essential to policymakers and politicians alike, economists were shunned by the Reagan administration. Reagan even threatened to leave open the Council of Economic Advisers' chair.[70] Whereas most economists blamed anti-intellectualism, or political expediency, for their declining status, Sir Alec Cairncross, delivering the Richard T. Ely lecture to the American Economic Association in 1985, blamed economists themselves for giving conflicting and often wrong advice.[71] Policymakers could

63 Flora Lewis, "A 'Star Wars' Cover-up," *New York Times*, December 3, 1985.
64 For a good summary of two major scandals that hit medical research in 1986, see Philip M. Boffey, "Major Study Points to Faulty Research at Two Universities," *New York Times*, April 22, 1986, p. C1. On economic forecasting, see Daniel S. Greenberg, "Alchemy," *Baltimore Sun*, January 13, 1988, p. A7.
65 *New York Times*, February 26, 1988, p. 1.
66 Feinstein, "Scientific Standards," pp. 1261–2.
67 *New York Times*, April 8, 1989.   68 Ibid.
69 *New York Times*, May 13, 1984.
70 *New York Times*, January 3, 1986, January 2, 1985, and September 27, 1987; *Wall Street Journal*, December 31, 1987, p. 1.
71 *New York Times*, January 2, 1985.

not trust economic theorists when theorists differed so widely, the distinguished British economist told his audience. Or, as the chief economist at Connecticut National Bank quipped, "What do you do? You've got 400 economists and 800 forecasts."[72]

Nowhere was the precipitous decline in expert authority more wrenching than among the thousands of Americans employed in the production of nuclear weapons. A *New York Times* article headline, "Fear Corrodes Faith at Atomic Plants," captured the dramatic evolution. Nuclear workers had always shared a special spirit, dating back to the Manhattan Project. "'Build your future with atomic energy,' the workers remember being told in the 1950s, as Government contractors moved through the country on recruiting drives."[73] Many of the 600,000 men and women who signed on did just that, affixing the symbol of the atom to their high school buildings and naming their teams "the Bombers." A series of disclosures about past accidents and unsafe practices and the inconclusive debate currently raging between experts have left these men and women with a crisis of confidence. "How much don't they know about the risks of their jobs? Can they afford to trust the Government with their lives? What choice do they have? Those are the kinds of questions many of the workers have been exchanging more than ever in recent weeks," the *Times* reported. "The Department of Energy assures these workers that they are no less healthy than other industrial workers," but "many of the workers remember the inconclusive language of the epidemiologists, who speak in probabilities instead of certainties," the article concluded.[74]

Experts themselves, though often oblivious to the longer-term cause, were among the first to recognize that their authority had declined. Some, including auditors, economists, physicians, and SDI project physicists, felt that the problem was simple: the public expected too much.[75] The solution lay in lowering these expectations. The American Medical Association (AMA), for instance, linked high patient expectations to soaring malpractice costs. It drew up plans to dampen "Marcus Welbyism" by holding seminars, distributing leaflets and sponsoring speeches in order to present a more realistic picture of the limitations of medicine.[76] What these earnest efforts to dampen expectations neglected was the source of those expectations. In the prominstrative state experts were called upon to create high expectations in order to create demand for their services sufficient to garner political resources – particularly funding and autonomy. The AMA official who complained that "we're too damned confident with our patients" was correct, as far as his analysis went. He failed to grasp, however, just how crucial that overconfidence was to the elaborate network of education, research, and applied technology that prominstrators had carved out.

72 *Wall Street Journal*, December 31, 1987.
73 *New York Times*, December 11, 1988, p. 36.   74 Ibid.
75 On auditors, see *New York Times*, May 13, 1984; on economists, see ibid., January 13, 1986; on physicians, see ibid., February 10, 1985; on SDI physicists, see ibid., December 16, 1985.
76 Ibid., February 10, 1985.

Although some experts decided that the public's opinion of their profession was irrationally high, most analysts continued to worry about those "opponents" of science who harbored irrationally low opinions. A group of science journalists sitting around a table on public television discussing the American Association for the Advancement of Science meeting they had just attended in January 1989 captured the consensus of the science community best. Barbara Culliton, a noted commentator on science, discussed Stanford president Donald Kennedy's impassioned speech. Culliton informally labeled it "science under siege." Kennedy talked about critics ranging from animal rights activists to neighborhood groups who, from the scientists' point of view, were trying to impede research. He also noted "that the public no longer seems to trust scientists even though there is a high level of appreciation of what science can do." It was not their competence but their motives that were suspect. Kennedy suggested that the scientific community go on the warpath and start defending itself more aggressively.[77]

Moderator Dave Marash intervened to elaborate on Culliton's point. "So it's really the same complaint that a lot of politicians are making, isn't it? That single issue constituencies are getting better and better at becoming roadblocks." Culliton confirmed Marash's hunch, but insisted that scientists were "probably less sophisticated than some other groups."

They want to speak to these activists rationally....And the activists, the special interest, single issue groups in this field as in any other are less interested in a rational dialogue. And the scientists have not yet learned how to fight back in a political sense, so that irrespective of who ends up being right or wrong in any given case there's an unequal footing politically. And I think Don Kennedy is saying, you know, let's get our act together.[78]

Before American prominstrators try to get their act together in the 1990s they would do well to study the history of their past performance. My interpretation of their fifty-year run as the nation's most dynamic new political force suggests that they overcame their antiscientific, anti-intellectual opponents long ago. To continue to rail against this enemy is to misjudge their critics and underestimate the deep reservoir of support for science. Today's most effective opponents are far more likely to be colleagues, perhaps working for a different political jurisdiction, possibly in a different subdiscipline, almost certainly highly educated and undoubtedly skilled in the ways of issue network politics. In other words, the prominstrative state has triumphed.

The victory has not been systematic. In fact, prominstrators who have learned how to adapt to the residue of a state once dominated by "courts and parties" in the late nineteenth century and the impulse to organize and coordinate characteristic of the 1920s and 1930s have fared the best. Victory has altered the

77 Transcript of *Science Journal* 2, no. 3, (January 19, 1989), Washington, D.C.: WETA, p. 8.
78 Ibid.

trajectory of American politics but has hardly eradicated previous patterns that remain deeply ingrained in the culture.

Once victory is acknowledged, the proministrative ensemble can perform its crucial second act. The script for that act has already been written by those who promised to employ expertise and administrative capacity to improve social conditions. For those who promised too much, the act will undoubtedly strain their audience's trust. But for other political actors, particularly those who read their reviews carefully and even solicit criticism at dress rehearsals, a promising career beckons.

Proministrators have not risen to such strategic positions of power in today's political system by eschewing politics. They should acknowledge the significant political agendas that they have always harbored. Only then can they begin to assess the long chain of reactions that has diminished their autonomy but made access to expertise a virtual precondition for effective politics. Although the final results of that chain reaction are far from predetermined, one lesson should be clear. The scope of debate, regardless of policy area, degree of insulation, or level of expertise, has been inexorably broadened. Although broader public participation can and has been in many instances consciously retarded, the absolute autonomy that may well remain the proministrative ideal cannot be sustained in America's pluralist landscape.

Rather than briefly postpone the inevitable, why not invite a full range of participation far earlier in the policy process, and certainly, long before implementation? Why not go still farther and test for demand rather than seek to create it artificially? No doubt these suggestions require a far longer attention span than most political actors are willing even to consider. But if the proministrative state cannot learn how to plan without excluding a broad cross section of political actors who ultimately will be involved in the implementation and consumption of services, can it master the increasingly complex programs it undoubtedly will seek to undertake? I think not. Should a few historically minded yet farseeing proministrators heed this advice, they may not only alter the subsequent chain of reactions, they may begin to harness it to achieve the social purposes that brought experts and administrators together in the first place.

Index